Research Reports ESPRIT

Subseries PDT (Product Data Technology)

Project 2195 · CADEX

Edited in cooperation with
the Commission of the European Communities and
the Product Data Technology Advisory Group (PDTAG)

H. J. Helpenstein (Ed.)

CAD Geometry Data Exchange Using STEP

Realisation of Interface Processors

 Springer-Verlag

Berlin Heidelberg New York
London Paris Tokyo
Hong Kong Barcelona
Budapest

Volume Editor

Helmut J. Helpenstein
Gesellschaft für Strukturanalyse mbH (GfS)
Pascalstraße 17, D-52076 Aachen

ESPRIT Project 2195 "CAD Geometry Data Exchange (CADEX)" belongs to the technological area "Computer Integrated Manufacturing and Engineering" of ESPRIT, the European Specific Programme for Research and Development in Information Technology supported by the European Communities.

The objective of this project was to support and further develop the ISO STEP standard for a neutral file format for CAD data exchange by:

- proving the practicality of the standard
- providing application protocols for the efficient use of the standard
- developing comprehensive implementations for CAD geometry exchange
- implementing state-of-the-art testing methodologies
- developing pre- and post-processors to interface a range of CAD equipments and Finite Element Modelling (FEM) systems.

The project continued the work of ESPRIT project 322, which led directly to the STEP standard.

The experience gained in the project will be used to maintain and strengthen European influence in the relevant international standardisation bodies.

Computing Reviews Subject Classification (1991): J.6, J.2, K.6.3, D.2.1, E.2, H.2.3, G.1.2

ISBN-13:978-3-540-56902-2   e-ISBN-13:978-3-642-78335-7
DOI: 10.1007/978-3-642-78335-7

Publication No. EUR 15380 EN of the Commission of the European Communities, Dissemination of Scientific and Technical Knowledge Unit, Directorate-General Information Technologies and Industries, and Telecommunications, Luxembourg. Neither the Commission of the European Communities nor any person acting on behalf of the Commission is responsible for the use which might be made of the following information.

Typesetting: Camera-ready by authors
45/3140 – 543210 – Printed on acid-free paper

Foreword

The Product Data Technology Advisory Group, short PDTAG, was established on 30 September 1992 under the auspices of the ESPRIT CIME Division of the Directorate General XIII of the European Commission. Its goals include promoting European cooperation and improving the European infrastructure in Product Data Technology, particularly in connection with the new standard STEP (ISO 10303). The dissemination of information on Product Data Technology and on European contributions to STEP is of crucial importance to this development.

The current volume is the first title in a new PDTAG subseries to Springer Publishers´ Research Reports ESPRIT. This new subseries intends to form a comprehensive repository of publications on Product Data Technology resulting from ESPRIT Projects and from European contributions to standardisation based on ISO/STEP. PDTAG welcomes the opportunity to make this information more accessible under the format of a coherent subseries within the established framework of Research Reports ESPRIT. Much valuable background on the new international PDT standard can thus be found in the same collection.

The CADEX project (ESPRIT Project 2195) has played an important role in the second generation of ESPRIT projects which was recently completed. It has made major contributions to the exchange of geometry data between CAD systems using STEP by providing practical tools for the implementation of STEP software, by developing numerous STEP processors and by contributing to the test and validation methodology for STEP. These are important prerequesites for the practical introduction to the STEP standard. This volume on the results of CADEX is thus an excellent opener for the new subseries.

Horst Nowacki
Chairman of PDTAG

Preface

This document was produced at the end of the project CADEX (CAD Geometry Data Exchange, ESPRIT 2195) running from July 1989 to October 1992. It comprises all outstanding deliverables i.e. the results of all workpackages including the final documentation. These workpackages are:

- Application specification and contribution to STEP (Application Protocols)
- Specification and development of common software tools
- Development of STEP data exchange processors
- Test and validation

The following partners contributed to this report:

Chapter 1: GfS
Chapter 2: CADDETC, DISEL, FEGS, HP, ITALCAD, SI
Chapter 3: BMW, CADDETC, FEGS, GfS, HP, ITALCAD, SI, SNI
Chapter 4: DISEL, DnV, FEGS, GfS, HP, ISYKON, ITALCAD,
          K3DP, SI, SNI
Chapter 5: BMW, CADDETC, FIAT

Acknowledgement

The CADEX consortium gratefully acknowledges the support given by the Commission of the European Communities, Directorate General XIII. Funding this pre-competitive work in research, development and standardisation has brought European Information Technology industry to top world level.

Table of Contents

1. Introduction

CAD data exchange has been the goal of many initiatives, investigations, research works, collaborations and projects for many years. A lot of effort has been invested in linking two dissimilar systems to each other. With increasing number of available CAD systems, with further desired links to analysis and machining systems, this practise was no longer feasible.

The idea of putting a neutral format between the different specific outputs and inputs has lead to significant success and to some more or less accepted standards. While IGES and VDAFS came from industrial applications and yielded good results in a limited scope, the CEC funded Esprit project CAD*I was the first to produce a neutral representation to cover an actually large range of product related data, putting emphasis on CAD and analysis data. The results of CAD*I included entity definitions in a formal description language, as well as a physical file specification, and a data base representation of all entities. The CAD*I members were the first to develope prototypes of data exchange processors using the CAD*I exchange format. With this they brought European development in that area to the top of the world.

With the appearing of an international standard for the exchange of product data (ISO 10303, called STEP), a purely European solution had little chance of becoming worldwide accepted. Therefore the project CADEX pursued three important goals:

- Ensuring the European influence on the new STEP standard

- Being early applicators of the new standard

- Offering processors for the new standard and a sufficient number of European CAD systems.

In this report the results of the project CADEX are described. They can be divided in four categories:

1. Data specification, Application Protocols, ISO relations

2. Software for reading, writing, handling, converting and checking data (Common Tool Kit)

3. Processors dealing with data (category 1) and using the tools (category 2) for CAD data exchange between dissimilar systems

4. Test and validation of data transfer

In principle these categories reflect the main workpackages of the CADEX project. In this report a main chapter was assigned to each of them. In this way the chapters show the CADEX approach to solve the problems of data exchange:

Chapter 2 deals with the data modelling results that came out of the project. Beginning with a table comparing possible data types that could be used for exchange, the mapping to STEP is described. The applications using these data types are described. From these result the Application Protocols of which CADEX has produced five:

- for manifold solid boundary representation models -> ISO 10303 Part 204

- for surface models -> ISO 10303 Part 205

- for wireframe models -> ISO 10303 Part 206

- for compound boundary representation models

- for constructive solid geometry models

Those marked with "ISO 10303" have been forwarded to ISO and are to become part of the emerging international standard ISO 10303 (STEP).

The implementation of these Application Protocols is regarded another main task of CADEX. Enabling a CAD system to do data exchange according to one or more Application Protocols means to provide it with an appropriate interface processor. Since all interface processors deal with STEP data have have similar functionality, the CADEX members have developed a set of software that can be used in any interface processor. The development effort and the usage of this package could be shared, therefore it is called the Common Toolkit. Its functions and components are described in chapter 3.

The toolkit is used in the processors of each CAD vendor within CADEX. The processors communicate on one side with STEP files, on the other side with native data repositories. In between there is intermediate data storage, checks, conversions and control. In chapter 4 each individual CAD vendor describes the data exchange processor developed in the project.

An important task in the development of software is its test and validation. Chapter 5 describes the test methodology, selected test parts and the results of individual tests.

The full text of the Application Protocols is available from the authors or from the Project Management. Those forwarded to ISO are also available from there.

The ongoing changes in the STEP standard have required a continuing adaption of the processors and the Common Tools supporting them. This report reflects the situation in June 1992, at the end of the CADEX project. Thus the results described in this report are regarded as the starting point for all further work in the field of STEP data exchange.

2. Application Protocols and STEP relations

2.1 Entity tables and mapping to STEP

This section is an elaboration of a detailed table of STEP entities for the CAD and FEM systems involved in CADEX. Each partner has analysed his own system in light of the entities available in Part 42 of IPIM, and made a list of the expected mapping from native to IPIM and from IPIM to native formats.

The entities that are not covered are listed as generic forms. These entities will be found in other schema within IPIM in due course.

2.1.1 Objectives for the table of entities

Each CADEX partner has specified the intended scope of his data exchange referring to the individual entities available in IPIM Part 42: Shape Information Model. Part 42 includes the entities needed to exchange the shape of a product: the design model schema, the topology model schema and the geometry model schema. Based on this study, a table has been made to reflect the intended data exchange of entities to and from the individual CAD system.

In addition to the entities which have a similar representation in Part 42, there are entities and attributes that are in use in each of the present CAD and FEM systems and need to be exchanged, but are not covered by Part 42. These entities and attributes should be identified.

There are also entities in the various CAD and FEM systems that are not covered in IPIM. These entities should be identified. They will be considered in detail in a later phase of the project since the corresponding IPIM models are not stable yet. A list of the requirements for these entities is worked out.

The entity table may be used to derive processor implementation specifications, especially the definition of Application Protocols (APs).

The entity matrix is a means to identify the common subset for STEP data exchange for the involved CAD/FE systems.

2.1.2 The table of native entities mapped to STEP

All the CAD and FEM partners have specified the intended mapping to and from STEP entities for their systems for both STEP pre- and post-processors. The result is shown in the STEP entity matrix in Table 2.1.

The table includes only the entities in Part 42. The next section covers the other CAD model entities and attributes for STEP model data transfer.

A first version of this entity matrix was presented at the first CADEX review in February 1990, and it has been further developed the project. The following table shows the status at the end of the project:

```
-------------------------------------------------------------------------
STEP entity map  |DISEL|HP ME|ITALC|ISPRO| K3DP| SI | SNI |FEGS | GfS |DNVR |
                 |     |     |     |     |     |    |     |     |     |     |
to= Pre; fr=Post |to|fr|to|fr|to|fr|to|fr|to|fr|to|fr|to|fr|to|fr|to|fr|to|fr|
-------------------------------------------------------------------------
                 |     |     |     |     |     |    |     |     |     |     |
GEOMETRY
  Coord. System  |M /M |     |M /M |     |     |Dm/Dm|Dh/Dh| /Dl|Dh/Dh| /Dl|Dl/Dl|
  Point          |     |     |     |     |     |    |     |     |     |     |
  Cartesian Pt.  |M /M |Mh/Mh|Mh/Mh|M /M |Dh/Dh|Dh/Dh|Mh/Mh|Mh/Mh|Mh/Mh|Dm/Dh|
  Point on Curve |D /D |     |     |  /  |     |     | /Dh|Dh/Dh|     |Dh/Dh|Dm/Dm| -/Am|
  Point on Surfac|D /D |     |     |  /  |     |     | /Dh| /Dm|     |Dh/Dh| /Dl| /Am|
Vector
  Direction      |D /D |Mh|Mh|Dh/Dh|D /D |Dh/Dh|Dh/Dh|Dh/Dh|Dh/Dh|Dh/Dh|Dl/Dm|
  Vector with mag|D /D?|     |Mh/Mh|D /D |     |     |     |     |  /  |     |Dl|     |
Axis Placement
  Axis 1 Placemen|D /D |Dm/Dm|Dh/Dh|     |     |     |     | /Dm|D /D |Dh/Dh|Dh/Dh|Dl/Dm|
  Axis 2 Placemen|D /D |Dh/Dh|Dh/Dh|Dh/Dh|Dh/Dh|Dh/Dh|Dh/Dh|Dh/Dh|Dh/Dh|Dl/Dm|
  Transformation |M /M |Dh/Dh|Mh/Mh|     |Dl|     |Dh/Dh|     |Mh/Mh| /Dm|Dl/Dl|
Curve
  Line           |M /M |Dh/Dh|Dh/Dh|Dh/Dh|Dh/Dh| /Am|Mh/Mh|Dh/Dh|Dh/Dh| /Am|
Conic
  Circle         |M /M |Dh/Dh|Dh/Dh|D /D |Dm/Dm| /Am|Mh/Mh|Dh/Dh|Dh/Dh| /Am|
  Ellipse        |M /M |     |Dh/Dh|Dh/Dh|D /D |Dm/Dm| /Am|Mh/Dh|Dh/Dh|Dh/Dh| /Am|
  Hyperbola      |M /M |     |Dl|Dh/Dh|D /D |Dm/Dm| /Am| /Dh|Dh/Dh|Dh/Dh| /Am|
  Parabola       |M /M |     |Dl|Dh/Dh|D /D |Dm/Dm| /Am| /Dh|Dh/Dh|Dh/Dh| /Am|
Bounded Curve
  Polyline       |D /D |Dm/Dm|Dh/Dh|D /D |     |     | /Al|Ah/Ah|Mh/Mh|Mm/Mm| /Al|
  B-spline curve |M /M |Dm/Am|D /A |Dl/Dl|Dh/Dh|Dh/Ah|Dh/Dh|Mh/Mh|Ah/Ah|Dm/Ah|
  (*polynomial3D |M /M |Dm/Dm|Dh/Ah|     |Dh/Dh|Dh/Dh|     |Dh/Dh|     |Dm/Dh|
  (*Bezier   3D  |M /M |     |Mh/Mh|     |     |     |Dh|     |Ah/Ah|     |Dh|
  Trimmed Curve  |M /M |     |Dl/Dl|     |Dh/Dh| /Am|     |Dh/Dh|Dh/Dh|Dl/Dm|
  Composite curv |M /M |     |Dl/Dl|     |     | /Al|Dh/Al|     |Mh/Mh| /Al| /Al|
Curve on Surface
  Pcurve         |M /M |     |     |     |     | /Dh| /Ah|     |Dh/Dh| /Am| /Am|
  Surface Curve  |D /D |     |     | /Al|     |Dh/Dh|Dh/Ah|     |Dh/Dh| /Al| /Am|
  IntersectCurve | /M |     |     | /Al|     | /Dh| /Ah|Am/Am|Mh/Mh|Mh/Mh| /Am|
  CompCurveOnSur |D /M |     |     | /Al|     |Dh/Ah|Dh/Al|     |Dh/Dh| /Al| /Al|
Offset curve
  D2 offset curv | /M |     |Dl/Dl|     |     |     | /Am|     |Dh/Dh| /Dl| /Al|
  D3 offset curv |     |     |Dl/Dl|     |     | /Dm| /Am|     |Dh/Dh| /Dl| /Al|
Surface
Elementary Surf
  Plane          |M /M |Dh/Dh|Mh/Mh|D /D |Dm/Dh| /Am|Dh/Dh|Mh/Mh|Dh/Dh|Dl/Dm|
  CylindricalSur |D /D |Dh/Dh|Mh/Mh|D /D |Dm/Dh| /Am|Dh/Dh|Mh/Mh|Dh/Dh|Dl/Dm|
  Conical Surfac |D /D |Dh/Dh|Mh/Mh|D /D |Dm/Dh| /Am|Dh/Dh|Mh/Mh|Dh/Dh|Dl/Dm|
  Spherical Surf |M /M |Dh/Dh|Mh/Mh|D /D |Dm/Dh| /Am|Dh/Dh|Mh/Mh|Dh/Dh|Dl/Dm|
```

```
--------------------------------------------------------------------------
STEP entity map  |DISEL|HP ME|ITALC|ISPRO| K3DP| SI | SNI |FEGS | GfS |DNVR |
                 |     |     |     |     |     |    |     |     |     |     |
to= Pre; fr=Post |to|fr|to|fr|to|fr|to|fr|to|fr|to|fr|to|fr|to|fr|to|fr|to|fr|
--------------------------------------------------------------------------
    Toroidal Surfa|M /M |Dh/Dh|Mh/Mh|   |   |Dm/Dh|    /Am|Dh/Dh|Dh/Dh|Dh/Dh|   /Am|
    Swept Surface
    Surface of Rev|D /M |Dh/Dh|Mh/Mh|   |/Dl|Dh/Dh|    /Am|Dh/Dh|Dm/Dm|  |Al|   /Am|
    SurfaceLinExtr|D /M |Dh/Dh|Mh/Mh|   |/Dl|Dm/Dh|    /Am|Dh/Dh|Dm/Dm|  |Al|   /Am|
    Bounded Surface
    B-spline surfa|M /M |Dm/Am|D /A |Dl/Dl|Dh/Dh|Dh/Ah|D /D |Mh/Mh|  |Am|Dm/Ah|

    (*B-spl.SurRatio|  |   |   |Dh/Dh|   |Dh/Dh|    /Ah|D /D |Mh/Mh|   |   /Al|
    (*B-spl.SurPolyn|M /M |Dm/Dm|Dh/Ah|   |Dh/Dh|Dh/Dh|D /D |Dh/Dh|   |Dm/Dh|
    (*Bezier surf  |M /M |   |   |Mh/Mh|   |Dh/Dh|    /Dh|    /D |Ah/Ah|   |   /Dh|
    RectangTrimSur|    /D |   |   |Mh/Mh|   |Dh/Dh|    /Am|   |Dh/Dh|Am/Am|   /Am|
    CurveBoundSurf|M /M |   |   |Mh/Mh|   |Dh/Dh|Dh/Ah|   |Mh/Mh|  |Al|Al/Am|
    Offset surface|    /M |   |   |Dl/Dl|   |Dh/Dh|    /Am|   |Mh/Mh|  |Dl|   /Al|
    RectaComposSur|M /M |   |   |Dm/Am|   |Dl/Dl|    /Al|   |Mh/Mh|  |Al|   /Al|

TOPOLOGY MODEL
Vertex          |D /D |Mh/Mh|Ml|   |Mh/Mh|Dl/Dl|Dm/Dm|Dh/Dh|Mh/Mh|Mh|Mh|Mm/Mh|
Edge            |D /D |Dh/Dh|Ml|   |Dh/Dh|Dl/Dl|Dm/Dm|Dh/Dh|Mh/Mh|Mh|Mh|Mm/Mh|
Curve LogicStruc|D /D |   |   |   |Dh/Dh|Dl/Dl|Dm/Dm|Dh/Dh|Dh/Dh|Dm/Dm|Dm/Dh|
Path
Edge LogicStruct|D /D |Dh/Dh|   |   |Dh/Dh|Dl/Dl|Dm/Dm|Dh/Dh|Dh/Dh|Dm/Dm|Dm/Dh|
Loop
  Vertex Loop   |   |Mh/Mh|   |   |Dh/Dh|Dl/Dl|Dm/Dm|M /M |Mh/Mh|Dm/Dm| -/Dm|
  Edge loop     |D /D |Dh/Dh|Ml|   |Dh/Dh|Dl/Dl|Dm/Dm|Dh/Dh|Mh/Mh|Mh/Mh|Dm/Dm|
  Poly Loop     |M /M |   |   |   |Dl|   |   |   |   |Mh/Mh|  |Dh|  |Dl|
  SurfaceLogicStru|   |Dh/Dh|   |   |Dh/Dh|Dl/Dl|Dm/Dm|Dh/Dh|Dh/Dh|Dm/Dm|Dm/Dm|
  Loop Logical Str|   |Dh/Dh|   |   |Dh/Dh|Dl/Dl|Dm/Dm|Dh/Dh|Dh/Dh|Dm/Dm|Dm/Dm|
  Face          |D /D |Dh/Dh|Ml|   |Dh/Dh|Dl/Dl|Dm/Dm|Dh/Dh|Mh/Mh|Dh/Dh|Dm/Dh|
  Subface       |   |   |   |   |   |   |   |   |   |D1/D1|   /Dl|   |
Shell
  Vertex Shell  |   |   |   |   |   |   |   |   |   |Mh/Mh|   |   |
  Wire Shell    |   |   |   |   |   |   |   |    /Dl|   |Mh/Mh|Dh/Dh|Dl/Dl|
  Open Shell    |   |   |   |   |   |   |Dl|Dl|Dl/Dl|Dh|Dh|Ml/Ml|   |Dm|Dl/Dm|
  Closed Shell  |D /D |Dh/Dh|Ml|   |Dh/Dh|Dl|Dl|    /Al|Dh|Dh|Mh/Mh|Mh/Mh|Dl/Dm|
  Face LogicStruct|   |   |   |   |Dh/Dh|Dl|Dl|Dm/Dm|Dh|Dh|Dl/Dl|Dm/Dm|Dm/Dm|
  Shell LogicStruc|   |   |   |   |Dh/Dh|Dl|Dl|Dl/Dl|Dl/Dl|Dh/Dh|Dm/Dm|Dl/Dm|
  ConnectedEdgeSet|   |   |   |   |   |   |   |    /Dl|   |Dh/Dh|Dm/Dm|Dl/Dm|
  ConnectedFaceSet|   |   |   |   |   |Dl|Dl|Dh/Dh|   |Dh/Dh|Dm/Dm|Dm/Dh|

Shape Model
  Solid Model
    Man.Solid BREP|D /D |Dh/Dh|Mm|   |Dh/Dh|Al/Al|    /Al|Dh/Dh|Mh/Mh|Dm/Dm|Dl/Dm|
    Facet(man)BREP|D /D |   |/Dh|   |   |/Dl|Al/Al|    /Al|   |Dl/Dl|   |   |
    CSG Solid    |M /M |   |/Dl|Dm|Dm|/Dl|   |   |   |   |   |   |
    Boolean Expression
      Boolean Term
        Union      |M /M |   |/Dl|Mm/Mm|/Dl|   |   |   |   |   |   |
        Intersection|M /M |   |/Dl|Mm/Mm|/Dl|   |   |   |   |   |   |
        Difference |M /M |   |/Dl|Mm/Mm|/Dl|   |   |   |   |   |   |
      Solid Instance|M /M |   |   |Dm/Dm|/Dl|   |   |   |   |   |   |
    CSG Primitive
        Sphere     |M /M |   |/Dl|Mm/Mm|/Dl|   |   |   |   |   |   |
        RightCircCyli|M /M |   |/Dl|Mm/Mm|/Dl|   |   |   |   |   |   |
        RightCircCone|M /M |   |/Dl|Dm/Dm|/Dl|   |   |   |   |   |   |
        Torus      |M /M |   |/Dl|Mm/Mm|/Al|   |   |   |   |   |   |
      Primitive with Axes
        RightAngWedg|    /D |   |/Dl|Dm/Dm|/Dl|   |   |   |   |   |   |
        Block      |M /M |   |/Dl|Mm/Mm|/Dl|   |   |   |   |   |   |
```

```
---------------------------------------------------------------------------
STEP entity map |DISEL|HP ME|ITALC|ISPRO| K3DP| SI  | SNI |FEGS | GfS |DNVR |
                |     |     |     |     |     |     |     |     |     |     |
to= Pre; fr=Post |to|fr|to|fr|to|fr|to|fr|to|fr|to|fr|to|fr|to|fr|to|fr|to|fr|
---------------------------------------------------------------------------
    Swept Area Solid
      Solid of Rev|M /M |     | /D1|Dm/Am| /D1|     |     |     |     |     |
      SolidLinExtr|M /M |     | /D1|Dm/Am| /D1|     |     |     |     |     |
      Half Space  | /A  |     |Mm/Mm| /A1|     |     |     |     |     |     |
    Box Domain
    Surface Model
      ShBasedSurfMod|   |     |     |     | /D1|D1/D1|D1/D1|Dh/Dh|Mh/Mh|D1/Dm|   /Dm|
      FaceBasedSurMo|   |     |     |     | /D1|D1/D1|D1/D1|Dm/Dm|     |Mh/Mh|D1/Dm|Dm/Dh|
    Wireframe Model
      ShellBasedWirM|   |     |     |     | /  |D1/D1| /A1|     |Mh/Mh|D1/Dm|     |
      EdgeBasedWireM|D  |     |     |     | /D1|D1/D1| /D1|     |Mh/Mh|D1/Dm|     |

    Geometric Set
      Geometric2DSet|   |     |     |     |     |     |     | /Dm|     |     |D1/Dm|     |
      Geom.ProjecSet|   |     |     |     |     |     |     | /D1|     |     |     |     |
      Geom.3DCurvSet|M /M |   |Dm/Dm|D1/D1|Dh/Dh| /Dm|     |Mh/Mh|D1/Dm|D1/D1|
      Geom.3DSurfSet|M /M |Dm/ |Dh/Dh|D1/D1|Dh/Dh|Dh/Dh|   |Mh/Mh|D1/Dm|D1/Dm|
      ProjectiveView|    |     |     |     |     |     |     | /D1|     |     |     |     |

    Product Information|Dm/Dm|
    Shape_def_rep  |Dm/Dm|
    Shape_repres   |Dm/Dm|
    Product_def_shap|Dm/Dm|
    Represent_cont |Dm/Dm|
    Prod_definition|Dm/Dm|
    Prod_def_context|Dm/Dm|
    Product_version|Dm/Dm|
    Product        |Dm/Dm|
    Apl_interpr_model|Dm/Dm|
    Apl_protocol   |Dm/Dm|
    Year_of_ap     |Dm/Dm|
---------------------------------------------------------------------------
```

In this table of mapping from native to STEP entities letters appear with the following meaning:

- The first letter indicates the mapping:

        M: perfect mapping

        D: derived entity,

        A: mapping by approximation

        a blank means not specified

- The second letter indicates priority for implementation:

        h: high

        m: medium

        l: low

- Entities marked '(*' are not independent entities in STEP, as they are mapped to b_spline_curve and b_spline_surfaces; this information is provided for clarification of mapping of supertype.

- Special case for DNVR: voids and holes are not permitted in closed_shell entities.

2.1.3 Other CAD model entities which require a STEP model transfer

A CAD system's data model is composed of shape model descriptions and of model descriptions which give the model a clear semantic meaning in specific applications of the CAD system. This information is also subject for model exchange by means of the STEP files.

Here is a list of the major entities which need further investigation and specification and implementation during the CADEX project:

This additional model information consists of entities which belong to several categories: The part hierarchy (e.g. assemblies and parts), and the associated model information.   The associated model information covers associated properties, element names, attributes for model representation (colours, line types, shading) etc.

2.1.3.1 Assembly Model Information and External File References

Assembly model structure is required as a part of state-of-the-art CAD system. The assembly model structure can contain the following:
- 3D-assemblies, 3D-parts, geometric-sets, 2D-assemblies, 2D-parts.
- 3D-assemblies can contain 3D-assemblies, 3D-parts, geometric-sets.
- 3D-parts can contain shape_models.
- Shape_models can contain solid_models, surface_models, wireframe models,
- Geometric_ sets(2D and/or 3D).

Some of these assembly mechanism may be available in anyone of the CAD systems present in the project. No system will have all the mechanisms. Typically, a CAD system X supports 3D_Assemblies, 3D_Parts,3D_geometric_sets, 2D_assemblies, 2D_parts; its 3D_part supports a shape_model with a b_rep_solid and 3D_geometric_sets (It does not support surface_models nor wireframe_ models).

There are entities in the current STEP definition to represent those assembly requirements. An assembly model as a core model (a STEP resource model) is not defined. Especially the integration or link to the shape model needs clarification. It is obviously a requirement to set up for STEP a self-contained assembly model in the IPIM which can be used in several applications (e.g. assemblies in design, production assembly simulations, robotics etc.)

2.1.3.2 Requirements on external references for STEP files

An exchange file will typically contain the model representation of an individually existing part or an assembly. In industrial applications there is a need for a mechanism to cope with complex model structures where dependent structures are stored in individual physical files. It is therefore necessary to refer files from within other files. For example external submodels (e.g a solid model) may have elements 'owned' by an assembly.

The file owning the submodels refer the external parts via a logical or a physical name. The 'owned' object in a STEP file indicates that it is referred externally. This mechanism is considered in the STEP physical file specifications.

2.1.3.3 Co-existence of different shape models within one STEP file

In some CAD systems it is necessary to support multiple shape models within one CAD model which is mapped to one neutral STEP file: a CSG model is co-existent with a facetted Brep (polyhedron) model (CATIA); e.g.a Brep model is co-existent with a facetted Brep model (ME30).

The current STEP file specifications seems not to support this concept; at least the administrative header_section does not indicate this.

CADEX is in contact with the persons responsible for the respective STEP specifications at ISO. We expect that a satisfying solution for industrial use of STEP will be available soon.

2.1.3.4 Associated model information

The Application Protocols for CADEX (AP_BREP, AP_SS, AP_CSG, AP_WF) have initially a restricted scope with respect to geometry and topology. This is a reasonable and necessary starting point, but not sufficient for a state-of-the-art model transfer, for for example BREP solid models. Typically in a CAD model there are associated data attached to individual entities which represent integral parts of the model and are subject to a complete model transfer. Not transferring this model information would mean to violate the product definition.

In order to re-build the complete transferred CAD model in the receiving system all the associated information has to be present in the STEP file. A list of requirements for associated model data is given below.

One could consider a further category of submodels: form features and tolerances. These areas are intentionally left out of this detailed study because not many CAD systems today do support those. They are nevertheless important in the near future since those model categories carry real application related information. It's intended to have some 'hook' mechanisms defined for CADEX pre- and post-processors in order to be able to integrate those future requirements later.

The model draughting aspects are not considered here; nevertheless a concept for integrating them into the shape definition is required.

2.1.3.5 Model precision information

All CAD systems have a precision of the model data that can be specified, often expressed in a number of significant digits based on the model units. This number of significant digit is typically mapped to a model space (e.g. a model sphere) which results in the model precision. This information should be associated with assembly model structures and with model parts.

2.1.3.6 Model units

A CAD model is based on a clearly defined length and angle unit : e.g. length unit: mm; angle unit: degree; The entity which can 'own' the unit is the part and/or the assembly model structure.

2.1.3.7 Product information block

This comprises administrative information for data exchange. For example important information on a data model:
- date of creation, creator, department, organisation,
- version, release status,
   The product information block should be specified for all objects on the assembly model structure.

2.1.3.8 Construction geometry

This is auxiliary geometrical model information (points,curves,surfaces) which belongs to a designed part,or to an assembly procedure, etc. It is associated to a part (e.g. a B_rep model). This geometry category could be mapped to geometric_sets of the shape model.

2.1.3.9 General data association mechanism

For certain applications there is a need to indicate that model entities are to be interpreted in a certain application dependent way. Individual model entities get attached application relevant information. Often a concept like association of text strings to model entities is applied, e.g.'labeling'(association of text and its visualisation).

2.1.4 A general association capability

A general association capability should be allowed to attach any kind of data to any entity. (e.g. INTEGERs, REALs, TEXT strings or data structures). The associated information should be directly accessible by the application through the owning entity (see special consideration below).
   A minimum requirement is to be able to associate INTEGERs, REALs, TEXTs to any element.
   Note: This general mechanism can be used to derive special 'agreed on' association entities to be applied in specific application areas. This might be a by-

pass to the current STEP definition but there is the urgent need to define an efficient solution for the outstanding issues. The following model information categories are candidates for model data associations.

Entity names. In many CAD systems all the individual entities carry names which are subject to data transfer. This is necessary for efficient use of STEP for file transfers as well. (It could be performed by means of associativity entities).

Visualisation attributes. There's a variety of visualisation requirements for product model data. Some basic ones which are required in CADEX are:

- for curves: colour(e.g.RGB),line type, line width

- for surfaces: colour, reflectance, transparency;

- on boundaries or intersections: colour & line types

Level information. Most CAD systems have a model which allows to categorise the individual entities of the model to a logical level. Sometimes the logical level is mapped to a numeric or an alphanumeric notation of the level.

Material. Material and density attributes are essential information in mechanical engineering and applications based on it. (e.g. Material properties: name or key; isotropic; composites; density; temperature expansion coefficient, etc).

Material attributes can be associated with physically existent parts; hence i.e. in a B_rep solid model a body may carry this associated material attribute. A minimum requirement is an associativity entity which indicates the material and its parameters

Physical properties. For models where the full 3D representation is replaced by an alternative model with reduced dimensionality, the use of physical properties will describe the missing information. for a shell, the thickness will be an attribute, for wire frame models, the cross sectional properties of the curve will represent the real 3D object. Among the properties are eccentricities, point masses, etc.

Analysis modelling specific properties. Analysis modelling with FEM will require a description of the environment the object is placed in. The description of the environment includes:

- boundary conditions (free, fixed, prescribed)

- initial conditions (displacement, velocity)

- loads (loadcase, volume load, temperature)

A minimum requirement is an associativity entity which indicates the physical properties.

Features. Many features can be characterised, at least from a shape point of view, as groups (sets) of geometry and topology with attached attribute information (which indicate feature functions) and names. This information could be treated as associated model information.

A minimum requirement is to be able to associate the entities which represent a feature in one associativity entity and assign it a name.

Tolerancing. Tolerances have various aspects; there are form-tolerances and positional tolerances. Examples are tolerances for manufacturing, distance, minimum and maximum length or diameter value, tolerance for an assembly, minimum and maximum distance for x-axis parallel fitting or concentricity, minimum and maximum offset.

Tolerances also deal with relations between parts within assemblies. A requirement for tolerancing is an associativity entity which indicates the type of tolerance, the objects referenced, and relevant parameters.

Parametric models. Very often in CAD modelling the concept of variational geometry is applied: In those cases the geometry is not of fixed size or values, but instead parameterised with parameter variables. This concept is not addressed in STEP. Even so there is a clear need from a CAD modeller/user point of view. An additional requirement is to associate restriction rules. Parametric models are relevant for features as well.

This topic is not in the focus of CADEX but should be considered seriously in STEP since this belongs to the near future major CAD requirements.

2.1.5 Considerations on the attribute association of STEP entities

To be able to represent these attributes, an associative relationship must be established between the geometry description and the associated properties. This can be achieved in three ways:
- the entity definitions are extended to cater for additional information, each entity having a set of new attributes
- a new entity is defined having all relevant properties as attributes. A list of the entities using this property entity instance is included as an attribute
- a new entity is defined having all relevant properties as attributes. An additional entity relates a particular entity to a particular property instance.

These three alternatives are detailed in the following using the geometry, surface, curve and point entities as an example. The properties included are an identifier and properties related to the visualisation of the entity. Each of the entities will have particular properties as well as common properties. The property entity for visualisation is divided into a hierarchy of properties migrating the common properties as far as possible up the entity hierarchy.

Alternative A: Additional attributes added to the existing definitions. This would be a natural extension of the pure mathematical representations used now, and will put the entities into a context. All the additional attributes may be optional. In this case the property entities do not know which entities they are associated with.

In EXPRESS this looks like:

```
ENTITY geometry
SUPERTYPE OF( XOR(coordinate_system,
                  point,
                  vector,
                  axis_placement,
                  transformation,
                  curve,
                  surface));
        identifier    :      OPTIONAL STRING(#);
(* other properties that are general, ie used by all geometry
   entities *)
END_ENTITY;
ENTITY point
        SUPERTYPE OF (XOR(cartesian_point,
                          point_on_curve,
                          point_on_surface));
        SUBTYPE OF (geometry);
        presentation  :      visualisation_information_point;
END_ENTITY;
ENTITY curve
SUPERTYPE OF (XOR(line,
                  conic,
                  bounded_curve,
                  curve_on_surface,
                  offset_curve));
SUBTYPE OF (geometry);
        presentation  :      OPTIONAL
                             visualisation_information_curve;
WHERE
        arcwise_connected(curve);
        arc_length_extent(curve)>0;
END_ENTITY;
ENTITY surface
SUPERTYPE OF (XOR(elementary_surface,
                  swept_surface,
                  bounded_surface,
                  offset_surface);
SUBTYPE OF (geometry);
        presentation  :      OPTIONAL
                             visualisation_information_surface;
WHERE
        area_extent(surface) > 0;
        arcwise_connected(surface);
END_ENTITY;.
```

```
ENTITY visualisation_information
        SUPERTYPE OF (XOR(visualisation_information_point,
                      visualisation_information_curve,
                      visualisation_information_surface));
        level           :       INTEGER;
        colour_number   :       INTEGER;
END_ENTITY;
ENTITY visualisation_information_point
        SUBTYPE OF (visualisation_information);
        (* anything special for points *)
END_ENTITY;
ENTITY visualisation_information_curve
        SUBTYPE OF (visualisation_information);
        line_font_pattern       :       INTEGER;
        line_weight_number      :       INTEGER;
        (* anything special for curves *)
END_ENTITY;
ENTITY visualisation_information_surface
        SUBTYPE OF (visualisation_information);
        line_font_pattern       :       INTEGER;
        line_weight_number      :       INTEGER;
        reflectance             :       INTEGER;
        transparency            :       INTEGER;
        (* anything special for surfaces *)
END_ENTITY;
```

Alternative B: The property entities define both the properties and a list of the entities making use of the property instance. The entities having properties are unchanged. In this case the entities having properties do not know which properties they have, if any.

```
ENTITY entity_identifier
        identifier      :       STRING(#);
        entity          :       geometry;
END_ENTITY;
ENTITY visualisation_information
        SUPERTYPE OF (XOR(visualisation_information_point,
                      visualisation_information_curve,
                      visualisation_information_surface));
        level           :       INTEGER;
        colour_number   :       INTEGER;
END_ENTITY;
ENTITY visualisation_information_point
        SUBTYPE OF (visualisation_information);
        (* anything special for points *)
```

```
            list_of_points :        LIST [1:#] point;
END_ENTITY;
ENTITY visualisation_information_curve
        SUBTYPE OF (visualisation_information);
        line_font_pattern       :       INTEGER;
        line_weight_number      :       INTEGER;
        (* anything special for curves *)
        list_of_curves          :       LIST [1:#] curve;
END_ENTITY;
ENTITY visualisation_information_surface
        SUBTYPE OF (visualisation_information);
        line_font_pattern       :       INTEGER;
        line_weight_number      :       INTEGER;
        reflectance             :       INTEGER;
        transparency            :       INTEGER;
        (* anything special for surfaces *)
        list_of_surfaces        :       LIST [1:#] surface;
END_ENTITY;
```

Alternative C: The property and the geometry entities are associated through an associativity entity. Both the property entities and the geometry entities are unchanged. The definition used in this case do not know what associations are established in the model.

```
ENTITY entity_identifier
        identifier      :       STRING(#);
END_ENTITY;
ENTITY property_association
        SUPERTYPE OF (XOR(property_association_point,
                      property_association_curve,
                      property_association_surface));
END_ENTITY;
ENTITY property_association_point
        SUBTYPE OF (property_association);
        point_entity    :       point;
        identifier      :       entity_identifier;
        presentation    :       visualisation_information_point;
END_ENTITY;
ENTITY property_association_curve
        SUBTYPE OF (property_association);
        curve_entity    :       curve;
        identifier      :       entity_identifier;
        presentation    :       visualisation_information_curve;
END_ENTITY;
ENTITY property_association_surface
        SUBTYPE OF (property_association);
```

```
        surface_entity :        surface;
        identifier      :       entity_identifier;
        presentation    :       visualisation_information_surface;
END_ENTITY;
```

The A approach allows directly reference to attributes from an entity via pointers. It has the advantage that the attributes can directly be accessed from the referring entity.

The references from an entity could generally be allowed for all entities in the same way by means of a list of references. This is basically extending the number of attributes for a given entity as shown.

The B approach allows the properties to refer the entities using them. It has the advantage that the definition contains the list of geometry entities making use of them. IT is basically the opposite relationship from alternative A. To establish the direct link between the geometry entity and its properties a search is necessary.

The C approach allows any geometry to be associated to any property, none of them will know the relationship. The relationships can only be established through a search. The solution is general, as any entity can be associated with any other entity.

The disadvantages with this alternative is the need for an associativity entity for each of the associations. This is the extreme case for alternative B, as each property entity refers to only one entity. Alternative A needs the fewest number of entities and would allow direct attribute associations from an IPIM entity. This alternative would require changes to the existing entity definitions in IPIM and subsequently is not possible even though it would reflect the functionality of existing CAD systems. Alternative C is available in STEP.(examples are shown in the Representation Schema).

The current STEP definition only allows alternative C, which is more general but will require more processing time during pre- and post-processing. It will be of lower performance than both alternative A and B.

2.2 Work in the ISO and the concept of Application Protocols

2.2.1 Overview on CADEX contributions and activities to ISO/STEP

When the project CADEX started in 1989 it was assumed that the STEP documentation and specification is almost complete and ready to use for a typical prototype implementation in a consortium like CADEX.

However, early in the project it was realised that the stabilisation of the STEP product model exchange definition will take still considerable time before it stabilizes. One reason for this was and still is that the ISO organisation is mainly driven by volunteer work with frequent changes in personel. It was observed that the STEP meeting participants are very often from Univeristities and research institutes with a lot of good new ideas, however this does not necessarily satisfy the need to quickly stabilize and release STEP in its first version. The lack of stability might also be caused by the fact that not many CAD/CAE system vendors did participate regularily in the STEP standardisation work in order to push for stabilization. The STEP definition scope is very broad and one can expect that the standardisation process takes very long time with this approach.

On the other hand the majority of users of the CAD/CAE systems are expecting very ambitious solutions by means of STEP.

This mismatch had to be addressed and therefore some CADEX representatives took the initiative to support the standardisation activity in STEP more than originally planned in the project.

CADEX partners took the approach to concentrate on some kernel product definitions which are very critical to be standardized and are of paramount importance to any product definition based on it.

The kernel product definition is considered in the geometrical and topological shape of the product. This is the basis on which other product definition data can be build on in a consistent and logical way.

The initiative was taken to define so called Application Protocols which are subsets of STEP for specific application areas. Within CADEX the focus was put on the mechanical design process and especially on the mechanical design data to be exchanged between dissimilar CAD systems via the standard STEP.

The basic design shape models in use within the CADEX project are a reflection of the different kind of CAD/CAE product models in use today in the market.

The design data model categories for Solid Modelling, Surface Modelling, and Wireframe Modelling were identified. Hence application protocols reflecting this data model categories were invented and defined within CADEX and also submitted to ISO.

Initially the Application Protocol for transfering Boundary Representation Solid Models (B-Rep Solids) was defined in its first version in 1989 which was the pathfinder for the other APs out of CADEX: The Surface Application Protocol, the Wireframe Application protocol and the Constructive Solid Geometry Application Protocol. An extension of the B-Rep Solids AP can be found in the Compound B-Rep Application Protocol which has its application in the field of Finite Element Analysis based on Solid B-Rep Geometry.

This buttom up approach of defining different shape oriented Application Protocols had to be structured in a more formal way. The concept of the Mechanical Design Application Protocol was invented. This Mechanical Design Application Protocol concept describes how the different shape oriented APs of

CADEX will fit together in one product data definition Application Protocol for Mechanical Design including non-shape descriptions like assemblies, annotations to the shape, tolerances, features etc. (See a more detailed description below).

The strategy to forward the CADEX developed APs to ISO was to submit the shape descriptions APs which have the broadest demand and support in the CADEX partner's customer base and which can be supported actively by attending the ISO meetings by CADEX partner representatives.

The following 3 APs are now official ISO STEP APs under definition for STEP:

a) AP204: Application Protocol for Mechanical Design Using Boundary Representation

b) AP205: Application Procotol for Mechanical Design Using Surface Representation

c) AP206: Application Procotol for Mechanical Design Using Wireframe Representation

These APs have been technically very stable during the last 18 months of CADEX, however the issue at ISO was and is that the ISO/STEP definition guidelines keep changing and a major effort had and has to be spend in keeping track on changes of guidelines and on stabilizing them.

Several company representatives from CADEX (CADDETC, FEGS, HP, PROCAD, SI) participated frequently on international ISO meetings to make sure that the technical contents of the submitted APs and the underlaying resource model definitions are stabilizing. A lot of detailed change requests were forwarded directly to ISO/STEP or indirectly to ISO via the national body representations like e.g. BSI, DIN, NVS. Up to April 1992 it looked very promising that the APs out of CADEX could be managed to be finalised for ISO Committe Draft by July 1992.

However, the latest change in April 1992 of the underlaying resource model Part 42 (Geometrical and Topological Representations) has caused a big dalay in finalisatioin of the APs 204/5/6 documentation. These drastic changes, caused by the overall STEP resource model integrations, force an update on the APs 204/5/6. The new Part 42 documentation release was planned to be ready by end June 1992. After this time the update of the APs 204/5/6 can be made. This is however beyond the project time frame of CADEX which ends in June 1992. It is planned that a follow-up ESPRIT project will continue some of this work.

2.2.2 The Concept of an Application Protocol

2.2.2.1 Introduction of STEP work

The section which follows provides the background within STEP which led to the concept of application protocols being formulated. The next section gives more detailed background to application protocol is taken from STEP Part 1, which gives a description of the contents of each clause.

Further documents are available which give more details; in particular, prototype application protocols are available within the ISO arena for review.

2.2.2.2 STEP background

At the ISO meeting held in Frankfurt am Main, West Germany, in June 1989, the results of the national ballot of STEP were presented. It had been submitted as a single document, although votes from each country were cast for each clause, It was clearly apparent that STEP needed to be divided into a series of parts, for two reasons:

- That clauses were at different stages of maturity.

- It was clearly inappropriate for CAD vendors to implement all of the standard and uncontrolled "subsets" of the whole were not desirable.

The concept of application protocols grew naturally from the solution to these two problems. Using work which had already been undertaken in the IGES testing project.

STEP was devided into Classes comprising a number of Parts. The description of the Clases is as follows:

- Introductory

  This contains only Part 1 of STEP and provides an introduction to the concepts of STEP and the structure of its Parts.

- Description methods

  Parts in this Class standardise the methods used when describing STEP entities. The data description language EXPRESS is the first example.

- Information models

  Parts in this Class describe the methods used in the developement of information models. Such parts include a model specification language (EXPRESS) and a framework for product data modelling. The framework defines an integrated product information model (IPIM) that specifies how information about products is modelled an application interpreted models which interpret the IPIM to provide required functionality for specific uses of product data.

- Implementation forms

  STEP provides a logically complete inforamtion model of a product which is capable of supporting multiple implementation methods. Physical file exchange is the first example; other forms will be added at a later date.

- Conformance testing methodologies

  These include the definition of the standard procedures and tools required to undertake conformance testing of products which claim to implement one or more STEP application protocol standards.

- Resource information models

  These models define the data content which provides the basis for the development of application protocols. They include models of almost universal applicability and those which support a particular application or class of application. The product data is encapsulated in an implemention independent form and is only implemented indirectly via an application protocol.

- Application Protocols

These are refined from the resource information models to provide a specific functionality. The logical data content of each is self-contained and complete. Application protocols state explicitly the information needs of a particular application, specify an unambiguous means by which information is to be exchanged for that application, and provides a basis for conformance testing.

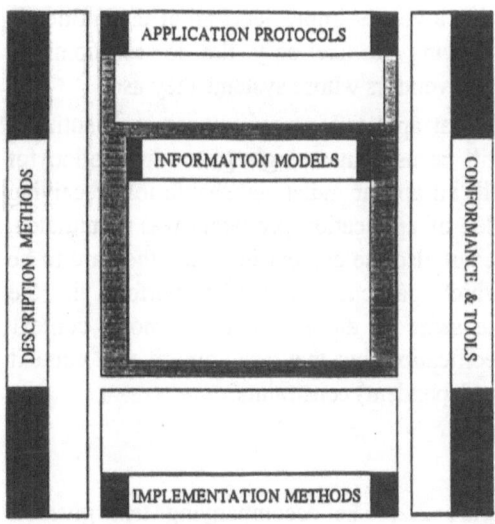

Fig. 2.1: Classification in STEP

2.2.2.3 Application Protocol background

All of the entities in the STEP resource information models (RIMs) are context independent; that is, they are general purpose and may be used in a wide variety of applications. For example, the polyline entity may convey two dimensional information in the context of technical publications, either as a line on a line drawing or as a small segment in a more complex curve or character. It could also be perceived as representing a three dimensional edge in a wire frame model, a pipe on a piping and instrumentation diagram, a connection in an electrical circuit, or an asis of symmetry important to machining applications, even though more appropriate entities may exist for such purposes.

Direct implementation of the STEP RIM constructs would produce a penalty: that CAD vendors would have a choice when implementing the specification in the context of the application which their own particular system supports. In practice, each vendor would implement those entities which are perceived to be relevant to his system, which would result in communication between different systems being hampered because each system would support a different set of entities.

Existing specifications use an application subset, which reduces this problem by presenting an enumerated list of entities which are to be used in a particular application. For example, MIL-D-28000 has provision for technical publications, engineering drawings, electrical and electonic and NC machining. By mandating that no entities other than those in the list shall appear in a file produced by a processor which claims conformance to the application subset, a common "subset" of entities may be used for exchange.

However, there is no requirement that a vendor implements all of the entities in the subset. This results in users having to use only the lowest common denominator of entities supported by the vendors whose systems they use.

Further, no guidance is giben on what application requirement each entity is intended to fulfil: a polyline may still be used in a single given application for many and varied purposes, which will all appear indistinguishable to a receiving post-processor. Consequently, the idea of application protocols was formulated, which give not only a list of entities, but also the context in which they are to be used and a mapping indicating which particular task they perform in the application. The entities are still the same as those (general purpose, context independent) entities in the main specification, but they are now given a context and probably some additional (context dependent) constraints.

2.2.2.4 The development of an AP

This process is encapsulated in figure 2.2. The accompanying text gives a commentary and also endeavours to provide examples to make the text more tangible.

Application protocols are based upon four main ideas and each AP has sections which reflect them:

* Scope and context possibly from an acitity model, described for example in IDEFO

* An application interpreted model which defines the requirements formally.

* An application interpreted model which uses STEP IPIM constructs in order to satisfy the requirements given in the ARM; the requirement and this information model are functionally equivalent and requires a one-to-one mapping to reflect this.

* Conformance requirements and test purposes for AP compliant processors are presented in a hierarchical from corresponding to the levels of detail added by stages one to three above.

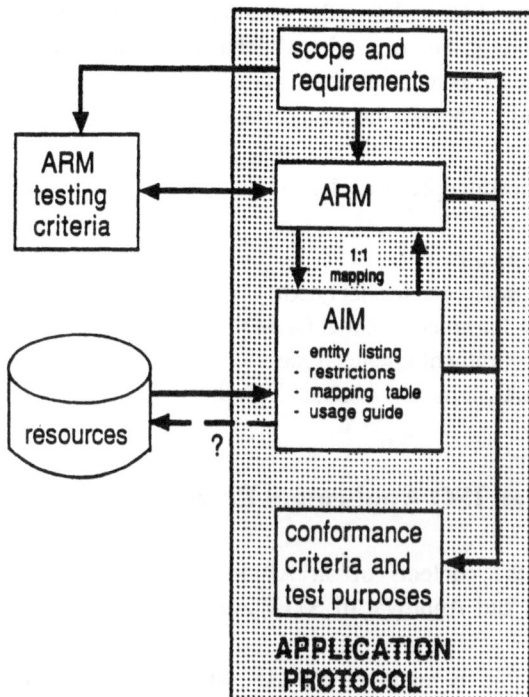

Fig. 2.2: AP development process

The development process is that of progressive detailing, moving from general requirements through a formal and specific presentation of detailed requirments, to a specific information model which satisfies those requirements.

The ARM undergoes a peer review to ensure that it is useful in the area which the AP addresses and that it is self-consistent. This is part of the process which integrates and produces the AP itself. Such a review will probably identify conformance requirements which are included in the AP. The review also ensures that the model satisfies the scope of the AP within the stated context.

The peer review is termed viability testing, and comprises both fitness testing and integrity testing. The fitness testing ensures that the ARM meets the scope and requirements laid down in the previous section; it is best achieved by taking instances of concepts which the AP is intended to address and ensuring that the example can indeed be encapsulated by the AP. Such example "test pieces" can be kept in an appendix to the AP but do not form an integral part of it. The integrity testing ensures that the model is self-consistent and self-contained.

The AIM then undergoes integrity testing and also a review to ensure that the ARM and AIM are transformationally equivalent.

The development of an application protocol is predicated on the fact that there are a number of general-purpose entities already available as "resources". The entities in the AIM are refinements of those drawn from the resources, with the AP adding context-dependent constraints. For example, a technical publications AP may constrain the polyline entity to have z-values of zero, making it two rather than three dimensional. It will also specify the single requirments in the ARM which such a polyline fulfils.

The development of an application protocol for STEP should follow these guidelines:

1. Definition of scope and requirements
2. Development of an Aplication Reference Model (ARM)
3. Development of an Application Interpreted Model (AIM)
4. Development of an Application Protocol Usage Guide
5. Developement of Conformance Criteria and Abstract Test Suite

2.2.3 Table of contents of an ISO Application Protocol

See a detailed description on the contents of an AP in the ISO document 'Guidelines for the Development and Approval of STEP Application Protocols , Version 1.0', Febr. 1992.

Scope: This clause defines the domain of the AP and summarizes the functionality and data that are accomodated by the AP. A description of the functionality and data that are specifically outside the scope of the application may be also defined to clarify the domain.

Normative references: All normative references are listed.

Definitions: This clause includes definitions of concepts necessary to understand the introduction, Scope and Information requirements clauses.

Information requirements: This describes the required functionality and information of the AP. References are given to the application activity model (AAM) into the Annex E and to the application reference model (ARM) in annex F.

Application Interpreted Model: This comprises the essential portion of the AP in computer readable form, in the EXPRESS language defined model of the AP. This model is composed of the

- Mapping Table, which documents the correspondence between the information requirements and the AIM in an unambigous form.

- AIM EXPRESS short form, which consists of USE FROM statements which select resource model entities or complete AICs (Application Interpreted Constructs, which are functional sets of application interpreted model entities which are shared between more than one AP).

Conformance requirements and Test Purposes: This section describes the conformance requirements, completeness requirements, and test purposes for conformance testing. A test purpose specifies an information requirement documented in the guidelines. Test purposes might be organized in a hierarchical structure of test groups.

Annexes:

- AIM Express long form:

  is normative and contains the complete list of the AIM. It replaces all USE FROM statements in the short form.

- AIM entity and type abbreviation:

  is normative and contains all applied abbreviations of entities and enumeration types.

- PICS (Protocol Implementation Conformance Statement) (normative):

  defines the flexibility of implementation of the AP. It is in form of a questionaire to be completed by the implementer before undergoing the conformance test.

- Implementation specific requirements (normative):

  may contain additional requirements for the implementation, e.g. required interpretation of values not specified in the normative clauses.

- Application Activity Model (informative):

  describes the application domain and the activities which use the product data described in the application context. The AAM shall be represented in the form of an IDEF0 diagrams.

- Application Reference Model (informative):

  describes the information requirements and constraints of the application context as defined in the scope and requirements of the AAM. It specifies so called Units of Functionality and ARM diagrams. The ARM uses application specific terminology and rules familiar to an expert from the application context. The model independent of any physical implementation. Formal data description languages like IDEF1X, NIAM, are used. The ARM together with the AAM and the scope sufficiently describe the AP domain to a person familiar with the application.

- AIM EXPRESS-G (informative):

  is a graphical representation of the data model modelled in the AP. It's purpose is for easy communication of model components and structures mainly used between human beeings.

- Application protocol usage guide (informative):

  explains AP usage rules, e.g. cardinality constraints over an AP compliant data set.

- Technical discussions (informative):

  contains resolutions of issues and valuable background informations for potential users of APs.

- Bibliography (informative):

  lists relevant references

- Resource entity definitional references:

lists thoses entities referenced within the AIM which are constrained not to occur in any instantiation of the AIM.

2.2.4 The Mechanical Design Application Protocol concept

The individual shape oriented Application Protocols of CADEX, like the B-Rep Solid Model AP, the Surface Model AP, the Wirefram Model AP, the CSG Solid Model AP and the Compound B-Rep Model AP had to be integrated conceptually in order to match the requirements of Product Model exchange which typically will go beyond pure geometrical/topological shape descriptions in the future.

The formal approch to this was the invention of the Mechanical Design Application Protocol (MD AP) which was initiated mainly by the CADEX partners BMW, Hewlett-Packard and PROCAD together in CADEX and in a DIN working group in Germany.

The purpose of this AP concept is to structure the underlaying product model data categories Solid models, surface models and wireframe models in a consistent and complementary way in one umbrella application protocol:

The elementary geometric shape building blocks of the MD AP are the sub-APs like B-Rep Solids, CSG-Solids, Surface Models, Wireframe Models and the Compound B-Rep Models.

On top of these elementary geometrical shape building blocks the other product data categories are build. These 'on top of' defined product model data categories are described in more detail below.

A Mechanical Design product model data set typically can be composed of the following data categories:

The categories are:

Product Information Description: This section covers basic administrative data for managing the product through the design process and to other departments in a manufacturing/engineering company. Information covered here is for example:

- identification of the product (name),
- ownership (company, department, designer,manager)
- history (date of design, updates, release status, revisions)

Assemblies: Assemblies are composed of individual parts or of other subassemblies. An assembly is typically represented as a tree of parts and assemblies. The relation between assemblies and subassemblies or parts contain positional and orientational information in 3D space.

Part description: This contains an aggregation of administrative data for a specific part. The administrative description has a at least a reference to the geometric shape description.

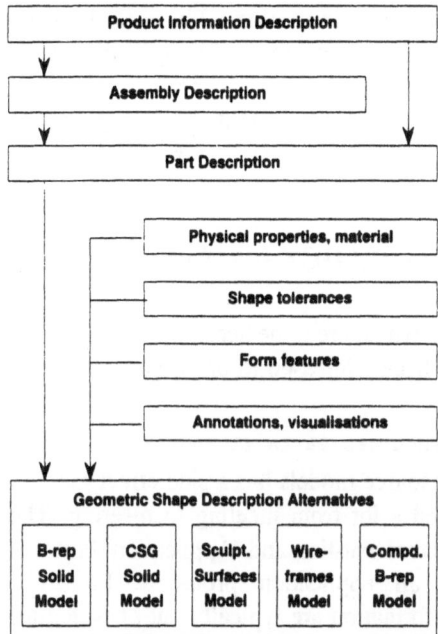

Fig. 2.3: Categories in the product model

Physical property description: There is a variety of physical properties attachable to a design part. One essential property is the material property. It's assumed that each model part has one material property. There might be other physical properties (e.g. surface roughness).

Shape tolerances: An individual design part has typically a shape tolerance which is relevant for manufacturing and for its functionality.

Form Features: Form Features here are typically design based form features which are composed of e.g. sets of faces representing a specific technical functionality like e.g. a slot for a fitting.

Annotations and Visualisations: This category of model data is mainly for visualisation purposes of shape dimensions or for clarification reasons for humans. Also colors, shading, reflections, transparency belong to this.

Geometric shape description (Geometry and topology): The shape description of mechanical design models can have different levels of completeness and constraints. Also different modeling techniques might be applied, especially in a heterogeneous design world where different supplying departments and tools are used. Geometric description alternatives are:

  - Volume based design, (e.g. B-Rep or CSG)
  - Surface based design,
  - Wireframe based design with curves

The different geometric descriptions typically do not exist all simultaneously for one product rather they are set up in the design in one of the above mentioned forms.

The product data categories Product Administration, Physical Property (material), Features, Tolerances and Annotations are not finally defined in detail yet in STEP. These categories are conceptual placeholders to be integrated later into the Mechanical Design AP when the STEP resource models allow this to do. Most of these data have references to the underlaying shape describing data.

2.2.5 List of open STEP issues from CADEX point of view

The following are some examples of issues which are to be handled still in a STEP implementation. Some of the issues have been addressed to be solved, and some are still to be addressed.

Issue 1:   Stability of Integrated Resources Parts 41 and 43

The integration of the different STEP resource models has a side effect that the integration part P41 and P43 gets changed a lot from meeting to meeting. The Parts P41 and 43 contain product structure entities and Application Protocol information which has an impact to all application protocols. Prototype implementations which have to use this information are typically out of date after another STEP meeting.

Issue 2:   The product structure definition in Part 43 lacks semantics

The various entities within P43 have a very complex relationship to each other, the underlaying semantics of the individual entities is not clear. The entities very often have a text string as a contents and the meaning of this text string is different to different people in different companies. The effect is these entities don't carry reasonable information accross when a STEP file is exchanged.

Issue 3:   Geometrical definition of Part 42 is not complete

There are special design shape definition requirements which cannot be represented yet in STEP. Examples are: an arbitrary extrusion of profile cannot be represented in STEP by means of the given 2 linear and rotational extrusion definitions. A design having this general requirement has to map the shape to another representation, e.g. to a B-Spline surface which doesn't allow the receiving system to re-capture the original design shape defined as a general swept surface.

Issue 4:   Accuracy of data models is hardly to convey and specify
           with the current STEP definition of the physical file
           in Part 21

Currently there is no data item which allows to specify the accuracy of the data conveyed in a STEP file. Even though this was submitted and requested to be integrated to the STEP physical file it was rejected by the working group. The implementations of CADEX did show that this kind of specification would really help in the practical data exchange.

Issue 5:   The current STEP definition needs more prototype
           implementations before it can be considered to be stable.

Some detailed conventions of the STEP definitions are not fixed yet, neither in the Resource Models nor in the existing Application Protocols.

As example we can take the transfer of a B-Rep Solid model containing a closed B-Spline surface as a geometrical surface. In this case it is not defined if at the closing edge topological data like vertices and an edge has to be transfered or not.

Issue 6:   Stability of STEP guidelines (e.g Application protocol
            guidelines)

Throughout the project runtime of CADEX it has been an issue that the STEP guidelines for Application Protocols have been changing. Also the resource models which have been considered to be very stable like e.g. the Part 42 geometry and topology, have been shaked through the very late integration process of the resource parts. This late and to the outside observer confusing scheduling of actions in ISO is very confusing and leads to the effect that STEP will be late and risks to lose credibility.

These kind of issues have to be addressed in further actions at ISO meetings otherwise STEP will not become accepted in industry.

2.2.6 Recommendations to STEP

The CADEX project has been represented in ISO TC184/SC4/WG2-9 through several participants, who have (during the last period) attended the following ISO meetings on behalf of the CADEX project:
- ISO TC184/SC4/WG2-9 meeting in San Diego, 6-12 April 1991
- ISO TC184/SC4/WG7 meeting in Abingdon, UK, 12-19 July 1991
- ISO TC184/SC4/WG2-9 meeting in Houston, 18-27 October 1991
- ISO TC184/SC4/WG2-9 meeting in Oslo, 2-8 Feb 1992
- ISO TC184/SC4/WG2-9 meeting in Seattle, 12-18 April 1992

2.2.6.1 The work undertaken by FEGS as ISO representative

The meeting in San Diego was spent looking at the concepts of Application Protocols, and to contribute to the definition of the Guidelines for the Development and Qualification of STEP Application Protocols. The ad hoc approach that dominates the first few years of STEP development is now being replaced by a more mature and quality aware approach, where quality assurance procedures are used to build quality into the solutions. The work on the AP Guidelines continued throughout the year and FEGS participated in the AP methods meetings to know what direction the document was given as well as feeding experience from making APs based on the document back to the committee.

The meeting in WG7 STEP Implementation Methods, concentrated on the review of comments to the Committee Draft of Part 21: 'Clear Text Encoding of the Exchange Structure' and the specification of the Standard Data Access Interface, SDAI. Both very relevant to the CADEX project: the definition of STEP exchange file format and the access routines to the Intermediate Data Structure.

The comments to Part 21 suggested editorial changes to the document, structural changes to the chosen approach or improvements which previously had been agreed for the next version of STEP. A resolution from the WG meeting to the workshops to define EXPRESS version 2.0 was put together to emphasise the need for changes to the EXPRESS language, ISO 10303 Part 11: 'Description Methods: The EXPRESS Language Reference Manual'.

Experience from the CADEX project was presented as contributions to the SDAI discussions and influenced the resulting conclusions to a great extent:

- the use of several data repositories addressed through SDAI in the same session was supported from the CADEX project
- the need for incomplete models in transition between stages of conforming product models was presented as a CADEX requirement and subsequently included in the specification of SDAI

The ISO meeting in Houston approved the project of progressing the wireframe Application Protocol from CADEX as ISO 10303 Part 206: Mechanical Design Using Wire Frame Representation. The content will be aligned with the other two APs from the CADEX project, so they together form a class of APs for geometry description of product models. A new version of the AP will be created for the last ISO meeting held before the CADEX project ends, April 1992.

The meeting in Oslo was concentrated on AP development and the inclusion of Application Interpreted Constructs, AICs, in the documents. Further feedback on the content of the AP Guidelines were given. By this time is was clear that the APs from CADEX would have no chance to reach Draft International Standard, DIS, status unless the CADEX project provided the resources necessary for the Qualification Project.

The preparation of an ISO document consists of a number of complementary activities. When the technical content is completed, the working group will release the document as a Committee Draft, CD, for further development towards an international standard. Additional work is required to satisfy formal criteria, as well as being compatible and consistent with the other parts in STEP , the documents must have the correct content and a pre-defined lay-out. A part of this activity is the Qualification Project, where the document is evaluated against a number of metrics for content to ensure a consistent high standard. This process is time consuming and requires people with the right back ground.

Every document, APs included, must make available to the Qualification Project a named person to contribute to the qualification process, not only the document in question, but also the other documents worked on at the time. Any AP project which does not provide such resources will be stone-walled until resources are made available.

To get the APs from CADEX out of this deadlock, Jon Aas volunteered to be the qualification resource for the CADEX APs until the end of the project. This enabled the Qualification Team to include the CADEX APs in their busy schedule, both in the Oslo meeting and in the Seattle meeting.

Time has been spent in the CADEX project to Qualify parts at ISO meetings, both in Oslo, in Seattle and at home. Jon Aas has taken part in the pre-qualification activity of the integrated version of Part 42, and will continue to participate also in the ISO meeting in London.

The development of an AP for wire frame models has been completed to a point where the wire frame AP document has reached the same level of maturity as the other two ISO APs, i.e. Part 204 and Part 205. The document is written in LATEX, and has all the sections defined for an AP according to the AP Guidelines. The document is based on Part 42 before the re-integration however, and substantial re-work has to be done as soon as the underlying Integrated Generic Resources are stable.

2.2.6.2 US/EC collaboration on AP integration

In addition, there has been established a collaboration between the CADEX project and PDES Inc. to progress the concepts of application protocol integration. The first meeting was held at CADDETC in Leeds, 10-13 December 1991. Several new meetings are planned for the remaining part of CADEX. A continuation beyond the CADEX project depends on the availability of funds to do the work, both to cover travel expenses, and to cover man time required to make progress.

The idea is to co-ordinate the description of product model data in the different application protocols by re-using the same constructs. A construct is typically a collection of entities that form a larger unit, for example the surfaces used to describe a sculptured surface geometry, the topology used in the topology hierarchy up to face, etc.

The application protocols from the CADEX project are aimed at describing different kinds of geometry models, adding very little of application specific information, see CADEX report D2. These descriptions may be constructs in their own right that other application protocols may use to define geometry shape, for example manifold solid boundary representation models, non-topological surface models, edge-based wire frame models, etc. The candidate application protocols for sharing data are:

- ISO 10303 Part 203: Configuration Controlled Design
- ISO 10303 Part 204: Mechanical Design Using Boundary Representations
- ISO 10303 Part 205: Mechanical Design Using Surface Representations
- ISO 10303 Part 206: Mechanical Design Using Wire Frame Representations

Commonality between these and the draughting APs, e.g. Part 201 and Part 202, may be identified subsequently. Typically, Part 203 will be using constructs from Part 204, Part 205 and Part 206 as far as geometry descriptions are concerned. Part

204 and Part 205 will share constructs used to define geometry and topology as far as they are common between the models. This will enable the conversion between the different formats without to much hassle, and to send product model data between systems of dissimilar type, e.g. from a solid modeller to a surface modeller.

A large number of application protocols will be developed in the near future, using the concepts defined within the ISO TC184/SC4/WG2-9 on the STEP standard. A structure for sharing product model data will significantly simplify the implementation of STEP for data exchange. The concept of sharable constructs will contribute to the concept of processor interoperability.

The collaboration between CADEX and PDES Ic is a follow up of the EC/US Workshops on Manufacturing Technologies that have been held in 1991, including one in Berlin 29-31 July 1991.

2.2.7 Other ISO work

In addition to the direct contribution to STEP by the development of the Application Protocol Parts 204, 205 and, more recently 206 the CADEX project has had an indirect affect upon other work within the TC184/SC4 Working Groups.

During the development and review of Part 42 (Geometric and Topological Representations) the practical implementation experience of the members of the CADEX project enabled them to provide many constructive criticisms which directly influenced the development of this Part. More recently the work of editing Part 42 in response to the CD ballot and of modifying it to incorporate the resolutions of the Leeds integration workshop would have been impossible without the direct support of the CADEX project.

In the development of APs, and in particular in the integration of APs the concept of Application Interpreted Constructs (AICs) is playing an increasingly important role. This concept was first introduced at the workshop which addressed the integration requirements for Parts 204 and 205. Most of the subsequent work in developing prototype AICs and in defining the requirements for an AIC library and how they should be documented has been done withgin the CADEX project. The CADEX project is also playing a leading role in the development of test purposes for APs.

2.3 ISO Application Protocols

2.3.1 Application Protocol for mechanical design using boundary representation

2.3.1.1 Introduction

This application protocol is one of a group of APs developed within the CADEX project. As such it was designed from its first conception to inter-operate with the other CADEX APs. In parallel with its development within the project it has been developed as one of the APs selected for the first release of ISO 10303 (STEP) and will eventually be published as ISO 10303 Part 204. The technical content of this AP has been stable for some time but the document itself continues to evolve as the ISO procedures develop and the requirements for Application Protocol documentation are finalised.

Within the working groups of ISO TC184/SC4 the development of APs has been an evolutionary process and the CADEX APs, particularly Parts 204 and 205, have played a key role in this process. In particular the requirement for data sharing between APs has now been recognised within the ISO community and has introduced the concept of AICs (Application Interpreted Constructs) which are essentially shareable data models between a number of APs. The latest ISO version of this AP (ISO TC184/SC4/WG3/P1 N124) is documented using several AICs. As a result of this development it is likely that in the very near future the support within Part 203 (AP for configuration contrlled 3D design) will be fully compatible with Part 204.

This Application Protocol has particularly close relationships to two other parts of STEP. All the geometric and topological entities required are provided by Part 42. Part 21 provides the detailed specification for the physical file which is essential for the implementation of this AP.

2.3.1.2 Scope

This AP contains the definition of conforming Boundary Representation (BRep) models and the mechanisms to transfer them via a physical file. The application reference environment where these BRep models are used is the generation and exchange of volume based data in the Computer Aided Mechanical engineering design process.In this Part the BRep's are characterised by the fact, that they can represent:

- Models with only planar surfaces (so called polyhedrons),
- Models with only analytical surfaces (elementary Brep)
- Models including sculptured surfaces and curves.

These different categories of BRep models recognise the reality that not all systems provide the same level of support for BRep models and enable the implementation and conformance requirements to be more precisely specified. Further details of these 3 levels, based purely on geometric complexity, and their potential applications are given in the following paragraphs.

2.3.1.3 The Brep AP levels

BREP level 1:

For models with purely planar surfaces, referred to as Polyhedron models.

The basic level of geometric complexity is given by shapes of parts which have only planar surfaces as the bounding surfaces. Part examples are: Boxes, and facetted simplified/approximated models of more complex shapes.

At level 1 much of the topological information is implicit. Edge and vertex information is not given and the shells consist of faces bounded exclusively by polygons. The face geometry is implicitly planar. The complete part shape is represented by the polyhedron.

Applications of these parts can be seen
- in rapid prototyping manufacturing (e.g.stereolithography),
- for visualization purposes
- for collision checks of parts
- for kinematic studies
- for robot programming and simulations

* Level 1 models can either be an exact model of a simple part or a simplified model of a more complex part which is suitable for a selected range of applications (e.g. stereolithography, or Finite Element Analysis).

* Level 1 models can be represented in a more compact form than models from level 2 or level 3: edges and curves are not required to be explicitly defined since those are always straight lines. The connecting points are sufficient.

* Explicit topology is not included at Level 1.

BREP level 2:

For models with elementary surfaces; for parts whose geometric shape can be represented by means of analytic (elementary) surfaces (planes, cones, cylinders, spheres, tori).

Part examples:
- bolts(ignoring the thread)
- screws(ignoring the thread)
- piston of a simple piston-engine
- motor housings

* All topological information is included at Level 2, all geometry is unbounded and the bounding information is given by the topology data.

* For the same part there can exist different model representations, which correspond to different application requirements.

* For a part there might exist a representation of level 1 and of level 2.

* Level 2 is a proper subset of Level 3.

BREP level 3:

For models with advanced surface (or curve) descriptions.

This level will be used for modelling of parts whose geometric shape is representable with elementary and/or sculptured surfaces and/or swept surfaces with linear or rotational extrusions. The generator curves for the extrusion can be analytic and/or free-form curves. The sculptured surfaces or curves will be B-Spline based. Surfaces of revolution and linear extrusion are also included.

Application examples:
- any parts which require 3 to 5 axis NC machining for their manufacturing
- dies for moulding or forming
- ergonomically designed consumer products
- car surface parts like fenders

* Level 3 is a superset of level 2.

* Level 3 surfaces are surfaces which can have any shape.

Other information is included at all 3 Levels of implementation. The Application protocol supports the definition of assemblies, the attachment of presentation attributes and name assignments.

2.3.1.4 Assembly structures

Products in the real world of design and manufacturing are composed of individual parts and of collections of parts which form so called assemblies. Assemblies may consist of sub-assemblies and of individual parts. Individual parts are represented by specific geometrical shape descriptions. Assemblies have specific geometric relationships between each other and to individual parts.

These relationships are given by the following properties:

A reference from an assembly to another assembly or to a part.

A geometrical relationship, typically described with a transformation matrix, which allows translation, rotation and, possibly, mirroring and scaling.

The assemblies and parts carry names which typically are unique within one product assembly.

2.3.1.5 Presentation properties and assigned names

Each of the Functional Levels comprises a small set of features for the visual presentation ofthe shape models. They enable the user to assign pre-defined colour, line-style (dashed, dotted etc.) and line-width to any geometric or topological entity. Annotation text can be positioned in 3D-space and be combined with leaders. A layer mechanism is provided. Layers may contain geometric and topological entities as well as complete geometric models. An entity may be part of several layers.

Each of the Functional Levels allows for the preservation of entity names as they have been defined by a user. The user-defined name of an item will be used as an alias in the model in addition to possible implementation dependent identifiers.

2.3.1.6 Applications

This AP provides the capability to produce and exchange a complete desciption of the shape and size of a part. This description is done essentially in terms of the boundaries of the part, both internal and external. As such it is able potentially to support any application which has a requirement for such a precise and complete description. The target application for this Application Protocol is Mechanical Design using the CAD modelling technique Boundary Representation Solid Modelling.

The given application area places fundamental requirements on model exchange and the neutral representation of models. A requirement of this application is the exchange of CAD models at different stages of the design and engineering processes. This results in data exchange requirements between design and engineering and manufacturing companies. The transfer and archiving of such models requires the following to be maintained:

- the completeness of mapping
- the correctness of semantics
- the accuracy of relationships between model entities.
- the structural model data of assemblies and parts and shapes
- the geometrical relationship between assemblies, parts, shapes (transformation matrices)
- the associated attributes to assemblies, parts, shapes (names of parts/assemblies)

It is intended that the archiving and exchange of this information will be achieved by using a STEP file and the development of this AP has paid particular attention to the mapping from the conceptual schema to the physical file and to prototype implementations.

Applications which are beyond the direct scope of this Application Protocol but which could potentially make use of the information contained here would include dimensioning, toleranbces, manufacturing and inspection, finite element meshing, NC tool path generation.

2.3.1.7 Implementation and testing

The only implementation form supported by this Application Protocol is the physical file, the basic structure for this is fully described in Part 21. The AP contains detailed specific information on how the entities represented in this AP are mapped to the physical file. The concepts have been thoroughly tested during the CADEX project with implementations based upon a number of different systems. An informative annex to the AP contains examples of files corresponding to each of the 3 implementation levels. Within the AP itself it is an ISO requirement that conformance requirements and test purpuses shall be included, the conformance testing of implementations will be against the AP requirements. Still to be developed is the companion (Part 1204)to Part 204 which will contain the corresponding abstract test suites for this conformance testing.

2.3.2 Application Protocol for mechanical design using surface representation

2.3.2.1 History in CADEX

Application Protocols - processor specifications for CADEX Right from the start CADEX had a need for specifications for STEP- processors. Both information contents and implementation requirements needed to be settled for processor development. At that time, in the middle of 1989, the Application Protocols (AP) were determined by TC184/SC4 (STEP) to be the implementable Parts of the standard (ISO 10303).

CADEX wanted to generate STEP conformant software. Most of its processor specifications were, therefore, launched as ISO Application Protocols. In the course of the project Application Protocols were produced covering the main CAD-modelling paradigms:
- Boundary Representation AP (Brep),
- Surface AP,
- Wireframe AP,
- Constructive Solids Geometry (CSG) AP,
- Compound Boundary Representation AP.

The first three of these are accepted ISO Application Protocols; this section deals with the second one, of which two versions were submitted.

Versions 1 and 2: In CADEX two versions of the Surface AP were used as the basis for implementation. These are hereafter called version 1 and version 2 which is not coincident with the real versioning of the document:

	date	official name
version 1	21.05.1990	SS-AP, version 1.1
version 2	01.05.1992	ISO TC184/SC4/WG3/P1 N132.

The Surface AP is in its first version restricted to sculptured geometry and, therefore, often called for SS-AP. The second version has been extended to also comprise elementary geometry as circle, cylindrical surface etc. .

Version 1 is based on the Paris-version of Part 42 (January 1990), version 2 on the Houston-version (October 1991) which is an update of the Part 42 CD-document.

ISO-representation: In June 1990 (Gothenburg-meeting) the CADEX Surface AP was adopted by ISO. Together with four other proposals (among them the CADEX Brep AP) it was considered both to cover major needs in todays industry and to be implementable by the initial release of STEP. The document got the Part- number 205, its official name being ISO 10303-205. Jochen Haenisch from SI (Senter for Industriforskning) in Norway was appointed project leader.

Since then the Surface AP was - for the life-time of CADEX - represented at all ISO/STEP-meetings. At these meetings the progress of the document was presented and discussed. Requirements arising from Part 205 and concerning STEP resource models as e.g. Part 42 were forwarded. The document was developed in accordance to the appropriate STEP-guidelines.

2.3.2.2 Summary of contents

The contents of the Surface AP is modelled in correspondance to the capabilities of the STEP resource model for shape, ISO 10303-42. This Part offers 3 alternative surface model representations. These have been adopted by Part 205 as Functional Levels (FL), each level being an alternative specification:

```
      Name in Part 205                     Main construct in Part 42
      ----------------------------------------------------------------
FL1   Geometrically Bounded Surface Model   geometric_3d_surface_set
FL2   Non-manifold Surface Model            face_based_surface_model
FL3   Manifold Surface Model                shell_based_surface_model
```

Processors for surface models shall satisfy one of the three Functional Levels. FL1 consists of geometrical entities only; the boundaries of shapes are represented by their natural geometrical boundaries. FL2 allows - to a limited extent - to model non-manifolds as e.g. surface models with more than two faces using the same edge as part of their boundaries. FL3 is the topologically most sophisticated surface model before entering into the domain of Brep solids. This basic structure of Functional Levels is maintained in both versions of the document.

Version 1: The initial implementation of the Surface AP covers sculptured geometry only. Elementary geometry can be modelled, but only in the representation of B-spline curves and surfaces. The main geometrical entities used by all three Functional Levels are cartesian_point, b_spline_curve and b_spline_surface. The boundary of FL1-models is the whole parameter space of their B-spline-entities. Of the three Functional Levels FL1 is the in CADEX mostly supported one; eight partners developed processors based on SS-AP FL1. Version 1 does not handle any other aspect of a product than its shape.

Version 2: Version 2 of the Surface AP maintains the principle ideas of all the Functional Levels. It is, however, essentially extended compared to version 1. Modelling capabilities have been added concerning geometry and other aspects of a product than shape (valid for all three FLs except other stated explicitly):

```
Geometry:          elementary curves and surfaces,
                   trimming capabilities in FL1;
Product structure: surface model assemblies,
                   shape embedded into the context of a product;
Presentation:      assignment of presentation attributes as line-
                   style and colour on an entity basis;
Units:             assignment of global units valid for a whole
                   model;
Names:             assignment of user-defined names on an entity
                   basis;
Groups:            general grouping mechanism on an entity basis.
```

For some of these enhancements the document only provides slots which shall be filled by contributions from other members of the ISO/STEP-community. These slots concern product structure, units and names.

The background for having slots and distributed AP-development is the requirement for interoperable Application Protocols. The AP information models have been modularised, the modules being called Application Interpreted Constructs (AIC) (see also chapter 2.5.2) . Some of the AICs of Part 205 are used by other Application Protocols. Thus, the respective APs can interoperate within the domain of these shared AICs.

Version 2 of the Surface AP requires conforming implementations to keep a fixed accuracy of numbers that describe geometry.

2.3.2.3 Differences between specification and implementation

Though version 2 of the Surface AP is the specification of version 6.1 of the Common Toolkit, the implementation does not satisfy all the requirements stated in Part 205. What is covered, are the AICs concerning shape, topology and geometry:

```
- geometrically bounded surface AIC,
- non-manifold surface AIC,
- manifold surace AIC,
   the latter two including:
    - topologically bounded surface AIC
      that includes:
       - topologically bounded elementary surface AIC.
```

Product Structure has been implemented conformant to annex 1 of this chapter. The corresponding AIC of Part 205 could - as mentioned - not be provided in time. The mapping given in Annex D of Part 205 is, however, correct and conformant to the model in annex 1.

Presentation, Units, Names and Groups have not been taken care of. The same is valid for the functions specified in Part 205. No mechanisms are provided for checking the required representation of numbers. The rest of Annex D of Part 205, i.e. rules for mapping of EXPRESS entity specifications to the physical file format, have been incorporated. Compared to Part 21, the STEP Part defining the physical file format, these rules add information for the optional use of the SCOPE-statement.

2.3.2.4 Outlook

Work on Part 205 will after the end of CADEX be taken further by other projects. The ultimate goal of these activities is the standardisation of a Surface Application Protocol that is a useful specification for industrial applications within surface modelling.

The current document will be reworked as its foundation, the STEP Parts 41, 42, 43 and 46 have changed considerably. It shall, however, still satisfy the same set of requirements. Conformance requirements, test purposes and EXPRESS-G diagrams are still missing as well as supporting documents (AP validation report, Issues log, etc.).

The layout of the document will be maintained as well as its structure concerning AICs. The collaborations with different ISO/STEP-communities for the development of AICs will be continued. PDES Inc. is a current partner; VDA and ProSTEP are relevant partners for the near future.

Implementation work will be done in several ESPRIT III projects as PRODEX, MARITIME and INTERROB.

It is considered important to keep close contact to groups that want to apply Part 205, either the whole specification or AICs, especially the surface model AICs. Only by real world applications the Surface AP can get a shape that makes it acceptable as a world-wide standard.

2.3.3 Application Protocol for mechanical design using wirefame representation

The wire frame AP is is designed to meet these overall product data requirements:
- to cover the product data used for the representation of wire frame models as required for the mecanical engineering activity as defined for the other APs developed in CADEX.
- to cover the wire frame subset of the geometry and topology used in surface and solid boundary representations models within the surface AP and the b-rep AP in a compatible way
- to cover the same product data as other existing file formats for wire frame transfer, but in a more structured way.
- to form the foundation to which other attributes related to specific areas of the manufacturing process within the life cycle may be attached.

The models are described geometrically and topologically using ISO 10303 Part 42 as a resource model. The types of models which may be transferred using this wire frame AP includes:
- models which represent the centre line column type structures
- models which represent the edge loops of surface models
- models which represent the edge loops of solid models
- models which represent the edges and edge loops of solid models
- models which represent 2D sections of solid models (2D geometry may be used if the null coordinate is set to zero).

This application protocol forms the basis for product data modelling of wire frame models in the following application areas of mechanical engineering:

- engineering design models

- engineering analysis models used in engineering simulation

- robotics

- graphical presentation in mechanical part design

Figure 2.4 shows the types of mechanical engineering application areas where these types of models may be found.

The transfer and storage of wire frame models requires the following to be maintained:

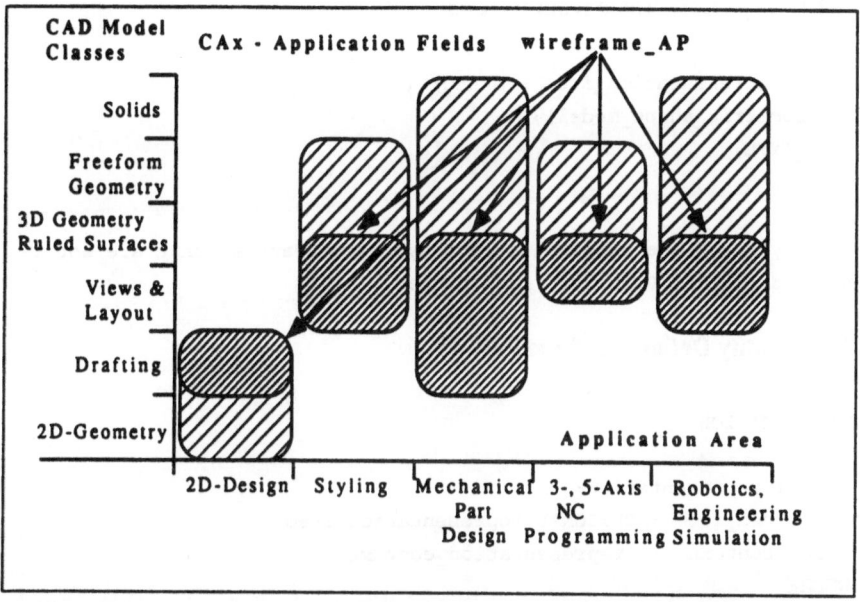

Fig. 2.4: The scope of the wireframe AP related to engineering activities

- the completeness of geometric and topological mapping

- the correctness of semantics

- the accuracy of relationships between model entities

The product data defined in this application protocol is intended to define the general product administration information and the basic shape description of a product. Other information such as materials, physical properties, presentation, etc is generally not completely transferrable across different areas of the mechanical engineering process. This information is classified as not mandatory. For the purose of this version only the generic foundation for assigning information to the model and one example test case, presentation, has been included. This application protocol will be the basis of other application protocols where specific attributes are addressed. It will also be a resource for application protocols which will use wire frame representation of shape as part of a larger product description.

The implementation in IDS is as specified in the ISO version of the document, apart from the presentation entities.

2.3.4 EXPRESS long form of the Product Structure module

```
(*
The following definitions are included in all three ISO Application
Protocols prepared by CADEX.
```

2.3.4.1 Type Definitions from Part 43

```
*)
  TYPE representation_item = SELECT (
    geometric_shape_model);
  END_TYPE;
(*
Remark
Those types of the original select type from Part 43 that are not
required in this Part have been pruned.
```

2.3.4.2 Entity Definitions from Part 43

```
REPRESENTATION
*)
  ENTITY representation;
    founded_item : OPTIONAL representation_item;
    rep_context  :  representation_context;
  WHERE
    wr1: SIZEOF (USEDIN (SELF, '') - USEDIN (SELF,
         'REPRESENTATION_SCHEMA.REPRESENTATION_RELATIONSHIP.' +
         'BASE_REPRESENTATION')) > 0;
(* As in this Part founded item is required, WR2 is superfluous.
    wr2: SIZEOF (USEDIN (SELF, 'REPRESENTATION_SCHEMA.' +
         'REPRESENTATION_RELATIONSHIP.BASE_REPRESENTATION')) > 0
         OR EXISTS (SELF.founded_item); *)
  END_ENTITY;
(*
*)
  RULE exists_founded_item FOR (representation);
  WHERE
    wr1 :  "where exists founded_item."
  END RULE;
(*
Remarks
Every representation shall have a reference to a geometric entity.
```

```
REPRESENTATION_CONTEXT
*)
  ENTITY representation_context;
    dimensionality : INTEGER;
  WHERE
    wr1: (SELF.dimensionality = 2) XOR (SELF.dimensionality = 3);
  END_ENTITY;
(*


SHAPE_REPRESENTATION
*)
  ENTITY shape_representation
    SUBTYPE OF (representation);
  END_ENTITY;
(*


TRANSFORMATION
*)
  ENTITY transformation
    SUBTYPE OF (geometry);
  END_ENTITY;
(*
```

2.3.4.3 Type Definitions from Part 41

```
*)
  TYPE shape_definition = product_definition_shape;
  END_TYPE;
(*
```

Remark

The other select type, shape_aspect, has been pruned. This Part treats the whole shape of a product as one identifiable element only.

```
*)
  TYPE characterized_product_definition = product_definition;
  END_TYPE;
(*
```

Remark

The other select type, product_definition_relationship, has been pruned. This Part is not capable of defining relationships between products.

```
*)
```

```
   TYPE part_status = ENUMERATION OF ( committee_draft,
                                       draft_international_standard,
                                       international_standard);
   END_TYPE;
(*

*)
   TYPE source =
      ENUMERATION OF ( make, buy, not_known );
   END_TYPE;
(*

*)
   TYPE year_number = INTEGER;
   END_TYPE;
(*

*)
   TYPE identifier = STRING;
   END_TYPE;
(*

*)
   TYPE label = STRING;
   END_TYPE;
(*

*)
   TYPE text = STRING;
   END_TYPE;
(*
```

2.3.4.4 Entity Definitions from Part 41

```
APPLICATION_INTERPRETED_MODEL
*)
   ENTITY application_interpreted_model;
     name                 : label;
     application_protocol : application_protocol;
   END_ENTITY;
(*

APPLICATION_PROTOCOL
*)
```

```
  ENTITY application_protocol;
    name                  : label;
    iso_10303_part_number : INTEGER;
    year                  : year_of_application_protocol;
  END_ENTITY;
(*


PRODUCT
*)
  ENTITY product;
    id                  : identifier;
    name                : label;
    description         : OPTIONAL text;
    frame_of_reference  : OPTIONAL SET [1:?] OF product_context;
  UNIQUE
    id;
  END_ENTITY;

  RULE not_exists_frame_of_reference FOR (product);
  WHERE
    wr1 :  "where not exists frame_of_reference."
  END RULE;
(*


PRODUCT_DEFINITION
*)
  ENTITY product_definition;
    documentation_id   : OPTIONAL SET [1:?] OF identifier;
    description        : OPTIONAL text;
    version            : product_version;
    frame_of_reference : SET [1:?] OF product_definition_context;
  END_ENTITY;

  RULE only_one_frame_of_reference FOR (product_definition);
  WHERE
    wr1 :  "where frame_of_reference is a set of max. 1
            product_definition."
  END RULE;
(*


PRODUCT_DEFINITION_CONTEXT
*)
  ENTITY product_definition_context;
    name                       : label;
```

```
      application_interpreted_models : SET [1:?] OF
                                   application_interpreted_model;
   END_ENTITY;

   RULE only_one_application_interpreted_model
    FOR (product_definition_context);
   WHERE
     wr1 :  "where application_interpreted_models is a set of max. 1
             application_interpreted_model."
   END RULE;
 (*


PRODUCT_DEFINITION_SHAPE
*)
   ENTITY product_definition_shape;
     definition            : characterized_product_definition;
   UNIQUE
     definition;
   END_ENTITY;
 (*


PRODUCT_VERSION
*)
   ENTITY product_version;
     id                    : identifier;
     description           : OPTIONAL text;
     of_product            : product;
     make_or_buy           : OPTIONAL source;
   END_ENTITY;
 (*


SHAPE_DEFINITION_REPRESENTATION
*)
   ENTITY shape_definition_representation;
     representation_model : shape_representation;
     representation_of    : shape_definition;
   END_ENTITY;
 (*


YEAR_OF_APPLICATION_PROTOCOL
*)
   ENTITY year_of_application_protocol;
     year                  : year_number;
     status                : part_status;
   END_ENTITY;              (*
```

2.3.4.5 Part 204/205/206 - product classification structure

The classification structure comprises entities from Parts 41 and 43:

```
application_interpreted_model
application_protocol
product
product_definition
product_definition_context
product_definition_shape
product_version
representation
shape_representation
representation_context
shape_definition_representation
transformation
year_of_application_protocol


*)
END_SCHEMA; -- part_XXX_product_structure_schema
(*
```

The diagram shows the relationship between the product structure entities and the shape entities that are pointing to topology and geometry.

```
SHAPE_DEFINITION_REPRESENTATION
   |          |
   |        PRODUCT_DEFINITION_SHAPE
   |            |
   |          PRODUCT_DEFINITION
   |              |          |
   |              |        PRODUCT_DEFINITION_CONTEXT
   |              |              |
   |              |            APPLICATION_INTERPRETED_MODEL
   |            PRODUCT_VERSION     |
   |              |              APPLICATION_PROTOCOL
   |            PRODUCT              |
   |
YEAR_OF_APPLICATION_PROTOCOL
   |
   |
   SHAPE_REPRESENTATION
      |           |
      |         REPRESENTATION_CONTEXT
      |              |
      |            (3);
      shape_entity
```

An excerpt from a STEP file shows the connection:

```
#20001=PRODUCT('Example ','STEP file',$,$);
#20002=PRODUCT_VERSION('1.0',$,#20001,.MAKE.);
#20003=YEAR_OF_APPLICATION_PROTOCOL(1992,.COMMITTEE_DRAFT.);
#20004=APPLICATION_PROTOCOL(     /* Adjust AP in next line !!! */
'Mechanical Design using Boundary Representation',204, #20003);
/* BREP=204, SF=205, WF=206, CSG=250, CBR=299 */
#20005=APPLICATION_INTERPRETED_MODEL('BREP_SCHEMA',#20004);
#20006=PRODUCT_DEFINITION_CONTEXT('as planned',(#20005));
#20007=PRODUCT_DEFINITION($,$,#20002,(#20006));
#20008=PRODUCT_DEFINITION_SHAPE(#20007);


#100=shape_entity;


#20009=REPRESENTATION_CONTEXT(3);
#20010=SHAPE_REPRESENTATION(#100,#20009);
#20011=SHAPE_DEFINITION_REPRESENTATION(#20010,#20008);
```

For shape_entity the following are possible:

```
MANIFOLD_SOLID_BREP                - AP 204
SHELL_BASED_SURFACE_MODEL          \
FACE_BASED_SURFACE_MODEL           - AP 205
GEOMETRIC_3D_SURFACE_SET           /
SHELL_BASED_WIREFRAME_MODEL        \
EDGE_BASED_WIREFRAME_MODEL         - AP 206
GEOMETRIC_3D_CURVE_SET             /
```

Some other shape_entities are provided in other CADEX Application Protocols that are not part of STEP. They cover CSG and Compound-Boundary models and are described in the following section.

2.4 Other CADEX Application Protocols

There are two more APs generated by CADEX. They are by no means less important than the three that have become STEP parts. They have been set up and tested and they have been used extensively for data exchange. It was only for lack of manpower that they could not be pushed in the ISO community. Whereas the Brep, Surface and Wireframe APs are available to everyone as STEP parts 204, 205 and 206, the other two CADEX APs are basically internal documents. They cover the important applications:

`Constructive Solid Geometry`

and

`Compound Boundary Representation.`

The complete Application Protocols are available as separate documents.

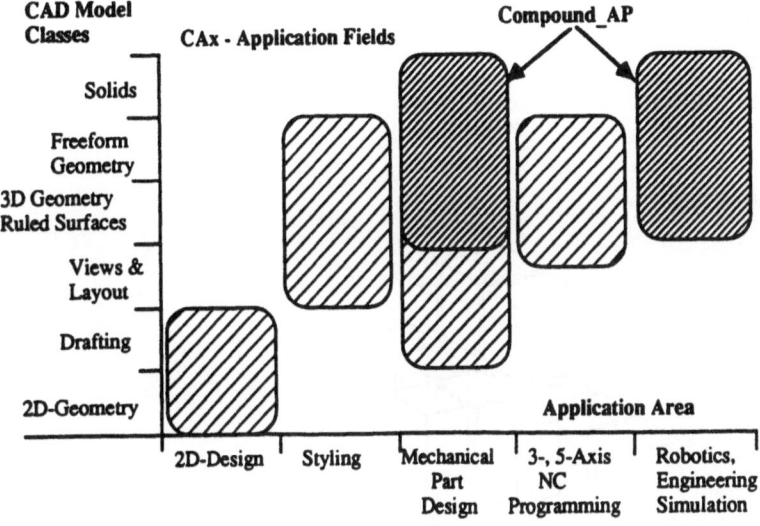

Fig. 2.5: The scope of the compound boundary representation AP related to engineering activities

2.5 Relationships between Application Protocols

2.5.1 Domains of the Application Protocols

The application protocols developed in the CADEX form a group of application protocols for geometry data exchange. Their coverage can be described in a three dimensional space spanned by the dimensions: Geometry, Connectivity and Model Type, see figure 2.6. The positions on the axes are described as domains with increasing complexity along the axes when going away from the origin:

Geometry:

- Facetted geometry
- Elementary geometry
- Sculptured geometry
- Offset geometry, etc

Connectivity:

- Geometry Bound, both a heap of points, line and surfaces and bounded geometry
- Topology Bound, both non-manifold and manifold topology
- Constructive Solid Geometry

Model Type:

- Wire frame models
- surface models
- Volume models
- Compound models

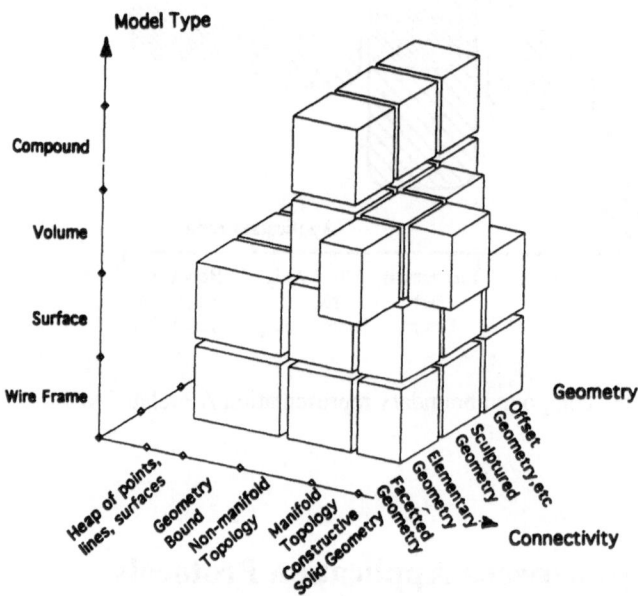

Fig. 2.6: The Application Protocols domain, a 3D view

In this space the APs from the project can be positioned as shown in figure 2.7.

The Constructive Solid Geometry AP is based on elementary geometry and covers volumes only.

The Boundary Representation AP has three geometry levels, facetted, elementary and sculptured. It covers manifold topology only, and addresses volumes only. The functional levels are:

- manifold topology of volume models with facetted geometry only
- manifold topology of volume models with elementary geometry only
- manifold topology of volume models with sculptured geometry as well as elementary geometry

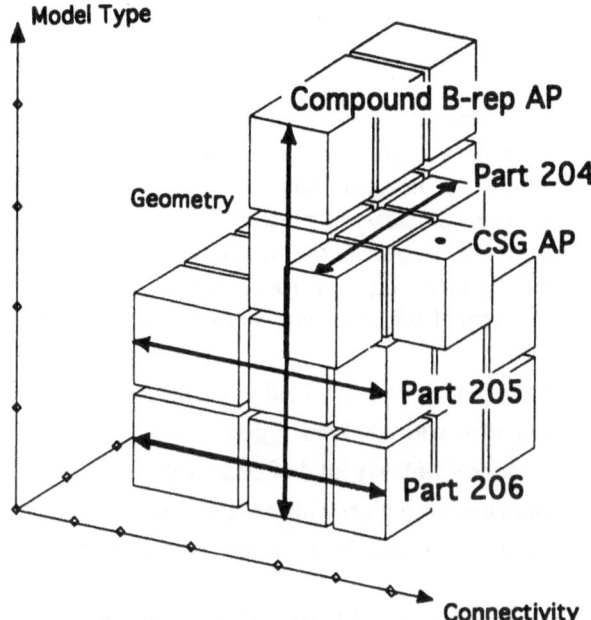

Fig. 2.7: The CADEX Application Protocols positioned in a 3D domain with functional levels given

The Surface AP has no faceted geometry, but covers the geometry axis for elementary, sculptured and offset geometry, etc. A model is connected by using either geometry bound and topology bound connectivity. The functional levels are:

- geometry bound surface model, where the geometry can be any kind of curves and surfaces, while there is no topology. The bounding is either not present, as for a heap of points, lines and surfaces, or present as trimmed curves and surfaces.
- non-manifold topology models where the topology bounds the model and no restrictions are imposed on the topology in terms of manifoldness, i.e. non-manifold topology.
- manifold topology models where topology bounds the model and restrictions for manifold topology is imposed.

The Wire Frame AP has no faceted geometry, but covers the geometry axis for elementary, sculptured and offset geometry, etc. A model is connected by using either geometry bound and topology bound connectivity. The functional levels are:

- geometry bound wireframe model, where the geometry can be any kind of curves and surfaces, while there is no topology. The bounding is either not present as for a heap of points and lines, or present as trimmed curves.
- non-manifold topology models where topology bounds the model and no restrictions are imposed on the topology in terms of manifoldness, i.e. non-manifold topology.

- manifold topology models where topology bounds the model and restrictions for manifold topology is imposed.

The Compound Boundary Representation AP has no faceted geometry, but covers the geometry axis for elementary, sculptured and offset geometry, etc. It covers only non-manifold topology, i.e. no restrictions are imposed. That does mean however, that the compound boundary representation AP will cover manifold topology as well, since this is a special case of non-manifold topology. The model type axis is covered by including wire frame, surface and volume models, as well as the compound model which is any combination of the three others. The functional levels are:

- non-manifold topology wire frame models with any kind of geometry

- non-manifold topology surface models with any kind of geometry

- non-manifold topology volume models with any kind of geometry

- non-manifold topology compound models with any kind of geometry

From the diagrams in figure 2.6 and 2.7 it is possible to deduce the following information:

- the wire frame and the surface APs have similar functionality, as they share both geometry and connectivity domains. The functional levels are along the connectivity axis. The surface AP includes both curves and surfaces, while the wire frame AP only includes curves. The wire frame AP is a true sub-space of the surface AP in all functional levels.

- there is overlap between the wire frame and the compound b-rep APs in the non-manifold wireframe model, i.e. the edge based wire frame models

- there is overlap between the surface and the compound b-rep AP in the non-manifold surface model, i.e. the face based surface model.

- the boundary representation AP is limited in functionality compared to the wire frame and surface APs, as no provisions are made to cover geometry bound or non-manifold models. The functional levels are in the geometry axis direction with limited geometric coverage, the offset geometry, etc is not included as possible geometry.

- the faceted boundary representation for volume models has an interface to manifold solid boundary representation for elementary geometry only. There are no other kinds of explicitly defined faceted models. However, the information can be stored in a more elaborate structure for both wire frames, surface, volume and compound models using boundary representation.

- the constructive solid geometry AP is on its own, connected to the manifold boundary representation for elementary geometry only. The significance of the manifold solid boundary representation is therefore clear, it has a number of neighbours in other formats and is central for communication of these models after conversion.

- the manifold boundary representation is a special case of non-manifold topology, but is still kept separately because of the significance of manifold topology in existing CAD modellers.
- the compound boundary representation AP adds another dimension to the definition of the application protocol domain, as the functional levels span the model type axis. It introduces the concept of non-manifold solid models, a domain not included in STEP Part 42 so far. The span in the geometry axis direction coincides with the wire frame and surface APs, and allows the definition of volumes bounded by offset curves and surfaces.
- the compound boundary representation AP introduces the concept of non-manifold compound models, i.e. models consisting of a mixture of wire frame, surface and volume information. This is extensively used in analysis modelling for discretisation methods like finite element analysis.

2.5.2 Conversion between different Application Protocols

Data exchange between different systems will be of two kinds:
- exchange of a particular class of model between functionally equivalent modelling systems, see figure 2.8.
- exchange of models between dissimilar systems, with conversion between different classes of models as part of the transfer, see figure 2.9.

The amount of data exchange between similar systems will be limited to the cases where the same kind of modelling system is used, i.e. people doing the same kind of job within the organisation or in other organisations. The data exchange between dissimilar systems will dominate the scene, as most product development activities need different kinds of modelling systems and different kinds of models. Conversion of models between different representations will be a key to successful data exchange in an industrial environment.

The definition of application protocols with identified functional levels simplifies the conversion process, as the different formats are well defined. Each different modelling system must conform to one or more of the functional levels in these application protocols, and the conversion between them defined. With this structure in place, data exchange between dissimilar systems will be possible to automate. The number of conversions needed for conversion between all models is high. A conversion path from a particular representation to another using intermediate formats will reduce the number of necessary conversions. An example will cast some light over the principles:

A system for surface modelling will be used to produce the mould for a plastic bottle, but it can only hold the information as a collection of individually defined bounded surfaces using polynomial spline representation. The original model is held in a manifold solid modeller as a manifold solid model bounded by elementary surfaces.

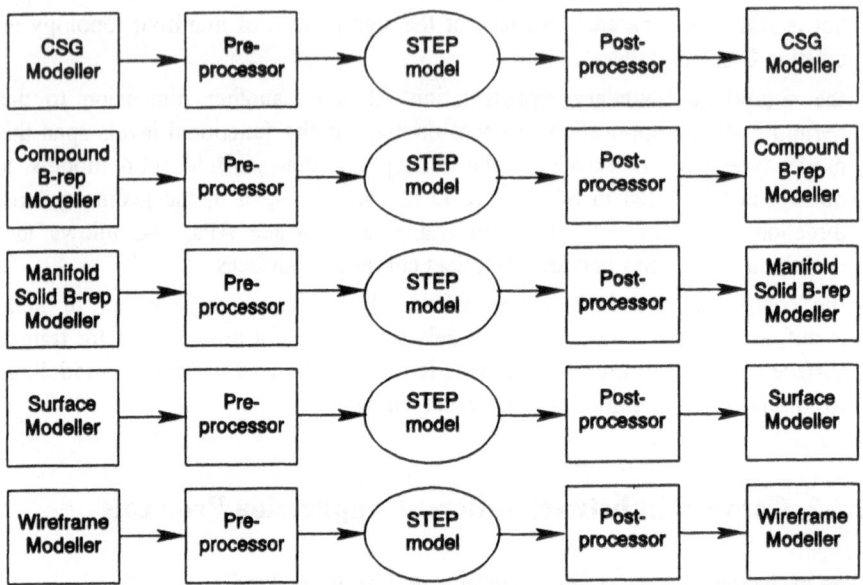

Fig. 2.8: Geometry exchange between similar systems

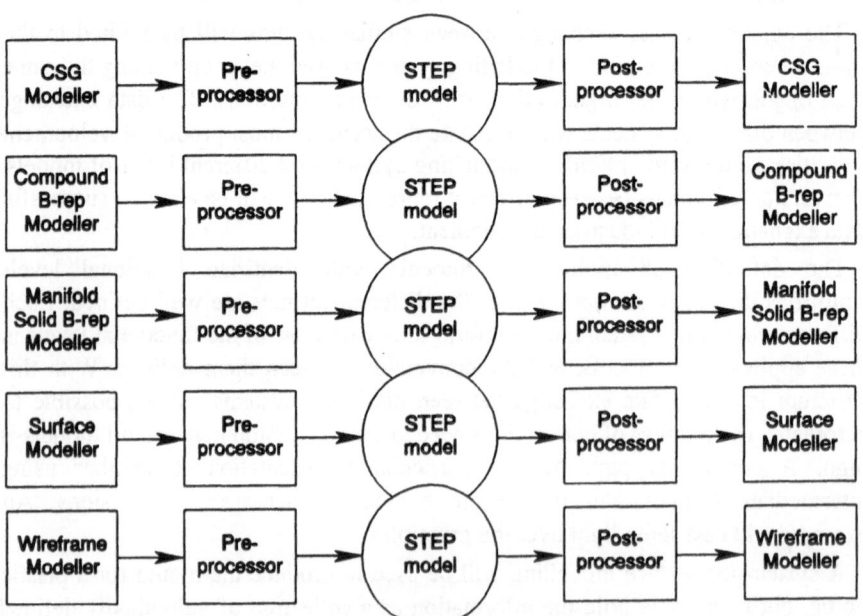

Fig. 2.9: Geometry exchange between dissimilar systems

The conversion path will typically be as follows:

- the model starts off conforming to Part 204, functional level 2, elementary b-rep model
- The closed shell of the manifold solid model is converted to conform to the closed shell used in the Part 205 functional level 3
- the geometry representation based on elementary surfaces is converted to a general NURBS representation for curves and surfaces and then to a format which is mathematically compatible with polynomial splines.
- the need for geometry subdivision to maintain model accuracy may require additional model modifications by introducing cuts through the surfaces while he model remains a shell based surface model.
- at the end of this process, the model is still a shell based surface model.
- the model is then converted to a geometry bound surface model using the geometry 3D surface set in Part 205 functional level 1. The conversion should ensure that all surfaces are bound, either by their natural boundaries or the boundary used by the bounding curves from the original model representation.
- the model is then sent to the receiving system as a surface model conforming to Part 205 functional level 1.

The model will conform to a particular functional levels from the existing APs at all stages in the conversion process. This can be tested using verification tools on the data repository holding the intermediate model representations. The conversion tools introduced in the CADEX projects operate both on the native data repositories and the intermediate data structure, IDS. The conversions between native representations and the STEP representations will ensure compatibility between systems and STEP, while conversion on the IDS will ensure conversions between systems of different kinds.

A number of tools have been developed in the project to operate directly on the IDS for conversion between different STEP representations. The area of conversions is very large and the implementation in CADEX is just scratching the surface of it. The use of the IDS as a working form, level 2, implementation (PDES Inc. terminology) has been proven functional.

2.5.3 Shape entites from multiple APs in single files

2.5.3.1 The problem

On the basis of the Cadex assumption on support of the Application Protocol concept, any shape model transferred between different CAD systems via a Step file must consist of entities belonging to a single AP. The fact that the file stores one model implies that all entities appearing in the file belong to the same AP. On the other hand, product models can be variously complex from the geometrical and topological point of view, so that they could be composed by several shape models, possibly from different APs (examples can be easily shown of models made by, e.g., surfaces, solids and wireframes). In this case, the need arises that a single Step file contains entities from multiple APs.

Moreover, since a same (shape) entity can belong to several APs, possibly being defined in each of them with different constraints, a semantic-based scanning algorithm for Step files requires the knowledge of the AP each entity belongs to. It should be noted that this fact is not covered in Step syntactical specifications on physical file format, as defined in ISO 10303-21, "Clear text encoding of the exchange structure".

As a consequence, some related problems originate. Can a same entity be referenced by entities from different APs (e.g., a cartesian point as a part of the definition of a wireframe model and a surface model) ? In that case, which AP the defining entity belongs to? Which AP-based constraints should the entity satisfy?

2.5.3.2 The adopted solution

In order to reach a workable solution to the presented problem, a rather conservative approach has been taken, assuming that any entity in a file can only be referenced in the definition of entities belonging to a same AP. This solves all application problems related to the possible semantic inconsistencies in the way entities are defined in different APs. However, a mechanism must be defined to identify the AP any entity in a file belongs to.

A first solution has been to define the construct, not present in ISO 10303-21:

> ...
> USE_AP <name of an AP>;
> ...
> END_AP;
> ...

to be inserted in the Data Section of the Step files, interpreted according to the rule that all entities defined between USE_AP and END_AP belong to the AP whose name is indicated in the USE_AP line.

This solution has been implemented and tested in its suitability. Although actually working, it has been eventually rejected, mainly because relying on syntactical constructs not part of the ISO standard.

The current draft of ISO 10303-41, "Integrated resources: fundamentals of product description and support" provides a mechanism for inserting in a same physical file several shape models, considered as constituting a global product model. In this view, each shape top entity is indirectly associated with an APPLICATION_PROTOCOL entity via a SHAPE_DEFINITION_REPRESEN-TATION entity, conceptually identifying a product structure and defined by both shape and product information.

In this line, the current version of the Cadex Common ToolKit has been designed, implementing a subset of the entities defined in ISO 10303-41. This allows to represent and transfer not only purely geometrical and topological models, but also their structure as (parts of) products, and therefore the information on the AP each entity belongs to.

2.5.4 Application Interpreted Constructs

2.5.4.1 Summary

This chapter presents the concept of AICs (Application Interpreted Constructs) and their impact on the CADEX work. The role of AICs for the interoperability of information models is discussed. The current STEP proposal for implementation of AICs is introduced. The chapter concludes with a list of unresolved issues.

2.5.4.2 The challenge of interoperability

The term interoperability is not well-defined. In the context of STEP the meaning of interoperability can be limited to communication among STEP conformant applications.

The STEP-community allows STEP-conformant implementations only by Application Protocols. A new Application Protocol must be initiated for each new application with requirements for database access or file exchange that are based on the STEP standard. The desired information model must be put into the frame of an Application Protocol and get the approval of the ISO-community. This has been done for the CADEX APs 204, 205, and 206.

However, what happens if a new application wants to communicate with the above mentioned ones? Shall a Mechanical Design AP - which certainly requires shape descriptions as provided by 204, 205 and 206 - redefine product shape? Obviously the reuse of APs - or parts of APs - by other APs is desirable. A product aspect should be modelled once; the resulting model should be shared by all applications that depend on this aspect. This would allow interoperability concerning overlapping parts of information models that are based on STEP.

The mechanism envisaged for this is the Application Interpreted Construct (AIC). The term construct refers to both types and entities. An Application Protocol may among others consist of one or several AICs. These parts of the product model can be accessed by applications that support the same AICs.

2.5.4.3 Characteristics of AICs

The AIC concept was first introduced to STEP at the Houston-meeting in autumn 1991. The implementation of the mechanism is still going on. In a collaboration between CADEX and PDES Inc. the first AICs have been defined and written. AICs will be part of the initial release of STEP.

An AIC is a construct - likely consisting of more than one entity - that is common to at least two Application Interpreted Models (AIM), which in practice means two Application Protocols. An AIC first comes into being, when a new AP requires a logical unit that already exists in another AP. An AIC must satisfy a certain naturally bounded application requirement. Under this condition it is possible to find among existing AICs suitable ones for new APs which have their requirements defined only.

The main body of an AIC is the EXPRESS representation of its information model. These EXPRESS constructs must be derived from the STEP Integrated Resources and must include their textual description as well. The original construct definitions may be additionally interpreted for the use by an AIC. Interpretation means that they may be adapted to application specific requirements by

- removal of optional attributes;

- addition of attributes to "abstract supertypes";

- addition of population constraints (e.g. radius < 5).

How far additional constraints may be applied to the use of the same AIC in different APs is not clarified, yet (see: Issues). Possibly further population constraints will be legal. AIC-constructs will then be sharable, AIC-instances not always.

In addition to the EXPRESS specification an AIC consists of:

- scope,

- mapping table from Integrated Resource construct to AIC-construct,

- references to other AICs,

- conformance testing that is independent of the use of the AIC,

- short form description,

- EXPRESS-G diagram of the information model.

As stated in the list, AICs may be nested.

APs using AICs will both reference these in a short form and include all the AIC EXPRESS-code with textual description (an AP shall be self-contained). The AP must also document those aspects of conformance testing that are specific to the use of the AICs in the AP½s application context.

All STEP AICs will be collected and maintained in a library.

2.5.4.4 AICs in CADEX

The CADEX project has actively participated already in the early AIC discussions. CADEX has contributed essentially to the AIC-library. CADEX representatives took part in the Leeds-meeting in autumn 1991 and the pre-Oslo-meeting in winter 1992 together with the STEP integration committee and PDES Inc. representatives to solve the interoperability problem. The collaboration with PDES Inc. showed the AIC-potential in Parts 204, 205 and 206, with Part 203 being a first user of them. The CADEX APs cover a domain of paramount importance to many coming APs: i.e. product shape. The information models that are implemented by CADEX will be largely spread due to the AIC-concept.

In particular the collaboration resulted in the following AICs:

Shape:
- topologically bounded elementary surfaces
- topologically bounded surfaces
- facetted Breps
- elementary Breps
- advanced Breps
- geometrically bounded surfaces
- non-manifold surfaces
- manifold surfaces
- geometrically bounded curves
- topologically bounded curves
- edge based wireframes
- shell based wireframes.

Presentation:
- basic surface presentation
- basic curve presentation.

AICs developed by PDES Inc.
- mechanical design context
- geometric measures
- name assignment.

2.5.4.5 Issues

The AIC-concept gains more and more acceptance. However, many issues concerning its practical implementation are still unresolved. Some of them are listed below.

1) How can software implementations exploit the modularity of AICs?

2) How does a query refer to AIC conformant information?

3) How do entities on a file reflect their AIC-background?

4) Is it necessary to interpret AICs in APs?

5) Who is going to maintain the AIC-library?

6) Who will make future AP-developers use the AIC-library?

7) What happens to an existing AP, parts of which have been identified to become AICs?

AICs are one tool to achieve interoperability of STEP-applications. It only covers overlapping parts of an information model. There are needs left e.g. for communicating both almost identical and supplementary product information. To solve these challenges possibly different implementation methods need to be invented.

3. Common Tool Kit

3.1 Introduction

The main goal of the CADEX project is the development of STEP related processors. The CAD vendors produce software that links STEP to their native systems. A great part of this software is the same in all processors. This part is called Common Tool Kit.

The specification of this Common Tool Kit was published as KfK report in January 1991 (KfK-PFT 159). This report was the Deliverable D3 according to the CADEX Technical Annex. The main development of the Common Tool Kit was described in the final report on workpackage 4. This was Deliverable D6 and was completed in August 1991.

Throughout the project all workpackages with close relations to STEP have been influenced significantly by delayed decisions and ongoing changes in ISO/TC184/SC4. Especially workpackage 4, the development of Common Tools, was affected severely. The CADEX goal to produce processors applying the international standard ISO 10303 (STEP) requires that data structure and basic tools (especially in the neutral interfaces) conform to formats and definitions given in the standard. Therefore the workplan for the last half year was modified such that workpackage 4 (Common Tools) was extended until the end of the project.

Common Tools means that this is a software package developed from many partners for many partners. A good collaboration and an intense exchange of code and other information, a lot of meetings, phone calls and written notes were necessary to achieve a consistent package meeting the requirements of the partners as well as the boundary conditions of the ISO. Some inhomogeneity could not be avoided, but the whole package has been integrated in several processors (see chapter 4) and has produced good results.

In the following sub-chapters first the architecture and then the different software pieces are described technically. A user's guide (chapter 3.4) is added and deals with the parts in the same order as the technical description. Emphasis in all descriptions was put on explaining how to integrate the modules in larger contexts.

The Conversion Tools Library is a large part of the Common Toolkit and contains a rather heterogenous collection of modules needed at different places in the processors and developed by different partners. Additional information on the structure of the conversion tools is given in a separate introduction at the beginning of chapter 3.3.6.

3.2 The architecture of the CADEX pre- and postprocessors

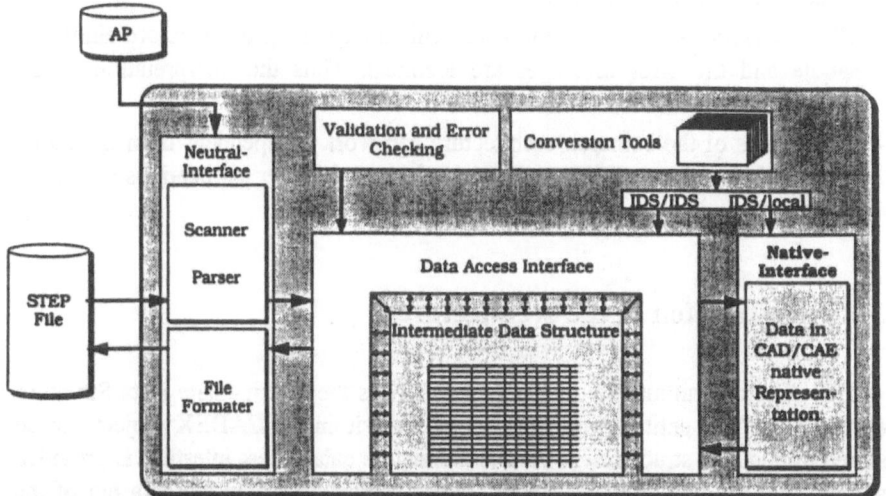

Fig. 3.1: CAx-interface reference processor

3.2.1 Overview

In order to obtain a powerful and efficient processor development tool kit the CADEX consortium has designed a functional overall processor architecture.

The reasons for this development are detailed below:

- The development of processors requires a great effort in terms of manpower and time. But most of the modules in all pre- and postprocessors are identical and only the systems interface to read or to write the data into and out of the system are specific to the database of the CAD system. Out of this a great part of the development effort can be shared out to the different partners. Therefore the use of common software in the project is strongly facilitated and the task of processor development becomes easier.

- Because STEP is not stable yet, the developed interface processors use the same STEP dialect by incorporating the same modules for reading and writing STEP files. Thus is guaranteed, that the pre- and postprocessors of the different systems fit together.
- To provide a useful mapping between the STEP entities and the system specific entities as well as to adapt incompatible model descriptions to the CAD systems (e.g. from Brep to Wireframe) conversion tools have been identified as an area of great importance. So as not to reinvent the wheel a common architecture enables the CADEX project to pool the conversion tools in a library and to share the effort of development.
- The validation of the STEP data after scanning the neutral file and before the writing of the STEP file can be done on the internal data structure of the processors by a common tool.
- All processors use the same modules. This means all processors are similar to handle and the error messages are identical. Thus the interpretation of the messages for the user is much more easier.
- All modules of the common architecture can work independent from the others and may be used as a STEP application platform in other applications within the different partner companies.

3.2.2 Description of the architecture

IDS: The central module of the architecture is the Intermediate Data Structure (IDS). All STEP entities which are supported within the CADEX project can be stored in this data structure. To handle this data a data access interface is provided which gives the possibility to create, delete, modify and read the data out of the IDS, see figure 3.2. Furthermore, tools for selection of subsets and navigation in the data are provided. All the other common modules which are described below work in cooperation with the IDS.

Scanner/parser: The scanner/parser software is a lex and yacc based module to read STEP physical files and to store the STEP file entities in the IDS. Syntactical checks and AP specific validation is supported. Semantic checks of the STEP file are not provided.

File formatter: This module transfers the data from the IDS into a physical STEP file. The output of the file formatter is controlled by an option file.

Validation and error checking: This tool provides semantic checks of the data stored in the IDS. It checks the AP specific WHERE clauses of the entities. The reason for extracting this task from the scanner and parser software was to enable the data to be checked in the IDS after scanning and parsing and also before the file formatter is called.

Conversion tools: This part of the architecture is a library of programs to convert and adapt entities to different representations. All of the software can be used on the IDS platform as well as for conversions "on the fly" in the systems interface.

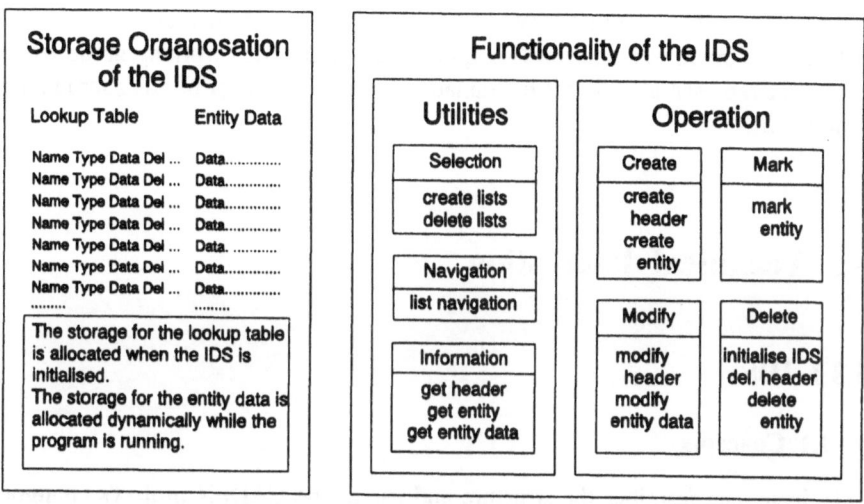

Fig. 3.2: Overview and functionality of the IDS

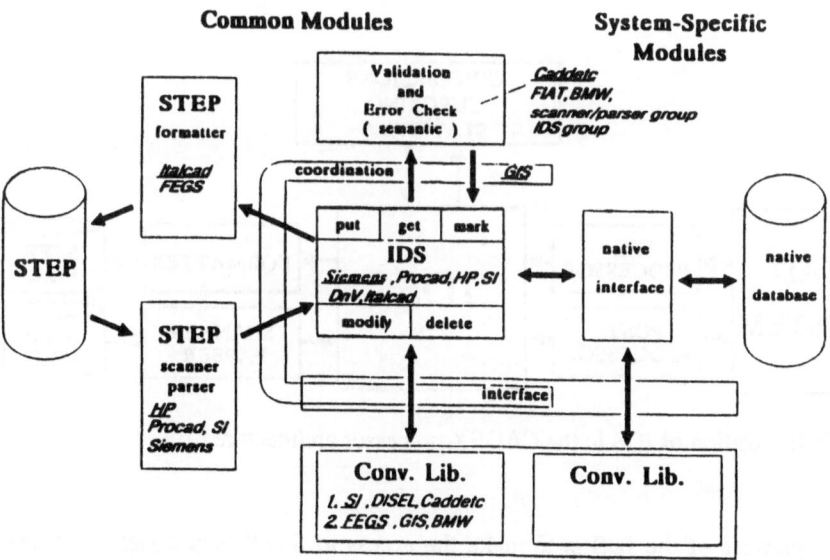

Fig. 3.3: Working groups for common tools

3.2.3 Organisation of the development

For the development of the common tool kit as described above a working group for each of the different modules was established. The underlined partners in figure 3.3 are the working group leaders. For the definition of the interfaces between the modules a coordination group was set up. Each working group leader is also the representative in the coordination group. The leader of this coordination group is GfS.

3.3 Technical description

3.3.1 IDS

3.3.1.1 Concepts

This document describes the structure and use of the CADEX project's common tools software called IDS (Intermediate Data Structure).

In the following figure 3.4 you can see the central role IDS plays in the STEP processors' architecture within the CADEX project. IDS will be used by most of the partners in both preprocessing and postprocessing for temporary data storage.

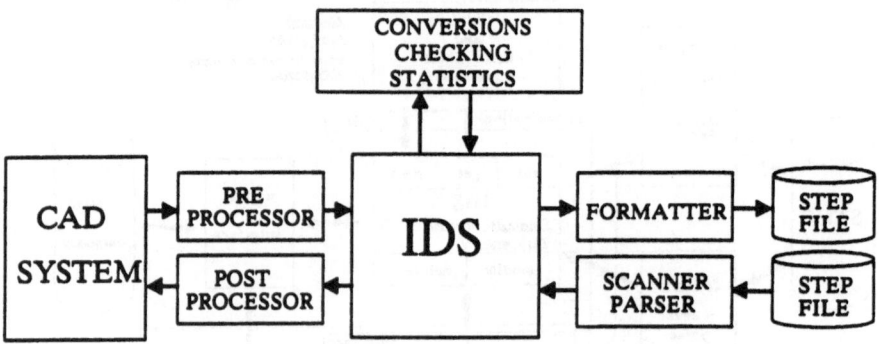

Fig. 3.4: Position of IDS in the CADEX processor architecture

The purpose of the IDS is to make the access to STEP data easier and more efficient than by direct access to the STEP physical file (random access instead of sequential access).

The IDS software is written in the programming language C because this programming language provides some advantages to store data structures.

However, as the IDS has to be accessible by all common tools, some of which are written in FORTRAN, there has to be an interface which enables FORTRAN programs to access the IDS. This "conflict" of writing the software in C, but also being able to access it from FORTRAN, has been solved by introducing a set of interface routines which can be called from both languages. They are the user's view of the IDS and perform the actions on the IDS internally.

3.3.1.2 Internal data structure

The internal design of the IDS has been done in a quite straightforward way:

The storage of data within the IDS is done with structures of the programming language C which are as similar as possible to the corresponding STEP entity. Example:

```
STEP entity:                          IDS struct:
ENTITY edge
SUBTYPE QF (topology);                struct edge_struct {
  edge_start:  vertex;                entity_reference  edge_start,
  edge_end:    vertex;                                  edge_end,
  edge_curve:  OPTIONAL curve_log_struct;               edge_curve;
WHERE                                                 }
  ...
END_ENTITY;
```

The inheritance of attributes resulting from SUBTYPE/SUPERTYPE expressions is an exception to this direct correspondence.

It was decided that the IDS should be optimised for top-down access. This means that it should be easy to identify the children of an entity, but not its parents. This in fact corresponds to the structure in STEP itself.

For more detailed information and examples of the IDS data structure please see file.

3.3.1.3 IDS access routines

The routines to access the data of the IDS are entity type specific: There is one put routine per entity type to store the data of an entity in the IDS, one get routine per entity type to retrieve the information which is stored in the IDS and one modify routine per entity type to modify the attributes of an entity already stored in the IDS.

The routines' names are composed as follows:

The 1st character is a 'c' indicating CADEX, the 2nd parameter is 'p' for put routines, 'g' for get routines or 'm' for modify routines, the following up to 4 characters identify the entity type.

Most of the routines' parameters can directly be derived from the IDS structure. The data types of the attributes of STEP entities are represented in the IDS as

follows (the limited set of used data types is a consequence of the requirement that
the IDS access routines should be callable within C and FORTRAN programs):

```
STEP                          IDS
--------------------          --------------------------------------------
REAL                          double (8 bytes)
INTEGER                       long (4 bytes)
LOGICAL                       long
                                       see file "enum.h":
                                       #define TRUE_BOOLEAN 2
                                       #define FALSE_BOOLEAN 1
enumeration                   long (values 1,2,3,...)
reference to entity           long (values 1,2,3,...)
string (e.g. header)          2 Parameters:
                                 1st:  long (length of string,
                                                incl. trailing \0)
                                 2nd:  char[] (string)
arrays (LIST, SET,...)        2 Parameters:
                                 1st:  long (length of array)
                                 2nd:  long[], double[] (array elements)
OPTIONAL REAL                 2 Parameters:
                                 1st:  long (indicates if attribute exists)
                                       see file "cdefin.h":
                                       #define OPT_PAR_EXISTS 1
                                       #define OPT_PAR_NOT_EXISTS 0
                                 2nd:  double (attribute value)
OPTIONAL INTEGER              2 Parameters:
                                 1st:  long (indicates if attribute exists)
                                       see file "cdefin.h":
                                       #define OPT_PAR_EXISTS 1
                                       #define OPT_PAR_NOT_EXISTS 0
                                 2nd:  long (attribute value)
OPTIONAL enumeration          long (value 0 indicates that enum
                                        doesn't exists)
OPTIONAL reference            long (value 0 indicates that there
                                       isn't a reference)
OPTIONAL arrays               2 Parameters as usual:
                                 1st:  long (array length 0 represents not
                                                existing arrays)
                                 2nd:  long[], double[] (array elements if
                                                existing)
```

There are some conventions:

The parameters are called by reference not by value. Input parameters come
before output parameters in the parameter list. In addition to the STEP/IDS
attributes there is an entity identifier as first function argument (i.e. the name

which appears in the STEP physical file, e.g. the integer number 127 for entity #127) and a status flag as the last argument. The status flag is also returned as the function value.

Example (store edge in IDS):

```
long cped ( long *name, long *start_vx, long *end_vx, long *culs,
            long *istat);
```

For more detailed information and examples of IDS data access routines see the files "ids_put.man" (IDS put routines), "ids_get.man" (IDS get routines) and "ids_mod.man" (IDS modify routines). These files hold in alphabetical order the headers of the put, get and modify routines. If you want to look at the code itself, you should examine the files "cdtput.c", "cdtget.c" and "cdtmod.c".

The IDS put, get and modify files have become quite large in the meantime. In order to avoid object files which are greater than necessary, you can decide at compile time which parts of the code you are interested in (i.e. which Application Protocols you will support): This "interest" is documented by providing some defines which are evaluated by the C preprocessor. You can provide them either by some compiler options or by including some #defines directly in the source code files. The relevant constants are the following ones:

CC_BREP_AP for the BRep Application Protocol,
CC_SF_AP   for the Surface Application Protocol,
CC_WF_AP   for the Wireframe Application Protocol,
CC_CSG_AP for the CSG Application Protocol
CC_CBR_AP for the Compound Brep Application Protocol.

3.3.1.4 Some other routines of the IDS package

The headers of the following routines are described in alphabetical order in the file. If you want to look at the code please look into the source file mentioned with each of the following paragraphs. A chapter called "File Organisation of IDS" will tell you where to find each group of routines.

Header of STEP Physical File: The header of a STEP physical file can be stored in the IDS with a set of header put routines. The data can be retrieved from the IDS with header get routines. The implementation approach is a little bit different to the access routines of the geometric entities described above: For each attribute of the header entities there is a separate put and a get routine. For details of the code see the files "chdput.c" (put header) and "chdget.c" (get header).

Initialisation of the IDS: Before you do anything with the IDS you have to initialise it. This is done with the routine 'cinids'. It has two relevant parameters: the length of the lookup table (i.e. the maximum number of entities which can exist within the IDS) and the number of top entities (i.e. the length of an array in a so-called world entity to store these entity ids). As an additional action the structs to store the STEP physical file header information are initialised. For more information see the file "cinit.c".

Leave the IDS: When you intend to leave the IDS and want to free the memory which has been used by the IDS, you have to call the routine 'cfrmem'. See the file "cidsut.c".

Scope information: There are two routines in the IDS package to handle scope information. 'cpscpe' stores the information that entity 'x' is defined within the scope of entity 'y', 'cgscpe' retrieves this information. See the file "cscope.c".

Set and read flags: In the IDS there are some flags which are stored together with entities (AP information, DELETE flag, some extra flags). The corresponding routines to set and read this information are in the file "cidsut.c".

Other utilities: In the IDS package there are various routines which enable navigation through the IDS.

These are the routine 'cgtype' which returns the entity type of a specified entity, the routine 'cgnumb' which returns the number of entities which lie below a specified entity and are of a requested type and the routine 'cglist' which works similar to cgnumb, however returns additionally an array with all ids entity found. There are routines which return the number or a list of entities which are referred by a specified entity (i.e. the children) and routines to return the number or a list of entities which refer to a specified entity (i.e. the parents).

There are a lot of other utility functions in the IDS. For more detailed information please see the files "ids_rest.man" and "cidsut.c".

3.3.1.5 Function prototypes

In order to guarantee consistent calls to the IDS routines, there is a possibility for a C programmer to check at compile time the parameter list of an IDS routine. For each IDS routine there exists a corresponding function prototype describing the parameters. It is therefore recommended to include the necessary ".h" files. Please look at the chapter "File Organisation of the IDS" to find the ".h" files which are relevant for you.

FORTRAN programmers don't have this possibility.

3.3.1.6 Define statements for constants

The IDS software uses constants which are defined by '#define' statements in several files. The relevant files are "cdefin.h" and "cerror.h" of the IDS package itself, "y_tab.h" and "enum.h" of the Scanner/Parser package.

It is recommended not to use the values of the constants explicitly. You should include the ".h" files and use the symbolic names. Then you can be sure that with each software release you get (IDS or S/P), a simple recompilation of your software will be sufficient to be up to date with the newest version of these constants. The reason for this recommendation is, that it is not guaranteed that the (mostly integer) values representing the constants will remain unchanged for all versions of the common tools (some of them are automatically generated by YACC).

The direct way to include the ".h" files is not possible for the FORTRAN users. They should transfer the '#define' statements to PARAMETER statements in FORTRAN include files. This could be done automatically; in the moment only tools for keyword conversion (and partly for enumerations) are available (see chapter on conversion tools).

3.3.2 Scanner and parser

3.3.2.1 Introduction

As part of HP's contribution to CADEX, it has developed a set of tools called "The scanner/parser" (S/P) to allow the importation of STEP files into CADEX's intermediate data structure (IDS) environment. The purpose of this document is to describe the software's functionality, structure, and implementation. Refer to the figure 3.5 to understand how the (S/P) package fits into the processor architecture.

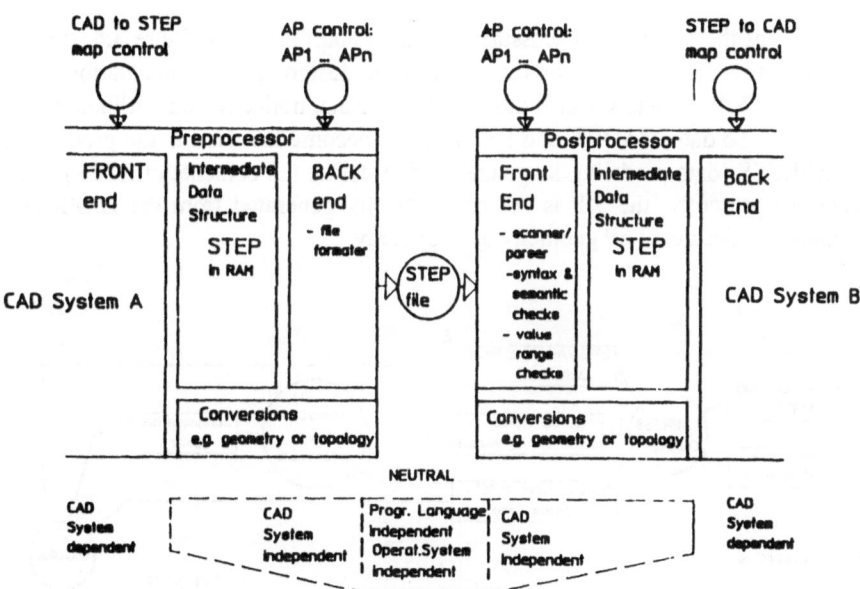

Fig. 3.5: Reference model for STEP pre/post processors

3.3.2.2 Specification

The development of HP's Post-Processor was driven by the following require-ments:

    * The S/P, or so-called "front-end" of the post-processor, whose responsibility is to scan and parse a physical STEP file, should be written as a tool, in the sense that

it is portable and usable by the other partners as HP's contribution to the CADEX toolkit. In particular, the S/P should support the multiple Application Protocols (AP)s used in the CADEX project.

* The S/P should be fully integrated with the IDS, a set of data storage and access routines created by CADEX to maintain the data embedded in a STEP file.

* In order to allow evolutionary support for future AP's, both the S/P and entire processor should be modular in nature, allowing for the easy addition of additional grammar and data in the future. Also the addition of higher software layers should be possible to decrease the amount of work which is necessary to build new AP's and allow a semi automatic generation of the S/P software.

* The S/P should be written in "C" to support the member platforms, but the back-end of the processor may be written in other languages. An example is C++, the language in which HP's new SolidDesigner CAD system is written. In addition, the software interface should be callable from FORTRAN, to support members of CADEX that work in this environment.

3.3.2.3 Scanner/parser software design

For much of the following discussing, it will be helpful to note figure 3.6, since it was and has remained the basis for much of the design and implementation work done on the S/P. So let's cover it in some detail. First, notice region outlined by the dotted line. To date, CADEX and HP has NOT become involved in the processing of EXPRESS (data modeling language used to define the contents of STEP) files. Therefore presently, the S/P is not automatically generated from the EXPRESS schemas that define STEP geometry and topology.

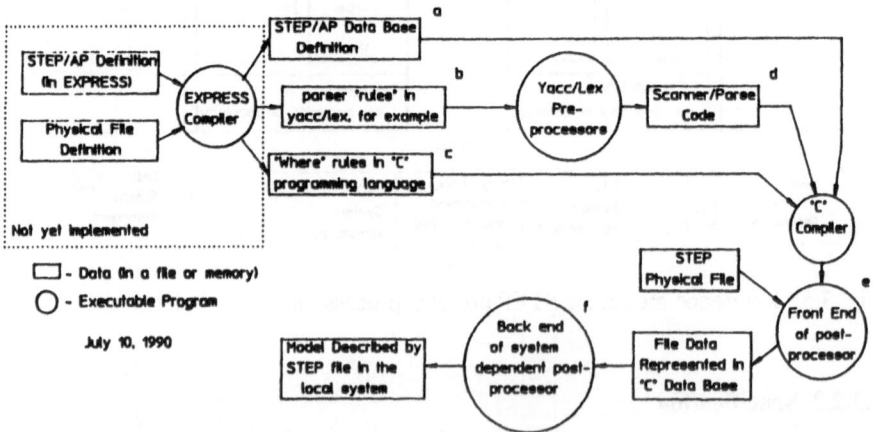

Fig. 3.6: CADEX's post-processor generation and execution

What has been accomplished though, is still quite substantial. The STEP definitions (or corresponding application protocols) have been modeled with "C" data structures and routines that install, access, and delete instances of these structures. This corresponds to box "a" in the drawing. For more information on this intermediate data structure (IDS) refer to the relevant documentation.

Next, the "rules" about how a STEP physical file looks, along with what kind of structures one is allowed to model in STEP or a corresponding AP, have been specified in a grammar specification language called YACC (this will be further discussed in the implementation section. The low level tokens that appear in the file (for example, floating point numbers) are specified in a token specification language (loosely defined) called LEX (also discussed later). This information corresponds to box "b" in the figure.

The where rules (contextual constraints on the data in a step file; for example, that the floating point number corresponding to the radius of a circle must be greater than zero) that are part of the STEP and/or AP definition have for the large part not been implemented as part of the S/P work. The group's goal is to implement such constraints as modules that will work on the IDS data structure after a file has been successfully installed into the IDS.

Box "d" corresponds to the following. The rules regarding the step physical file and AP's are then processed in order to convert them into "C" code that will be able to scan a file, read in low level tokens, collect these tokens to fit the rules that govern how STEP files can be written, and in the case when these rules are fulfilled, a particular action taken. The typical action taken is that correct information found in the file is "inserted" as an instance in the memory via calls to IDS' so-called "put" routines.

Box "e" corresponds to the executable file that when called, carries out the scanning, parsing, and installation in the IDS of a particular STEP file. The translation of the data within IDS into the user's native CAD system (box "f") has not impacted the S/P development; therefore it is not covered in this document.

3.3.2.4 Lex and Yacc for software generation

The scanning or reading in of the STEP file is accomplished by a C program that is generated from a specification that the UNIX tool "lex" can understand. Essentially all the basic token types (reals, integers, entity references, strings, etc.) are defined as tokens. In addition, actions (fragments of C code) to take place when such a token is found are also defined. The UNIX command "lex" is then executed as part of the makefile with this specification as input, generating a C file that contains the function "yylex()" which a calling program can use to get tokens from an input (in this case STEP) file. The file containing the "lex" specification is "lex.l". The file generated by "lex" is "lex.c". Lets look at a fragment of the "lex.l" specification to understand it better:

```
%{
/*  An integer. Note that leading zeros are allowed. */
%}
[+\-]?[0-9]+
    { yylval.int_val = atol((char *)yytext);
      log_token(yytext);
      return TOK_INTEGER;}
```

In this example, the token INTEGER is specified. It says that an integer may start with a "+" or "-" sign, followed by a series of digits with range 0-9. At least one digit must be supplied. In addition, if the routine generated from this specification, "yylex()", finds such a pattern, it converts the string containing the pattern to an integer, and saves it in the union structure "yylval". It then logs the found token into the log file, and returns a flag that implies that yylex() found an integer this time.

The parsing of these tokens according to a STEP Schema or AP is defined in a specification that the UNIX tool "yacc" can understand. All grammar rules regarding geometry, topology, shape, and AP variations are defined in so-called "yacc" files. They have as their suffix ".y". "yacc" is then run in the Makefile on such ".y" files, and a C file with the function "yyparse()" is generated which repeatedly calls "yylex" to get tokens, and then carries out the actions specified in the "yacc" file as each grammar rule is matched. Again, lets look at an example from the file "brep_ap.y". As you will see in the next section the example rule for a CIRCLE can also be found in sf_ap.y because a subset of their entities is used in both AP's:

```
circle : TOK_CIRCLE scoped_body tok_LP TOK_REAL TOK_COMMA
                                    TOK_REFERENCE
        { if (cpcicu(POP_ID(), &nodata, &$4, &$6, &stat) != CNOERR)
            logerror("Warning: Failure in installing CIRCLE in IDS",
                CDX_IDS_ERROR);
        }
        ;
```

This "rule" says that according to the BREP AP grammar, a circle must contain the token circle (a string defined elsewhere), a scoped body (defined by another rule in the file), an open parenthesis, a floating point number (this is the circle's radius), a comma, and an entity reference (a reference to the circle's local coordinate system). If these things are found, the routine "cpcicu()" (the IDS routine that creates a circle entity and stores its parameters) is called, and a error message is given if the routine returns a failure code.

At the end of parsing (either by receiving an EOF token from yylex() or by deciding that it has parsed enough of the present file (for example, the complete BREP section) the parser returns with either a value of 0 (ok) or non-zero (not ok). In addition, the parser writes into the global data structure "parse_results" to give more detailed information about the parsing session. This structure's contents and how to interpret the results will be discussed in the chapter "User's guide to the Scanner / Parser".

Note that in the current implementation, there are six different ".y" files. This means that there will be six different "yyparse" functions generated (one for the BREP AP, SS AP, WF AP, CSG AP, CBR and one for AP independent information - containing the header section definition, for example). To prevent a naming conflict, these three parsers are given the names yybrep_apparse(), yysf_apparse(), yywf_apparse(), yycsg_apparse(), yycbr_apparse() and yystepparse(), respectively. Although it is up to the user to call them and interpret their results accordingly, which parser to call (according to what comes next in the file being parsed) is always told to the user via the "parse_results" structure mentioned above. Refer to "main.c" for an example of how the calling of these different parsers can be managed. This issue will also be dealt with in more depth in the chapter "User's guide to the Scanner / Parser".

Once the last parser has returned the EOF signal, this means that the file has been completely parsed and stored in the IDS. The user is then free to manipulate the information in the IDS. Typically she would either do tests on the data, or call the back-end of her post-processor in order to translate the information into her native CAD environment.

3.3.2.5 Module and file organisation

According to the requirements specified in section 2 (3) the Scanner / Parser software demonstrate the evolutionary process of both the addition of new AP's and the S/P software itself. Whereas he first release supported only the BREP and SF AP, now STEP files containing all five CADEX APs can be handled. The original file structure is still valid but additional software generation tools are now part of the scanner parser to build some parts of the files automaticly.

For instance the used STEP keyword and STEP enum tables and matrices are build automaticly from a simple description of the used tokens and enumeration class dependencies. The generated tables are then included in the software by standard C "#include" statements. Therefore it is easy to change or add new enumeration types and additional token needed to support new AP's. The section "System Generation" describes the used tools. Instead of a manual and error prone handling of files like enum.h or the tables originally included in lutils.c only additional and easy to handle files like enum.hh and yacc.hhh must be touched.

Also the yacc rules to describe an AP specific parser can still be found in .y files. For each supported AP there exists a specific .y file. Please note, that there also existed an experimental version of the S/P which merged all entities used in all CADEX AP's into one large yacc grammar (described in "Esprit Project 2195 CADEX, Interim Report, 17 Jan 1992"). This concept was not only revised because such a large grammar is difficult to maintain and does not support any modularity. Such a concept is only oriented on the STEP resource models but not on AP's and cannot include constraints which are valid for a specific AP. Also the consequences for other supported AP's are not transparent if a rule in such a large grammar has to be changed.

Because of this evaluation the original "AP oriented" structure is still used in the S/P software. The decision to test the experimental structure was driven from the experience that the content and style of all CADEX AP's have overlapping components and the maintenance of different yacc files containing a lot of shared rules was error prone and does not guaranteed that the same rules specified for different AP's (e.g by the aic concept) is implemented in the same way. A "large" grammar would have solved this but introduced the other problems mentioned above.

The goal to have a more common structure and to share definitions of entities between different APs was addressed by using the new AIC concept. This simplifies maintenance and reflects the structure given by the APs itself. So for example the entities defined in the "topology bound elementary surface aic (tbes.aic)" can be shared between the SF and the BREP AP grammar. It should be stated that the AIC concept itself is experimental with open issues as for example missing rules for combining entities of different APs especially if they are named unique. For the current implementation an AIC is interpreted as a common subset of entities. If an AIC grammar is included to build an AP all entities of this AIC should be used and there should be no naming conflicts.

The whole S/P software is distributed over three directories: All S/P include files, which are needed by other parts of the CADEX toolkit or by a post processor are located in the shared "include" directory. All generating parts e.g. lex and yacc files are in the directory "sp/YACC" which contains a makefile to create C-source code from the description files by lex, yacc and other generation tools. The generated C code, C utility functions to simplify lists handling, and include files only used for the S/P module are located in the "sp" directory. The task of the makefile of this directory is to trigger the generation in the "sp/YACC" directory and to compile both generated files as well as utility C code to build an archive (libstep.a) which contains the whole S/P functionality. Because IDS and formatter are structured in the same way the build of a STEP processing program can be done by linking all toolkit module archives together with a "main" control program.

3.3.2.6 Tabular description of all files

Recognition of physical token

```
File / Module Name: sp/YACC/lex.l
Author, Owner:      Andrew Kutter, Hewlett-Packard GmbH, 1990
Bytes:              3890
Functionality:      Lex definition for all STEP token
Include files:      cdefin.h lutils.c
```

File / Module Name: sp/YACC/lex.c
Author, Owner: Andrew Kutter, Hewlett-Packard GmbH, 1990
Bytes: 14336
Functionality: Output of the lex processor on lex.l.
Relevant procedures: yylex.
Include files: cdefin.h lutils.c

Syntactical analysis of a STEP file

File / Module Name: sp/YACC/step.y
Author, Owner: Hewlett-Packard GmbH, 1992
Bytes: 28432
Functionality: Yacc definitions for the STEP grammar
                     that is independent of any particular AP
Include files: cdtput.h, cscope.h, yacc.h, yacc.hh.
Caution: Do not change this file. It is created automaticly
                by step.yy through the inclusion of "aic"-files

File / Module Name: sp/YACC/step.c
Author, Owner: Hewlett-Packard GmbH, 1992
Bytes: 51246
Functionality: Output of the yacc processor on step.y
Relevant procedures: yystepparse.
Include files: cdtput.h, cptput.h, cscope.h, yacc.h.

File / Module Name: sp/YACC/brep_ap.yy
Author, Owner: Peter J. Schild, Hewlett-Packard GmbH, 1992
Bytes: 2814
Functionality: Yacc definitions for geometry, topology
                     and shape used by the BREP AP.
Include "aic" files: ab.aic, fb.aic, pd.aic

File / Module Name: sp/YACC/brep_ap.y
Author, Owner: Hewlett-Packard GmbH, 1992
Bytes: 47532
Functionality: Yacc definitions for geometry, topology
                     and shape used by the BREP AP.
Include files: cdtput.h, cscope.h, yacc.h, yacc.hh
Caution: Do not change this file. It is created automaticly
                by brep_ap.yy through the inclusion of "aic"-files

File / Module Name: brep_ap.c
Author, Owner: Hewlett-Packard GmbH, 1992

Bytes: 77826
Functionality: Output of the yacc processor on brep_ap.y
Relevant procedures: yybrep_apparse
Include files: cdtput.h, cptput.h, cscope.h, yacc.h.

File / Module Name: sp/YACC/sf_ap.yy
Author, Owner: Peter J. Schild, Hewlett-Packard GmbH, 1992
Bytes: 2456
Functionality: Yacc definitions for geometry, topology
                          and shape used by the Surface AP.
Include "aic" files: tbs.aic, sf_aic.add, pd.aic

File / Module Name: sf_ap.y
Author, Owner: Hewlett-Packard GmbH, 1992
Bytes: 65589
Functionality: Yacc definitions for geometry, topology
                          and shape used by the Surface AP.
Include files: cdtput.h, cscope.h, yacc.h, yacc.hh.
Caution: Do not change this file. It is created automaticly
                  by sf_ap.yy through the inclusion of "aic"-files

File / Module Name: sf_ap.c
Author, Owner: Hewlett-Packard GmbH, 1992
Bytes: 102766
Functionality: Output of the yacc processor on sf_ap.y.
Relevant procedures: yysf_apparse.
Include files: cdtput.h, cptput.h, cscope.h, yacc.h.

File / Module Name: sp/YACC/wf_ap.yy
Author, Owner: Peter J. Schild, Hewlett-Packard GmbH, 1992
Bytes: 2178
Functionality: Yacc definitions for geometry, topology
                          and shape used by the Wire frame AP.
Include "aic" files: wf_all.aim, pd.aic

File / Module Name: wf_ap.y
Author, Owner: Hewlett-Packard GmbH, 1992
Bytes: 49594
Functionality: Yacc definitions for geometry, topology
                          and shape used by the WF AP.
Include files: cdtput.h, cscope.h, yacc.h, yacc.hh.
Caution: Do not change this file. It is created automaticly
                  by wf_ap.yy through the inclusion of "aic"-files

File / Module Name: wf_ap.c
Author, Owner: Hewlett-Packard GmbH, 1992
Bytes: 82781
Functionality: Output of the yacc processor on wf_ap.y.
Relevant procedures: yywf_apparse.
Include files: cdtput.h, cptput.h, cscope.h, yacc.h.

File / Module Name: sp/YACC/csg_ap.yy
Author, Owner: Peter J. Schild, Hewlett-Packard GmbH, 1992
Bytes: 8873
Functionality: Yacc definitions for geometry, topology
                    and shape used by the CSG AP.
Include "aic" files: pd.aic, tbes.aic

File / Module Name: csg_ap.y
Author, Owner: Hewlett-Packard GmbH, 1992
Bytes: 41178
Functionality: Yacc definitions for geometry, topology
                    and shape used by the CSG AP.
Include files: cdtput.h, cscope.h, yacc.h, yacc.hh.
Caution: Do not change this file. It is created automaticly
                by csg_ap.yy through the inclusion of "aic"-files

File / Module Name: csg_ap.c
Author, Owner: Hewlett-Packard GmbH, 1992
Bytes: 73615
Functionality: Output of the yacc processor on csg_ap.y.
Relevant procedures: yycsg_apparse.
Include files: cdtput.h, cptput.h, cscope.h, yacc.h.

File / Module Name: sp/YACC/cbr_ap.yy
Bytes: 6862
Functionality: Yacc definitions for geometry, topology
                    and shape used by the Compound B-rep AP.
Include "aic" files: pd.aic, ab.aic, fb.aic, sf_aic.add, wf_all.aim

File / Module Name: cbr_ap.y
Author, Owner: Hewlett-Packard GmbH, 1992
Bytes: 98566
Functionality: Yacc definitions for geometry, topology
                    and shape used by the Compound B-rep AP.
Include files: cdtput.h, cscope.h, yacc.h, yacc.hh.

Caution: Do not change this file. It is created automaticly
                   by cbr_ap.yy through the inclusion of "aic"-files

File / Module Name: cbr_ap.c
Author, Owner: Hewlett-Packard GmbH, 1992
Bytes: 139895
Functionality: Output of the yacc processor on csg_ap.y.
Relevant procedures: yycbr_apparse.
Include files: cdtput.h, cptput.h, cscope.h, yacc.h.

"AIC" related and general rules

File Name: sp/YACC/ab.aic
Author, Owner: Peter J. Schild, Hewlett-Packard GmbH, 1992
Bytes: 3026
Functionality: Mapping of the advanced_brep_aic
Include "aic" files: tbs.aic

File Name: sp/YACC/eb.aic
Author, Owner: Peter J. Schild, Hewlett-Packard GmbH, 1992
Bytes: 3250
Functionality: Mapping of the elementary_brep_aic
Include "aic" files: tbes.aic

File Name: sp/YACC/fb.aic
Author, Owner: Peter J. Schild, Hewlett-Packard GmbH, 1992
Bytes: 2492
Functionality: Mapping of the facetted_brep_aic

File Name: sp/YACC/generr.hlp
Author, Owner: Peter J. Schild, Hewlett-Packard GmbH, 1992
Bytes: 6910
Functionality: General rules / abbreviations used in all AP's

File Name: sp/YACC/generr_h.hlp
Author, Owner: Peter J. Schild, Hewlett-Packard GmbH, 1992
Bytes: 1027
Functionality: type declarations for generr.hlp

File Name: sp/YACC/pd.aic
Author, Owner: Peter J. Schild, Hewlett-Packard GmbH, 1992

```
Bytes:                  9133
Functionality:          Mapping of the product_definition_aic


File Name:              sp/YACC/scope.hlp
Author, Owner:          Peter J. Schild, Hewlett-Packard GmbH, 1992
Bytes:                  2110
Functionality:          General rules for entity occurrence and scope


File Name:              sp/YACC/scope_h.hlp
Author, Owner:          Peter J. Schild, Hewlett-Packard GmbH, 1992
Bytes:                  122
Functionality:          type declarations for scope.hlp


File Name:              sp/YACC/sf_aic.add
Author, Owner:          Peter J. Schild, Andrew Kutter,
                        Hewlett-Packard GmbH, 1992
Bytes:                  17625
Functionality:          All Yacc rules that implement the surface AP
                        without the rules imported from the
                        top_bound_surface_aic


File Name:              sp/YACC/sf_aic.hlp
Author, Owner:          Peter J. Schild, Andrew Kutter,
                        Hewlett-Packard GmbH, 1992
Bytes:                  6387
Functionality:          YACC rules which are used by the following
                        surface models:
                        geo_bound_surface_aic
                        non_manifold_surface_aic
                        manifold_surface_aic


File Name:              sp/YACC/tbes.aic
Author, Owner:          Peter J. Schild, Andrew Kutter,
                        Hewlett-Packard GmbH, 1992
Bytes:                  8101
Functionality:          Mapping of the top_bound_el_surf_aic


File Name:              sp/YACC/tbs.aic
Author, Owner:          Peter J. Schild, Hewlett-Packard GmbH, 1992
Bytes:                  7001
Functionality:          Mapping of the top_bound_surf_aic
Include "aic" files:    tbes.aic
```

File Name:	sp/YACC/wf_all.aim
Author, Owner:	Peter J. Schild, Hewlett-Packard GmbH, 1992
Bytes:	23196
Functionality:	YACC rules that implement all entities from the WF AP

Utility functions

File / Module Name:	sp/putils.c
Author, Owner:	Andrew Kutter, Hewlett-Packard GmbH, 1990
Bytes:	15553
Functionality:	Reset and initialize routines, log file handling, conversion of lists to (FORTRAN-)arrays.
Relevant procedures:	cd_list_to_long_array, cd_2xlist_to_double_array, cd_2xlist_to_long_array, cd_list_to_double_array, cd_free, cd_free_list, cd_str_llist_to_str_array, cd_free_str_list, init_parse_record, reset_parser, fetch_parse_results, init_parser, yyerror, logerror, markerror, log_token, log_char, cp_put_header, cd_2xlist_to_long_array_reverse, cd_2xlist_to_double_array_reverse
Include files:	cdtput.h, lists.h, lists.h2, parse.h.

File / Module Name:	sp/lutils.c
Author, Owner:	Andrew Kutter, Hewlett-Packard GmbH, 1990
Bytes:	6701
Functionality:	Correspondence between STEP keywords, enums and used tokens.
Relevant procedures:	lookup_keyword_token, lookup_token_keyword, lookup_enum_token, lookup_token_enum.
Include files:	lists.h, y_tab.h, enum.h, key.tab, enum_tm.tab

File / Module Name:	sp/lists.c
Author, Owner:	Andrew Kutter, Hewlett-Packard GmbH, 1990
Bytes:	2776
Functionality:	Linked list handling
Relevant procedures:	cd_add_node, cd_start_string_list, cd_add_string_node, cd_get_object, cd_get_next_node, cd_reverse_list, cd_del_head_from_list, cd_del_list, cd_count_nodes.
Include files:	lists.h

Include files

File / Module Name: include/enum.h
Author, Owner: Andrew Kutter, Hermann Ruess, Peter J.Schild
                     Hewlett-Packard GmbH, 1991
Bytes: 1532
Functionality: #defines for STEP enums, automatically generated
                     from enum.hh.
Caution: Do not change this file.

File / Module Name: sp/enum.hh
Author, Owner: Andrew Kutter, Hermann Ruess, Peter J. Schild
                     Hewlett-Packard GmbH, 1992
Bytes: 3523
Functionality: Class and Member declaration of STEP enumeration
                     types.

File / Module Name: sp/enum_tm.tab
Author, Owner: Andrew Kutter, Hermann Ruess, Peter J. Schild
                     Hewlett-Packard GmbH, 1991
Bytes: 3965
Functionality: Definition of step_enum_table and step_enum_mat
                     rix. Automatically generated
                     from enum.hh, included in lutils.c.
Caution: Do not change this file directly!

File / Module Name: sp/key.tab
Author, Owner: Andrew Kutter, Hermann Ruess, Peter J. Schild
                     Hewlett-Packard GmbH, 1991
Bytes: 8179
Functionality: Definition of step_keyword_table. Automatically
                     generated from yacc.hhh, included in lutils.c.
Caution: Do not change this file directly!

File / Module Name: include/lists.h
Author, Owner: Andrew Kutter, Hewlett-Packard GmbH, 1990
Bytes: 616
Functionality: Types used in lists.c

File / Module Name: include/lists.h2
Author, Owner: Andrew Kutter, Hewlett-Packard GmbH, 1990
Bytes: 813
Functionality: Function declarations used in lists.c

```
File / Module Name: include/lutils.h
Author, Owner:       Andrew Kutter, Hewlett-Packard GmbH, 1990
Bytes:               746
Functionality:       Function declarations for lutils.c


File / Module Name: include/parse.h
Author, Owner:       Andrew Kutter, Hewlett-Packard GmbH, 1990
Bytes:               901
Functionality:       Declaration of global data structure
                     parse_results


File / Module Name: include/putils.h
Author, Owner:       Andrew Kutter, Hewlett-Packard GmbH, 1990
Bytes:               1494
Functionality:       External variable and function declarations
                     for putils.c


File / Module Name: include/y_tab.h
Author, Owner:       Hewlett-Packard GmbH, 1990
Bytes:               4650
Functionality:       Output file of yacc that declares all token
                     types it expects to get back from yylex()


File / Module Name: include/yacc.h
Author, Owner:       Andrew Kutter, Hewlett-Packard GmbH, 1990
Bytes:               733
Functionality:       Set of external and static definitions used
                     by all .y files.


File / Module Name: sp/YACC/yacc.hh
Author, Owner:       Andrew Kutter, Hermann Ruess, Peter J. Schild,
                     Hewlett-Packard GmbH, 1991
Bytes:               5496
Functionality:       Set of token definitions used by all .y files.
                     Automatically generated from yacc.hhh.
Caution:        File must be sorted. Do not change this file.


File / Module Name: sp/YACC/yacc.hhh
Author, Owner:       Andrew Kutter, Hermann Ruess, Peter J. Schild
                     Hewlett-Packard GmbH, 1991
Bytes:               5496
Functionality:       Set of token definitions used by all .y files.
```

System generation

```
File / Module Name: sp/makefile
Author, Owner:       Andrew Kutter, Peter J. Schild
                     Hewlett-Packard GmbH, 1992
Bytes:               6444
Functionality:  The UNIX Makefile executed to compile
                the scanner/parser. Generates libstep.a, which
                contains all modules directly related to the
                s/p software. Forces the generation of all
                lex/yacc related C source code
Relevant subcall:   ap/YACC/makefile
Used files:     mkenumh.awkm, mkenumt.awk, mkenumm.awk,
                mkkeytab.awk, yacc_awk


File / Module Name: sp/YACC/makefile
Author, Owner:       Andrew Kutter, Peter J. Schild
                     Hewlett-Packard GmbH, 1992
Bytes:               5498
Functionality:  The UNIX Makefile executed to create all
                lex / yacc related C source.


File / Module Name: main/Makefile
Author, Owner:       Andrew Kutter, Peter J. Schild
                     Hewlett-Packard GmbH, 1992
Bytes:               3116
Functionality:  The UNIX Makefile executed to trigger the compilation
                of all toolkit modules (sp/ids/format)
                Uses the created libraries and compiles main.c
                to create one executable example program


File / Module Name: main/main.c
Author, Owner:       Andrew Kutter, Peter J. Schild
                     Hewlett-Packard GmbH, 1991
Bytes:               7136
Functionality:  Example program of how to call the scanner/
                parser and interpret its results


File / Module Name: sp/mkenumh.awk
Author, Owner:       Peter J. Schild
                     Hewlett-Packard GmbH, 1991
Bytes:               212
```

Functionality: Program file for awk. Generates from the
                       enumeration declarations in enum.hh the
                       #defines used in enum.h
Used files: stdin (e.g. enum.hh)
Generated files: stdout (e.g. enum.h)

File / Module Name: sp/mkenumm.awk
Author, Owner: Peter J. Schild
                       Hewlett-Packard GmbH, 1991
Bytes: 1412
Functionality: Program file for awk. Generates the body part
                       of the step_enum_matrix.
Used files: enum_tab.tmp (sorted structure definition of
                       used enums, created by "nawk -f mkenumt.awk
                       enum.hh |sort >enum_tab.tmp")
                       stdin (e.g. enum.hh)
Generated files: stdout (e.g. enum_mtx.tmp, temporary used enum
                       matrix) defines.tmp (Dimensions of generated
                       enum matrix)

File / Module Name: sp/mkenumt.awk
Author, Owner: Peter J. Schild
                       Hewlett-Packard GmbH, 1991
Bytes: 372
Functionality: Program file for nawk. Generates the body part
                       of the step_enum_table.
Used Files: stdin (e.g. enum.hh, see mkenumm.awk)
Generated files: stdout (e.g. enum_tab.tmp)

File / Module Name: sp/mkkeytab.awk
Author, Owner: Peter J. Schild
                       Hewlett-Packard GmbH, 1991
Bytes: 1413
Functionality: Program file for nawk. Generates step keyword
                       table:
                       "awk -f mkkeytab.awk yacc.hh >key.tab"
Used files: stdin (e.g. enum.hh)
Generated files: stdout (e.g. ">key.tab")

File / Module Name: sp/YACC/rec_incl.c
Author, Owner: Peter J. Schild, Hewlett-Packard GmbH, 1991
Bytes: 1116
Functionality: C-Program file to allow recursive inclusions

```
                        of "yacc_include" statements. Example:
                        Build up of brep_ap.y from brep_ap.yy
                        by including tbs.aic....tbes.aic...
Used files:             stdin (e.g. "step.y")
Generated files:        stdout (e.g. ">temp.y")
```

3.3.2.7 Tabular documentation of utility procedures

```
Procedure:       cd_add_node
Module:          sp/lists.c
Functionality:   Add a node of a particular size to the head of a
                 linked list. Note that the node is allocated
                 and attached- no object is copied into the
                 space allocated, however.
Parameters:
 in:             listptr list    List where object is to be added
                 unsigned int object_size  Number of bytes
                 needed for an list node
 out:            -
 return:         listptr   Expanded list


Procedure:       cd_add_string_node
Module:          sp/lists.c
Functionality:   Add an object to a list that is
                 assumed to be a string
Parameters:
 in:             listptr list
 out:            char *string
 return:         listptr
Relevant calls:  cd_add_node


Procedure:       cd count_nodes
Module:          sp/lists.c
Functionality:   Returns number of nodes in the list
Parameters:
 in:             listptr list
 out:
 return:         long


Procedure:       cd_del_head_from_list
Module:          sp/lists.c
Functionality:   Deletes the first node in a linked list and
                 returns the new head of the list
```

```
Parameters:
 in:            listptr list
 out:           -
 return:        listptr


Procedure:      cd_del_list
Module:         sp/lists.c
Functionality:  Deletes a list and returns NULL when successful
Parameters:
 in:            listptr list
 out:           -
 return:        listptr (NULL!)


Procedure:      cd_free
Module:         sp/putils.c
Functionality:  free allocated memory
Parameters:
 in:            char *ptr    Pointer to memory
 out:           -
 return:        -
Relevant calls:    free


Procedure:      cd_free_list
Module:         sp/putils.c
Functionality:  Free a linked list and its related
                dynamically allocated array.
Parameters:
 in:            listptr list       Linked list to be deleted
                char* new_array    Related array
 out:           -
 return:        -
Relevant calls:    cd_del_list, cd_free


Procedure:      cd_free_str_list
Module:         sp/putils.c
Functionality:  Delete a list of strings and all the structs
                that cd_str_llist_to_str_array() builds up
Parameters:
 in:            listptr list     List of strings
                char *str_block  Single string of all strings
                long *start  Index array
 out:           -
 return:        -
Relevant calls:    cd_free, cd_del_list, cd_free
```

```
Procedure:      cd_get_next_node
Module:         sp/lists.c
Functionality:  Return the next element of a list
Parameters:
 in:            listptr object
 out:           -
 return:        listptr  Next object of the list


Procedure:      cd_get_object
Module:         sp/lists.c
Functionality:  Return the object of a node
                (usually to be casted)
Parameters:
 in:            listptr node
 out:           -
 return:        char *  Pointer to memory containing the object
Relevant calls:


Procedure:      cd_list_to_long_array
Module:         sp/putils.c
Functionality:  Generate a dynamically allocated array of long
                integers from a linked list.
Parameters:
 in:            listptr list      Linked list (see list.c)
 out:           long *length_ptr  Length of allocated array
                long **new_array  Allocated Storage containing
                                  the list elements
 return:        CDX_NO_ERROR
Errors:         CDX_ALLOC_ERROR   Could not alloc enough memory


Procedure:      cd_2xlist_to_long_array
Module:         sp/putils.c
Functionality:  Generate a dynamically allocated array of
                integers from a linked list of lists of
                integers.
Parameters:
 in:            listptr list      Linked list (see list.c)
 out:           long *length_ptr  Length of allocated array
                long **new_array  Allocated Storage containing
                                  the list elements
 return:        CDX_NO_ERROR
Errors:         CDX_ALLOC_ERROR   Could not alloc enough memory
```

```
Procedure:       cd_2xlist_to_long_array_reverse
Module:          sp/putils.c
Functionality:   Same as cd_2xlist_to_long_array but new_array
                 is created in a way that u and v parameters
                 are exchanged. This function enables the
                 S/P to accept a AP conforming B_SPLINE_SURFACE
                 but have the old style ids interface


Procedure:       cd_list_to_double_array
Module:          sp/putils.c
Functionality:   Generate a dynamically allocated array of double
                 from a linked list.
Parameters:
 in:             listptr    list       Linked list (see list.c)
 out:            long    *length_ptr  Length of allocated array
                 double **new_array  Allocated Storage containing
                                 the list elements
 return:         CDX_NO_ERROR
Errors:          CDX_ALLOC_ERROR    Could not alloc enough memory


Procedure:       cd_2xlist_to_double_array
Module:          sp/putils.c
Functionality:   Generate a dynamically allocated array of
                 doubles from a linked list of lists of double.
Parameters:
 in:             listptr list           Linked list of doubles (see
                               list.c)
 out:            long    *length_ptr  Length of allocated array
                 double **new_array    Allocated Storage
                                 containing the list elements
 return:         CDX_NO_ERROR
Errors:          CDX_ALLOC_ERROR    Could not alloc enough memory


Procedure:       cd_2xlist_to_double_array_reverse
Module:          sp/putils.c
Functionality:   Same as cd_2xlist_to_double_array but new_array
                 is created in a way that u and v parameters
                 are exchanged. This function enables the
                 S/P to accept a AP conforming B_SPLINE_SURFACE
                 but have the old style ids interface


Procedure:       cd_reverse_list
Module:          sp/lists.c
Functionality:   Reverse the connections in a list
```

```
Parameters:
 in:              listptr list    List to reverse
 out:             -
 return:          listptr         Reversed list
Relevant calls:


Procedure:       cd_start_string_list
Module:          sp/lists.c
Functionality:   Begin a list by creating a node, attaching a
                 string as its object, and returning a pointer
                 to this node.
Parameters:
 in:              char *string    First string to be inserted
 out:
 return:          listptr  New created list


Procedure:       cd_str_llist_to_str_array
Module:          sp/putils.c
Functionality:   Take a list of strings, and convert them into a
                 single string with the number of characters in
                 the block, an index array pointing to the
                 beginning of each sub string  and the number
                 of strings originally in the list
Parameters:
 in:              list_ptr str_list   List of strings
 out:             char **str_block    Generated single string
                  long *num_chars     Number of characters in
                             the block
                  long **start        Index array
                  long *num_strs      Number of strings in the
                                      original list


Procedure:       cp_put_filename
Module:          sp/putils.c
Functionality:   Take the header information supplied by the
                 yacc parser, convert into the intermediate data
                 structure format, and then insert it into an
                 instance of this data structure
Parameters:
 in:              char *file_name
                  char *time_stamp
                  char *step_version
                  char *pre_proc_version
                  char *originating_system
```

```
                    listptr author
                    listptr organization
 out:               -
 return:            -


Procedure:          fetch_parse_results
Module:             sp/putils.c
Functionality:      Retrieve the results of the last parse
Parameters:
 in:                -
 out:               long *error       General result of parsing
                    long *use_ap        Last AP used
                    long *error_count    Number of errors detected
                    long *warning_count Number of warnings detected
 return:            -


Procedure:          init_parse_record
Module:             sp/putils.c
Functionality:      Reset error, use_ap, error_count and
                    warning_count
Parameters:         -


Procedure:          init_parser
Module:             sp/putils.c
Functionality:      Initialize the parser. Open the file pointers
                    to the step file and the log file (if asked to),
                    set the parse debug variable, and initialize
                    global variables
Parameters:
 in:                long write_log   If !=0: generate logfile
                    long debug    yacc debug flag
                    char *fname       Stepfile to parse
 out:               -
 return:            -


Procedure:          logerror
Module:             sp/putils.c
Functionality:      Print error message to the logfile and remember
                    error in parse results structure
Parameters:
 in:                char *s          Error message
 out:               long error_code
 return:            -
```

Procedure:	log_char
Module:	sp/putils.c
Functionality:	Write a character to logfile
Parameters:	
in:	char s Character to write

Procedure:	log_token
Module:	sp/putils.c
Functionality:	Write a token string to logfile
Parameters:	
in:	char *s String to write

Procedure:	lookup_enum_token
Module:	sp/lutils.c
Functionality:	Search the enum lookup table to determine if a text string recognized by Lex is STEP enum. If it is, return its value of it.
Parameters:	
in:	unsigned char *text String recognized by lex
out:	int *value Value of found enum member
	int *enum_class Value of found enum class
return:	CDX_ERROR Text not found

Procedure:	lookup_keyword_token
Module:	sp/lutils.c
Functionality:	Search the keyword lookup table to determine a text string recognized by Lex is STEP keyword. If it is, return the token number of it. Note this routine comes from McDonnel Douglas's express compiler.
Parameters:	
in:	unsigned char *text String recognized by lex
out:	-
return:	long Token number of text
Errors:	CDX_ERROR Text no valid token

Procedure:	lookup_token_enum
Module:	sp/lutils.c
Functionality:	This routine returns a pointer to the string related to the STEP enum. Note that a malloc() routine is called, and therefore the free() routine must also be called to release the allocated memory to the heap.

Parameters:
 in: long value   Enum for which text is needed
                long enum_class Enum class for which member
                                 value text is needed
 out: -
 return: char *   Text to enum value
 Errors: NULL   No text to enum found
 Relevant calls: malloc

Procedure: lookup_token_keyword
Module: sp/lutils.c
Functionality: This routine returns a pointer to the string
                related to the STEP token. Note that a malloc()
                routine is called, and therefore the free()
                routine must also be called to release the
                allocated memory to the heap.
Parameters:
 in: long token Token for which text is needed
 out:
 return: char *   Corresponding text
 Errors: NULL   No text to token found
 Relevant calls: malloc

Procedure: markerror
Module: sp/putils.c
Functionality: Set "^" at appropriate error position in logfile
Parameters: -

Procedure: POP_ID
Module: sp/pushpop.hlp
Functionality: This function is necessary to support the new
                scope section revised in ISO TC184/SC4/WG1
                April 23-27, 1990. Every time when an
                identifier is detected, its value is pushed
                to the identifier_stack. POP_ID() is used
                every time a valid token has been accepted
                to restore the (pushed) identifier id.
Parameters: -

Procedure: PUSH_ID
Module: sp/pushpop.hlp
Functionality: This function is necessary to support the new
                scope section revised in ISO TC184/SC4/WG1 in
                April 23-27, 1990. Every time when an

```
                    identifier is detected, its value is pushed
                    to the  identifier_stack.
Parameters:         long id         -


Procedure:          process_string
Module:             lutils.c
Functionality:      Copy a string to another string. Included "\n"
                    were removed
Parameters:
 in:                char *str        Original text
 out:               char *new_str_ptr  Text with removed "\n"
 return:            -


Procedure:          yy_error
Module:             putils.c
Functionality:      Record the column in which an error was found.
                    The parameter is required because the yacc
                    generated code looks for this routine with the
                    one parameter.
Parameters:
 in:                char *s  Error message
 out:               -
 return:            -
```

3.3.2.8 Generating the CADEX scanner / parser software tool

The following steps must be executed by the Makefile to generate an updated scanner/parser.

1. yacc.hhh -> yacc.hh
   1.1 Extract the %union declared for use with yacc from yacc.hhh.
   Create yacc.hh with the extracted %union.
   1.2 Extract the %tokens declared for use with yacc, sort them and append the sorted tokens to yacc.hh.

2. step.yy -> step.o
   2.1 Inclusion of yacc.hh, aic stuff and other utilities to create with tool rec_incl.cmd, create step.y
   2.2 yacc creates y.tab.c, y.tab.h.
   2.3 rename y.tab.h to y_tab.h.
   2.4 Change identifiers in y.tab.c to unique identifiers in the whole system:
   All identifiers beginning with "yy" were changed to identifiers beginning with "yystep".

The exception to this rule are yyerror, yylex, yylval and yydebug.
Creates step.c.

2.5 Compile step.c to step.o.

3. brep_ap.yy -> brep_ap.o
   < repeat the steps 2.1 - 2.5, exchange "step" with "brep_ap" >

4. ss_ap.yy -> ss_ap.o
   < repeat the steps 2.1 - 2.5, exchange "step" with "ss_ap" >

5. wf_ap.yy -> wf_ap.o
   < repeat the steps 2.1 - 2.5, exchange "step" with "wf_ap" >

6. csg_ap.yy -> csg_ap.o
   < repeat the steps 2.1 - 2.5, exchange "step" with "csg_ap" >

7. cbr_ap.yy -> cbr_ap.o
   < repeat the steps 2.1 - 2.5, exchange "step" with "cbr_ap" >

8. yacc.hhh -> key.tab
   8.1 Nawk script to build the step_keyword_table data structure.
   8.2 The dimensions needed for the keyword table were inserted at the
        beginning of the file key.tab.

9. enum.hh ->enum.h
   Nawk script extracts the class and member declaration in enum.hh
   to produce a valid c-include files with the #defines for the used enums.
   See 9.3 because further #defines were added to enum.h.

10. enum.hh -> enum_tm.tab
    10.1 Build temporary file enum_tab.tmp which contains the body of the
          step_enum_table data structure.
    10.2 Build temporary file enum_mtx.tmp which contains the
          step_enum_matrix data structure.
    10.3 The execution of 9.1 and 9.2 produces a file "defines.tmp" whic contains
          the dimensions needed for the enum matrix and enum table.
          "defines.tmp" is added to enum.h.
    10.4 "enum_tm.tab" is created which contains the complete enum table and
          enum matrix data structure.

11. lex.l -> lex.c

    Use lex to create lex.yy.c and rename it to lex.c

12. Compile lex.c, lists.c, lutils.c, putils.c and main.c

13. Archive all produced object files in library libstep.a

To create an example post processor also the compilation and archiving for the ids and formatter must be triggered. Together with a "main" program all archives have to be linked to build an executable.

3.3.3 Formatter

3.3.3.1 Design overview

The task of the Formatter is basically straightforward: from the contents of IDS to generate a Step file. Beyond Step specifications on mapping to the physical format (which, of course, must be fulfilled), during the Formatter design phases several further goals have been taken into account, mainly aimed to define some specifications on file format so that the work of the processors is eased (and possibly the readability of the file is enhanced). Each of these goals is related to an aspect in which Step specifications, as currently released, are not mandatory, nor explicit, nor clear.

* The definition of Step entities is relational in nature, several values of entity attributes being references to other entities. Although not formally required, the usefulness of a strict backward referencing (i.e., that each entity is defined always before it is used in the definition of another entity) has been pointed out, since this way a single-scan processing of the file is greatly eased. The implemented algorithm (ref. Section 2.3.) assures that this condition is always satisfied.

* Each entity in a Step file is uniquely identified by a positive integer number. Such identifiers are also used within IDS, so that in principle the Formatter can directly read them from it. Since the only requirements on these identifiers are related to their range and uniqueness, they can appear in the physical file in any order; since the Formatter always writes only backward referenced entities, in general the identifiers would not appear as increasingly ordered in the file. However, it could be useful that entities in the file are increasingly ordered, while strictly maintaining backward referencing. In general, this requires a renumbering of the entities. To this goal, a suitable, optionally activable, algorithm has been designed and implemented.

* Let "top entity" be an entity never used in defining another entity. A Step file (and IDS) can store more than one top entities at the same time. On the other hand, it is reasonable to assume that each top entity identifies a different (sub-)model.

Therefore it could be useful to handle each top entity and its defining entities separately from any other top entity. To this goal, two different options have been defined and implemented, allowing to choose a single top entity and to write it with its defining entities in the Step file, or to write each top entity with its defining entities in a different file. Moreover, entities from more than one AP can appear in the same file.

* While a Step file could physically appear as a single line, it can be useful (e.g., for readability or for network transfer) that it is optionally written according to some layout. Therefore several optionally activable strategies have been defined and implemented, allowing to differently "pretty print" the file.

3.3.3.2 General outline of the implementation

The Formatter is implemented as a set of ANSI C language functions, all declared and defined in the files cform.h, cformx.h, and cform.c. The functions which can be directly called from a main program are cfcdef(), cfcfil(), cfcpar(), cfmt(), and cform(), all declared in cform.h (so that the main program has to simply include such a file). These functions realize two main tasks: the configuration of the environment, and the formatting.

- The function cfcdef() initializes the data structure holding all parameter values for the configuration (see the Section 3 in the User's Guide) and sets their default values. It must be called before any other Formatter function.
- The function cfcfil() reads a configuration file which contains the definitions overriding the default values of one or more parameters.
- The function cfcpar() modifies a single parameter value, overriding the corresponding default value.
- The function cfmt() controls all steps of the formatting (see next Section).
- Moreover, in order to allow to execute the formatting by just one function call, the function cform() has been defined, which calls cfcdef(), then cfcfil(), and finally cfmt().

3.3.3.3 The basic algorithm of formatting

The function cfmt() implements the basic algorithm of formatting. Its conceptual outline is the following.

Function cfmt():
  if the option is set for entity renumbering:
    initialize the corresponding data structure;
  call make_scope_list(); /* generates a data structure for handling scope */
  get the list of the top entities in IDS;
  if the option is set for writing only one top entity:
    remove from the list of top entities all entities but the one to be
    written

call clear_visited(); /* clears the flag for writing the entities just once */
for each "top entity":
>   if the option is set for writing each top entity in a different file:
>>   call clear_visited();
>   initialize the output stream (either a file or the screen);
>   if the option is set for writing each top entity in a different file
>   or the first top entity of the list is to be written:
>>   call cwr_head(); /*writes the Step file Header section */
>   call cwr_ent() /* writes a top entity (and recursively all its
>   defining entities) */
>   close the output stream;
free all temporarily used data structures.

3.3.3.4 The algorithm for writing the entities

The Formatter main algorithm is related to the writing of the Data section of the Step file; it is contained in the function cwr_ent(). Its conceptual outline is the following.

Function cwr_ent(ID_of_the_entity_to_be_written):
>   get the parameters and the defining entities of ID; /* call the proper
>   function cgXXXX */
>   for each defining entity ENT:
>>   if ENT has not been already written and it is not included in the
>>   scope of ID:
>>>   call cwr_ent(ENT);
>   write the entity id of ID;
>   if ID has a scope:
>>   write "&scope";
>>   for each defining entity ENT which is included in the scope:
>>>   call cwr_ent(ENT);
>>   write "endscope";
>   write the entity definition line for ID.

Note 1 - This algorithm requires that a (possibly empty) list is associated with each entity, containing the identifiers of the entities included in the scope of the former entity. The data structure storing such a set of lists is generated by the function make_scope_list(). Thanks to this data structure, Scope information is properly written in the output file, even in case of nested structures.

Note 2 - In order to verify that entities are written just once in the Step file, the IDS flag visited is adopted (the function clear_visited() is then called for initializing such flags).

Note 3 - The way the described algorithm is implemented does not guarantee the semantics correctness of the generated output file, which is fully dependent on the correctness of the IDS contents. Specifically, it is therefore (syntactically) allowed that: entities of any type have a scope; any entity name and any enumeration variable appearing in a parameter list is omitted (as if it were optional); and so on.

Multiple top entities: In case of more than one top entity within IDS, the Formatter optionally writes each top entity, together with its defining entities, on a different output file. Moreover another option allows to write a single top entity in the file.

3.3.3.5 The algorithm for renumbering the entities in the output file

The Formatter offers an option allowing to write the output file so that it contains increasingly ordered entities, while maintaining only backward references in the definition of the entities. In general, this requires a renumbering of the entities.

The outline of the related algorithm is the following:

> given a BASE value and a STEP value for the sequence (read from the configuration file);
> set CURR_ID = BASE;
> allocate an array NEW_NAMES of ints, with as many elements as the IDS lookup table;
> each time a defined entity ENT is to be written:
> > instead of ENT, write CURR_ID;
> > if ENT appears as the N-th element of the lookup table:
> > > set NEW_NAMES[N] = CURR_ID;
> > set CURR_ID = CURR_ID + STEP;
> each time a defining entity ENT is to be written:
> > if ENT appears as the N-th element of the lookup table:
> > > instead of ENT, write NEW_NAMES[N].

Note that this algorithm is essentially based on the fact that in the output file an entity is defined always before it is used to define other entities (i.e. no forward references appear in the output file).

3.3.3.6 List of the implemented functions

In this Section, the prototypes of the functions defined in the Formatter module are listed. For each function the prototype, according to ANSI specifications for C language, and some semantic notes are indicated.

```
long cform      ( char *file_n, long *istat );
```
is a "macro" function, for a default call to the Formatter

```
long cfcdef     ( long *istat );
```
initializes the data structure for storing configuration parameters and reads their
  default values

```
long cfcfil     ( char *file_n, long *istat );
```
reads the configuration file of the Formatter

```
long cfcpar     ( char *keyname, char *keyval, long *istat );
```
modifies one configuration parameter of the Formatter

```
long cfmt       ( long *istat );
```
is the "main" function of the Formatter

```
long make_scope_list   ( long *istat );
```
scans the IDS lookup table and creates a temporary data structure for handling
   scope

```
void scope_handler     ( long *name );
```
checks whether the entity being written has a scope, and in that case calls the
   function for writing all entities in the scope

```
void write_ref ( long *name, long *to_write);
```
is a utility function: given a defined entity, writes one defining entity

```
void cwr_head  ( long *istat );
```
writes the header section of the output file

```
void cwr_ent    ( long *named );
```
writes a single entity in the output file

```
void pretty_printer    ( long *length );
```
handles newlines and indentation (low level function)

```
void wCHAR      ( char *what );
```
writes a string, with no formatting tests (low level function)

```
void wTOK       ( long *type );
```
writes the string of an entity type (e.g. "FACE") (low level function)

```
void wINT       ( long *what );
```
writes an integer value according to standard format (low level function)

```
void wFLOAT     ( double *what );
```
writes a float value according to standard format (low level function)

```
void wFFLOAT    ( char *how, double *what );
```
writes a float value, with a specified format (low level function)

```
void wSTR       ( char *what );
```
writes a string, with formatting control (low level function)

```
void wCHR       ( char *what );
```
writes a single character (low level function)

```
void wENUM      ( long *what, long cls );
```
writes an enumeration variable (low level function)

```
void wNNAME     ( long *name, long *pos);
```
writes the name of a defined entity (low level function)

```
void wNAME      ( long *name );
```
writes the name of a defining entity (low level function)

```
void wBEG_L     ( void );
```
writes a "begin of list" character (low level function)

```
void wEND_L     ( void );
```
writes an "end of list" character (low level function)

```
void wDEL_L     ( void );
```
writes a list delimiter character (low level function)

```
void wEND_E     ( void );
```
writes an "end of entity" character (low level function)

```
void wSKIP_LN   ( void );
```
inserts a new line (low level function)

```
void wENTITY    ( long *name, long *type );
```
writes the "left side part" of a defining entity (low level function)

3.3.4 Checker

3.3.4.1 IDS data checker functionality

The main item of test software developed is the IDS data checker. The Requirement Specification can be found in [9]. This code checks the contents of the IDS against the Express [39] definition of an AP and also includes physical file scope checking [38]. In brief, it provides semantic checking of the data, above the syntactic checking performed by the Scanner/Parser software. Together with the Scanner/Parser, the IDS data checker provides the required functionality of a STEP file checker required for both user and conformance testing.

The development has taken place in phases and covers B-rep [11] and Sculptured Surface [12] APs. The software has been successfully integrated into the common toolkit version 5.4.

The functionality of this code is one element of the suite of tools required for a conformance testing service for STEP. Descriptions of the other elements, and an overall architecture for conformance testing of other neutral format file processors can be found in the CTS2 project report [7].

Whilst no such specific software specifications have yet been defined for STEP, it is anticipated that the basic architecture will not differ substantially to that already specified in the CTS2 project. Indeed the whole methodology and architecture was designed with knowledge of what would be required for STEP, with the intention of applying it to STEP before completion of the project. The project now has formal commitments to pursue the development of STEP conformance testing during its final year.

The functional specification of the software can be found in [21]. It is coded in C and performs the following checks on the contents of the IDS:
- attribute data type checking
- entity existence checking
- entity referencability checking (scope)
- where rule checking

3.3.4.2 Errors reported

(Here, `***` represents an integer number and `$$$` a character string.)

Entity name and entity type errors:

error 101: entity ``#***'' is defined but not used by top entity

error 102: entity ``#***'' is not defined

error 103: entity ``#***'', attribute ***, list-element *** -- referenced entity
``#***'' should be $$$

error 104: entity ``#***'', attribute ***, list-element *** -- referenced entity
``#***'' should be $$$ AP entity

error 105: entity ``#***'' is of unknown type

Simple and structured data type errors:

error 201: entity ``#***", attribute *** -- integer value should be $$$

error 202: entity ``#***", attribute *** -- real value should be $$$

error 203: entity ``#***", attribute *** -- string should be $$$

error 204: entity ``#***" is a zero vector --

error 205: entity ``#***", attribute *** -- boolean or enumeration value should be $$$

error 206: entity ``#***", attribute *** -- value is of unknown $$$ enumeration type

error 207: entity ``#***", attribute *** -- list should $$$

error 208: entity ``#***", attribute ***, list-element *** -- duplicate reference to list-element ***

Entity referencability errors:

error 301: entity ``#***", attribute ***, list-element *** -- referenced entity ``#***" should not refer to itself

error 302: entity ``#***", attribute ***, list-element *** -- referenced entity ``#***" should not be ancestor of entity ``#***"

error 303: entity ``#***", attribute ***, list-element *** -- referenced entity is defined in scope of entity ``#***"

Where rule violation errors (topology):

error 401: entity ``#***", where rule ``topology constraints" violates --
non-unique edge/logical entities: entity ``#***" and entity ``#***"
non-unique loop entities: entity ``#***" and entity ``#***"
non-unique face entities: entity ``#***" and entity ``#***"
non-unique shell entities: entity ``#***" and entity ``#***"

error 402: entity ``#***", where rule ``topology constraints" violates --
edge entity ``#***" should be referenced $$$ - at least one edge should be referenced exactly once

error 403: entity ``#***", where rule ``topology constraints" violates --
open shell should have at least one hole

error 404: entity ``#***", where rule ``topology constraints" violates -- genus should be $$$

error 405: entity ``#***", where rule ``topology constraints" violates -- Euler equation is not satisfied

Where rule violation errors (geometry):

error 406: entity ``#***", where rule ``geometry constraints" violates --
direction inconsistency in use of edges

error 407: entity ``#***", where rule ``geometry constraints" violates --
>>given sequence of edge/logical entities does not form a loop
given sequence of edge/logical entities forms more than one
loop

error 408: entity ``#***", where rule ``geometry constraints" violates -- all weights
should be greater than 0.0

error 409: entity ``#***", attribute ***, where rule ``param constraints" violates --
sum of knot multiplicities should be $$$ - knot multiplicity
should be $$$

error 410: entity ``#***", attribute ***, where rule ``param constraints" violates --
knots should be in non-decreasing order

error 411: entity ``#***", attribute ***, where rule ``param constraints" violates --
incompatible knot type with knots
incompatible knot type with knot multiplicities

3.3.4.3 Where rule checking

Where rule checking is performed in 4 phases:

- data value constraints checking
- B-spline parameterization constraints checking
- topology constraints checking
- geometry constraints checking

Phases (1) and (2) are completed, but phases (3) and (4) are not - due to incompleteness of topology and geometry constraints functions defined in STEP Part 42 at the time [37]. The missing part in the topology constraints definition is due to the lack of practical and efficient shell genus computation algorithms.

The major missing part in the geometry constraints definition is geometric intersection computation for curves and surfaces.

Topology constraints checking implemented:

- entity uniqueness checking
- number of reference checking
- graph genus checking
- Euler equation checking for inequality

Topology constraints checking not implemented:

- shell genus checking
- open shell checking
- manifold checking

Geometry constraints checking implemented

- edge use direction consistency checking
- loop closure checking
- B-spline weights checking

geometry constraints checking not implemented
- (self-) intersection/overlap checking
- connectedness/disjointness checking
- closedness checking
- geometric extent checking
- coordinate system checking

3.3.4.4 IDS checker interface specifications

```
/********************************************************************
Module:             ckini
Author, Owner:      X. Ni, CADDETC
Last Modification:  5 April 91
File:               check_u.c
Parameters:
  char *step_fname  empty or name of a STEP file
  char *err_fname   name of checker error file
  long *istat       returned status
Function:           initialise the checker
Calls:              cgvtol, cgdtol
Include Files:      chdget.h
Errors: 0           no error
        not 0       initialisation error
/********************************************************************
Module:             ckall
Author, Owner:      X. Ni, CADDETC
Last Modification:  5 April 91
File:               check_u.c
Parameters:
  long *istat       returned status
Function:           check whole IDS
Calls:              cgwrld, cgapfl
Include Files:      cdtget.h, cidsut.h, cdatas.h
Errors: 0           no error
        not 0       checker error
/********************************************************************
Module:             ckent
Author, Owner:      X. Ni, CADDETC
Last Modification:  5 April 91
File:               check_u.c
Parameters:
  long *name        name of entity to be checked
  long *mode        recursive/non-recursive (1/0) checking mode
```

```
   long *istat         returned status
Function:              check named entity
Calls:                 cgapfl
Include Files:         cdtget.h, cidsut.h
Errors: 0              no error
        not 0          checker error
/******************************************************************
Module:                ckbrep
Author, Owner:         X. Ni, CADDETC
Last Modification:     5 April 91
File:                  check_u.c
Parameters:
   long *name          name of entity to be checked
   long *mode          recursive/non-recursive (1/0) checking mode
   long *istat         returned status
Function:              check named BREP_AP entity
Calls:                 -
Include Files:         cdtget.h, cidsut.h
Errors: 0              no error
        not 0          checker error
/******************************************************************
Module:                ckss
Author, Owner:         X. Ni, CADDETC
Last Modification:     5 April 91
File:                  check_u.c
Parameters:
   long *name          name of entity to be checked
   long *mode          recursive/non-recursive (1/0) checking mode
   long *istat         returned status
Function:              check named SS_AP entity
Calls:                 -
Include Files:         cdtget.h, cidsut.h
Errors: 0              no error
        not 0          checker error
/******************************************************************
Module:                ckend
Author, Owner:         X. Ni, CADDETC
Last Modification:     5 April 91
File:                  check_u.c
Parameters:
   long *istat         returned status
Function:              end the checker
Calls:                 -
```

```
Include Files:        -
Errors: 0             no error
        not 0         checker error
/*********************************************************************
```

3.3.5 Statistics tools

3.3.5.1 Concepts of STEP statistics tools

This document describes the functional structure and use of the STEP statistic viewer, the STEP file comparator and the STEP structure viewer.

The STEP statistic viewer provides information about the header and a statistic overview about the entities contained within a STEP file.

The STEP file comparator computes two input files and generates a statistic report by comparing the entities contained in both files.

The task of the STEP structure viewer is to provide a graphical overview of the relations between the entities within a STEP File. The "tree" like structure of a STEP File is represented by a sequential list of only ASCII characters enabling a clear output for both an "alpha" terminal and a printer.

All three programs, the statistic viewer, the file comparator and the structure viewer are written in the programming language C. The programs have been implemented on a Silicon Graphics 4D workstation and a DEC-Ultrix 5000 work-station. To compile and link the programs the scanner/parser version 5.3 or later and the IDS version 5.31 or later are necessary.

The user interface is realized by the I/O program tool kit "iotools" which is written in C. These routines are able to draw boxes, to place and center text into these boxes, to set the cursor and to compute other I/O management functions. These programs are realized by escape sequences for VT100 terminals or by plain "printf" functions. While compiling the programs the user can select the desired method. Only standard ASCII characters are used. This enables the programs to be run on most of the standard user terminals.

3.3.5.2 Description of the STEP statistic viewer

After initialization of the IDS and scanner/parser the following calls can be carried out from the main program.

1) Read the STEP input file into the IDS:

```
/*******************************************************************/
void scann(hd_error,brep_error,ss_error,wf_error,csg_error,(file))
/*  Function : checks inputfile and write it in the IDS          */
/*  Receives : --                                                */
/*  Returns  :                                                   */
     int  *hd_error;    /* error_flag for errors in header-section*/
```

```
        int  *brep_error;   /* error_flag for errors in brep-section */
        int  *ss_error;     /* error_flag for errors in ss-section   */
        int  *wf_error;     /* error_flag for errors in wf-section   */
        int  *csg_error;    /* error_flag for errors in csg-section  */
        (char *file;)       /* only in scomp :
                               file in which the error occures */
/*  Specified Errors : --                                           */
/*******************************************************************/
```

2) Read the header information from the IDS and write it into the output file:

```
/*******************************************************************/
void headinf(file_name,file_ptr)
/* Function : getting header-informations of the STEP-File and      */
/*            writing them in STEP-outfile                          */
/*  Receives :                                                      */
     char *file_name;          /* pointer to STEP-Input-File        */
     FILE *file_ptr;     /* file-pointer of Structure - Output-File*/
/*  Returns: --                                                     */
/*  Specific errors:                                                */
/*******************************************************************/
```

3) Collect the number of entity occurrences and write them into the output file:

```
/*******************************************************************/
void get_ent(file_ptr)
/* Function :                                                       */
/*     get the names and the number of different entities in the   */
/*     STEP-file. For each entity his token name, his type, his    */
/*     number and the total number of the entities is written to   */
/*     the output file.                                             */
/*  Receives :                                                      */
     FILE *file_ptr;  /* file-pointer of Structure - Output-File    */
/*  Returns: --                                                     */
/*  Specific errors:                                                */
/*******************************************************************/
```

3.3.5.3 Description of the STEP file comparator

The STEP file comparator works analogous to the statistic viewer. Initialy, file 1 is scanned into the IDS and the header information is stored in the output file. The contents of the STEP file is then stored in the following structure :

```
static struct entity_stat {
         char **token_names;   /* the token-names of the entities*/
         long *entity_typ;     /* the types of the entities       */
         long *numb_of_typ;    /* the number of the types         */
         long sum_of_typ;      /* the sum of all types            */
       } stat_cont_of_file[2];
```

The storage of the structure elements "token_names", "entity_typ" and "numb_of_typ" is allocated by the C-function "malloc".

Afterwards the above procedure is repeated for STEP file 2.

Finally the function "prod_output" compares the contents of the two structures :

```
/******************************************************************/
void prod_output (file_ptr,file1,file2,cont_of_file,
                  num_of_file1,num_of_file2))
/* Function : compares the contents of the structure "cont_of_file"*/
/*            and produces a statistic report                     */
/*  Receives:                                                     */
      FILE *file_ptr;    /* file-pointer of Structure - Output-File */
      char *file1;       /* name of STEP-file 1                   */
      char *file2;       /* name of STEP-file 2                   */
      struct entity_stat *cont_of_file    /*structure with the    */
/*                                   contents of file 1 and file */
/*    int   num_of_file1    /* number of entities in file 1       */
/*    int   num_of_file2    /* number of entities in file 2       */
/*  Returns: --                                                   */
/*  Specific errors:                                              */
/******************************************************************/
```

3.3.5.4 Description of the STEP structure viewer program

The structure of an entity: In order to store the references, the following structure is set up for each entity in the STEP file.

```
struct entity_struct {
    long ent_name;    /* this is the name of each entity in the IDS  */
    long ent_type;    /* this is the type of the entity in the IDS   */
    long ent_col;     /* the column between parent and child entity */
    long next_col;    /* the next parent-entity with the same column */
    long first_post_ent;/*the first child of this entity in the list*/
    long last_post_ent; /* the last child of this entity in the list*/
    long first_pre_ent;/*the first parent of this entity in the list*/
    long last_pre_ent; /* the last parent of this entity in the list*/
```

Out of these structures the output file can be created.

The lists generating the output: The following lists are created in order to sort the structures and to write the output file. The size of the lists depends on the number of entities in the STEP file. The storage space for all lists and structures is allocated by using the C function "malloc".

Entity_List: The sequence of the entities is stored in the variable "entity_list": struct entity_struct *entity_list;

Each entity of the STEP File, stored in the structure entity_struct, is represented once in the entity list. The storage for this list is allocated by using the IDS function "n_entities()". A recursive function called "get_childs" fills the list in a "tree" like structure.

Example :

```
                    #1
                    I
            I------I------I
            I      I      I
            v      v      v
            #2---->#5     #7
            I
            v
        I---I---I
        I       I
        v       v
        #3      #4
        I
    I---I
    I
    v
    #6
```

Entity #1 has following childs : #2,#5,#7
Entity #2 has following childs : #3,#4,#5
Entity #3 has following childs : #6
Entity #6 has no childs.

The following list is created:
entity_list[0] = #1
entity_list[1] = #2   (first child of Entity #1)
entity_list[2] = #3   (first child of Entity #2)
entity_list[3] = #6   (first child of Entity #3)
entity_list[4] = #4   (second child of Entity #2)
entity_list[5] = #5   (third child of Entity #2)
entity_list[6] = #7   (third child of Entity #1)

Column_Index_List: This list contains the page column for each entity on which its reference must be printed.

For example :

The entity #5 starts its reference at column 3
The entity #17 starts its reference at column 9

The structure of the list is as follows:
column_index_list[3] = 5
column_index_list[9] = 17

Structure viewer routines callable from outside: The following calls are carried out from the main program of SSTRUC.

1) Read the STEP input file name, create the output file name and initialize the IDS.

```
/*****************************************************************/
void org_data_input(inp_file,out_file)
/*  Function : output of the headline and getting the org-datas   */
/*  Receives:                                                      */
      char *inp_file[];        /* pointer to STEP-Input-Filename   */
      char *out_file[];        /* pointer to Structur-Output-Filename*/
/*  Returns:   --                                                  */
/*  Specific errors:                                               */
/*****************************************************************/
```

2) Read the STEP input file into the IDS.

```
/*****************************************************************/
void scan_pars(head_er,brep_er,ss_er,wf_er,csg_er)
/*  Function : checks inputfile and write it in the IDS           */
/*  Receives:                                                      */
/*  Returns:                                                       */
      int *head_er;       /* error-flag for errors in header-section */
      int *brep_er;       /* error-flag for errors in brep-section   */
      int *ss_er;         /* error-flag for errors in ss-section     */
      int *wf_er;         /* error-flag for errors in wf-section     */
      int *csg_err;       /* error-flag for errors in csg-section    */
/*  Specific errors:                                               */
/*****************************************************************/
```

3) Read the header information from the IDS and write it into the output file.

```
/*****************************************************************/
void get_header(file_ptr,file_name)
/* Function : getting header-informations of the STEP-File and     */
/*            writing them in STEP-outfile                         */
/*  Receives:                                                      */
      FILE *file_ptr;    /* file-pointer of Structure - Output-File */
      char *file_name;   /* pointer to STEP-Input-File              */
/*  Returns: --                                                    */
/*  Specific errors:                                               */
/*****************************************************************/
```

4) Allocate the necessary memory, create the described lists, analyse the structure and write the result into the output file.

```
/*****************************************************************/
void cstruc(file_ptr,in_file)
Function :  This Function controlls the course of the structural   */
/*           output. It allocates enougth memory for the lists :    */
```

```
/*              entity_list , name_index_list and column_name_list.    */
/*              These lists are necessary to bring the "parent and     */
/*              children" entities in a correct sequenz.               */
/*              Then the function "graph_struct_output" will be called,*/
/*              to make a correct output of references of the entities.*/
/*              At the end of the function, memory must give free.     */
/*  Receives:                                                          */
      FILE *file_ptr;     /* File-pointer of Structure-Output-file */
      char *in_file;      /* pointer to STEP-Input-file            */
/*  Returns:  --                                                       */
/*  Specific errors:                                                   */
/********************************************************************/
```

5) Display the output file on the screen.

```
/********************************************************************/
void show_output(out_file)
/* Function : writing the structure report on the screen.22 lines */
/*            per page will be show                               */
/*  Receives:                                                     */
      char *out_file;      /* name of the output-file (*.stc)     */
/*  Returns: --                                                   */
/*  Specific errors:                                              */
/********************************************************************/
```

Internal structure viewer routines

```
/********************************************************************/
void make_list_of_entities(entity_list,nam_index_list)
/* Function : This Function writes all entities in a "sequential"  */
/*            list. It also makes a list to get the "index" of each*/
/*       entity with the name of this entity.                      */
/*  Receives:   --                                                 */
/*  Returns:                                                       */
      struct entity_struct *entity_list[];/* pointer to list with  */
/*                                          entity structure       */
      int *nam_index_list[];          /* pointer to list with     */
/*                                      index of entities          */
/*  Specific errors:                                               */
/********************************************************************/
/********************************************************************/
void get_childs
         (nam_of_ent,entity_list,rec_count,stat_list,nam_index_list)
/* Function : This Function writes all entities in a "tree-like"   */
/*            list.  It also makes a list to get the "index"       */
/*            of each entity with the name of this entity.         */
/*  Receives:   --                                                 */
```

```
/*   Returns:                                                    */
     struct entity_struct *entity_list;    /* pointer to list    */
/*                                         with entity structure */
     int *rec_count;                /* counter for entity_list   */
     long *nam_of_ent;              /* pointer to name of entity*/
     int *stat_list[];              /* pointer to list which     */
/*                                  secures, that each entity*/
/*                                  is only one time in list */
     int *nam_index_list[];         /* index of each entity      */
/*   Specific errors:                                            */
/******************************************************************/

/******************************************************************/
int get_pre_and_post_ent (entity_struct_list,entities,
                          nam_index_list,col_index_list)
/* Function: This Function gives for each entity the first and last*/
/*           referenced and referencing entity.                  */
/*   Receives:                                                    */
     int *nam_index_list;      /* list with index of each entity */
     long *entities;           /* number of entities             */
/*   Returns:                                                     */
     struct entity_struct *entity_struct_list[];  /* pointer to   */
/*                                  list with entity structure*/
     int *col_index_list[];                    /* pointer to list */
/*                    with the number of columns for each entity */
/*   get_pre_and_post_ent (function_value) -- number of all columns */
/*   Specific errors:                                            */
/******************************************************************/

/******************************************************************/
void graph_struct_output (fl_ptr, entity_list, num_of_ent, num_of_ed,
                          nam_index_list, col_index_list,in_file)
/* This Function makes a correct output of references between all  */
/* entities. The entities will be shown as a box, which includes   */
/* the token-name of the entity and its name in the STEP-file.     */
/*   Receives:                                                     */
     FILE *fl_ptr;              /* file-pointer of Output-file     */
     struct entity_struct entity_list[];/*list of entity structure*/
     long *num_of_ent;          /* number of entities             */
     int *num_of_ed;            /* number of edges                */
     int nam_index_list[];      /* list of "index" for each entity */
     int col_index_list[];      /* list of "columns" for each ent. */
     char *in_file;             /* pointer to name of Input-file   */
/*   Returns:  --                                                 */
/*   Specific errors:                                            */
/******************************************************************/
```

Structure viewer internal help routines

```
/*********************************************************************/
int div_and_one(divide,diviso)
/* This Function returns the quotient of two integer values.       */
/* If the rest of the result is not equal 0, one will be added.    */
/*  Receives:                                                      */
/*      int divide;       /*  divident                             */
/*      int diviso;       /*  divisor                              */
/*  Returns:  --                                                   */
/*  div_and_one (function_value) -- reusult of divison             */
/*  Specific errors:                                               */
/*********************************************************************/

/*********************************************************************/
void strfill(str1,chr1,len)
/* Function : Fills the String "str1" with "len-1" characters      */
/*            of charcter chr1 => str1[len] = '\0' !!              */
/*            Memory must be allocated from calling program.       */
/*  Receives:                                                      */
/*      char chr1;   /* charcter to fill str1                      */
/*      int len;     /* number of charcters to fill str1           */
/*  Returns:                                                       */
/*      char str1[]; /* String to fill                             */
/*  Specific errors:                                               */
/*********************************************************************/

/*********************************************************************/
void strnfill(str1,pos,str2,len)
/*  Function : The String "str1" will be filled from position "pos"*/
/*             with the first "len" digits of String "str2".       */
/*             Memory must be allocated by calling program         */
/*  Receives:                                                      */
/*      char str2[];        /* String to fill str1                 */
/*      int pos;            /* start-position to fill str1         */
/*      int len;            /* number of digits of str2 to fill str1 */
/*  Returns:                                                       */
/*      char str1[];        /* String to fill with str2            */
/*  Specific errors:                                               */
/*********************************************************************/

/*********************************************************************/
int refer_to_ent(entity1,entity2)
/*  Function : Checks, wether entity1 is a child of entity2         */
/*  Receives:                                                      */
/*      long entity1;    /* "child"-entity                         */
/*      long entity2;    /* "parent"-entity                        */
```

```
/*  Returns:                                                      */
/*  refer_to_ent (function_value) -- 0: true  1:false            */
/*  Specific errors:                                             */
/*****************************************************************/

/*****************************************************************/
void make_formats(format1,format2,format3,format4,format5)
/*  Function : Makes "formats" for output of Structure-report     */
/*             Memory must be allocated of calling program        */
/*  Receives: --                                                 */
/*  Returns:                                                      */
     char format1[];       /* format of row_between_box           */
     char format2[];       /* format of top_box_row               */
     char format3[];       /* format of bot_box_row               */
     char format4[];       /* format of rows_in_box1 for entity-nam*/
     char format5[];       /* format of rows_in_box2 for token-name*/
/*  Specific errors:  --                                         */
/*****************************************************************/

/*****************************************************************/
void error_message(error)
/*  Function : Output of an error-message                         */
/*  Receives:                                                     */
     long error;              /* Number of error (look sstruc.h)  */
/*  Returns:    --                                               */
/*  Specific errors:  --    (look sstruc.h)                      */
/*****************************************************************/
```

3.3.6 Conversion tools

3.3.6.1 General remarks on conversion tools

Summary and applicable documents. This chapter summarizes the outcome of the conversion tool activities of CADEX. The background for the conversion tools development is described as well as the chosen way of implementation. Finally the tools are presented sorted by contributing partner.

The following documents are recommended supplementary reading:

1 Specification of a Common Tool Kit for CAD Geometry Data Exchange, CADEX, H.J.Helpenstein, H. Heinrichs, GfS mbH, Aachen, January 1991

2 Development of a Common Tool Kit for CAD Geometry Data Exchange Processors, CADEX, H.J.Helpenstein, GfS mbH, Aachen, 29.8.1991

3 CADEX Conversion Tools Library, User's Guide, Nicole Bach, Wilhelm Kerschbaum, BMW AG, München, August 1992

Scope. The architecture of the CADEX Common Tool Kit comprises a package called Conversion Tools. The CAD model exchange scenario of CADEX consists of many different CAD- and FEM-systems. Each system has its own proprietary representation of geometry and topology. A circle might be represented by a centre and a radius in one system and by three points in another one. In addition there is the neutral representation of STEP. This various environment requires conversions of geometric and topological representations to enable the exchange of data.

Conversions are as well required due to the different usage of CAD-data. Analysis requires wireframe data and deals most often with non-manifolds. Car body design is the domain of surface modelling, clash checking of the interior space of a car the domain of solid modelling. These different appearances of the same shape such as a car should nevertheless be derivates of the same basic model. The derivates should be the results of mainly topological conversions from the master representation.

Though each conversion tool is made for and should be used by several partners, the over-all needs for Conversion Tools vary from partner to partner. This fact had impact on the organisation of the tools. It was no use to collect all tools in one application. A library was found to be the best implementation framework for the Conversion Tools.

Software architecture. The Conversion Tools Library is part of the CADEX processor architecture. The library is as well a potential stand-alone product. To achieve both, an integration into the CADEX environment and a product that is independent of the rest of the CADEX software the architecture shown in figure 3.7 has been implemented.

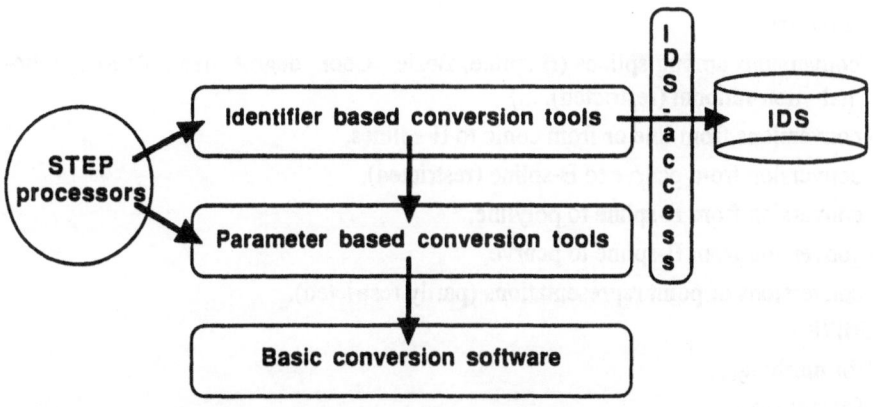

Fig. 3.7: Three versions exist of each conversion tool

Ideally there are three different versions of each conversion tool:
- one basic,
- one parameter and
- one identifier version.

Each version is represented by one C- or FORTRAN routine.

The basic version provides the basic conversion functionality. In many cases the basic routines existed already prior to CADEX (such as parts of the SI Spline Library and the spline conversion software by R.D.Fuhr from NIST) and were made available to all partners.

For the parameter version the input/output parameters of a tool were modified to be conformant to STEP. When converting a circle to a B-spline curve the STEP-attributes of the circle are input to the corresponding routine, the STEP-attributes of a B-spline curve are output.

The identifier version accesses the IDS and is, thus, dependent on the CADEX processor architecture. Only instances that reside in the IDS can be converted. The circle/B-spline routine mentioned above would have the IDS- identifier of the circle as input and the B-spline curve identifier as output. The circle instance would not be modified or deleted. The IDS would, however, be extended by a new B-spline curve instance.

Available Tools.   Mainly three of the partners contributed actively to the Conversion Tools Library: FEGS, GfS and SI. Some of their contributions are restricted in their use; i.e. conditions for distribution are subject to negotiations. Tools have been written in C and FORTRAN and are callable from both languages.

The tools belong to the following categories of functionality:

Topology
- generation of topology from geometry (e.g. *fltopo*),
- conversions between different levels of topology (e.g. *creawf, csi221, csi322*);

Geometry
- conversions among splines (B-spline, Bezier, Coon, degree manipulations, rational - non-rational (restricted), ...),
- conversions from line or from conic to B-splines,
- conversion from pcurve to B-spline (restricted),
- conversion from B-spline to polyline,
- conversion from B-spline to pcurve,
- conversions of point representations (partly restricted).

Utilities
- for meshing,
- for storage,
- for spline handling,
- for geometric calculations.

3.3.6.2 Conversion tools by SI

3.3.6.2.1 Module ccasbc

Name of File:	CSIA2S.C	
Function:	Converts b_spline_curves from aps to sisl format	
Input Parameters:	int idim	dimension in which the curve lies
	int ik	order of curve
	int in	number of vertices
	int ipknt	pointer to start of knot vector in escr
	int ipvert	pointer to start of vertices in escr
	double escr[ ]	array holding vertices and knots
	int kinxtfr	pointer to next free elem. in escr
Output Parameters:	SISLCurve **sisl_curve	b_spline_curve data in SISL curve-struct
	long *istat	status indicator
Calls:	-	
Include Files:	-	
Errors:	-	
Author/Owner:	Per Evenson, Senter for Industriforskning P.O. Box 124, Blindern, N-0314 Oslo 3	
Program Header:	void ccasbc(idim,ik,in,ipknt,ipvert,escr,kinxtfr,sisl_curve,istat)	

3.3.6.2.2 Module cccobc

Name of File:	CCCOBC.C	
Function:	Converts a conical curve in to a b_spline_curve; works for circles, ellipses, hyperbolas and parabolas	
Input Parameters:	long *type	type of conic
	double *capto	coordinates of the origin of the conic
	double *captx	coordinates defining the x-axis of the conic
	double *captz	coordinates defining the z-axis of the conic
	double *co_par_1	1st parameter of conic
	double *co_par_2	2nd parameter of conic
	double *epsge	geometry resolution
Output Parameters:	long *deg	algebraic degree of basis functions
	long *up_ind_ctrl_pnts	upper index of array of control_points (as IPIM-definition)
	long *n_ctrl_pnts	number of control_points (not in IPIM) (= *up_ind_ctrl_pnts + 1)

long **ctrl_pnts	array (of length *n_ctrl_pnts) with the names of the control_points defining the b_spline_curve
long *uniform	indicator of the type of knot set (OPTIONAL)
long *ex_up_ind_knots_data	flag is set if knots_data array is filled with data (OPTIONAL)
long *up_ind_knots_data	upper index of array of knots_data (as IPIM-definition) (OPTIONAL)
long *n_knot_mult	number of knot_multiplicities (not in IPIM; identical to *up_ind_knots_data)
long **kn_mults	array (of length *n_knot_mults) with the multiplicities of the knots (OPTIONAL)
long *n_knots	number of knots defining the b_spline (not in IPIM; identical to *up_ind_knots_data) (OPTIONAL)
double **knots	array (of length *n_knots) with the knot values used for defining the b_spline basis functions (OPTIONAL)
long *n_weights	number of weights (not in IPIM; identical to *n_ctrl_pnts) weights associated with the control points in the rational case (OPTIONAL)
long *form_no	indicates the analytical type of curve represented by the b_spline (OPTIONAL)
long *closed	indicates whether the b_spline is closed (OPTIONAL)
long *self_int	indicates whether the b_spline self intersects (OPTIONAL)
double **ca_pnts	array holding cartesian points
long *istat	status indicator

Calls:	ccsibc
Include Files:	-
Errors:	-
Author/Owner:	Per Evenson, Senter for Industriforskning P.O. Box 124, Blindern, N-0314 Oslo 3
Program Header:	long cccobc (type, capto, captx, captz, co_par_1, co_par_2, epsge, deg, up_ind_ctrl_pnts, n_ctrl_pnts, ctrl_pnts, uniform,

ex_up_ind_knots_data, up_ind_knots_data, n_knot_mults,
kn_mults, ca_pnts,istat)

3.3.6.2.3 Module ccicbc

Name of File:	CSICON.C
Function:	Converts a trimmed conic to a non-rational b_spline_curve

Input Parameters:	long *conic_name	name of conic
	long *pt_1	name of first trimming point in cartesian space (OPTIONAL)
	long *pt_2	name of second trimming point in cartesian space (OPTIONAL)
	long *ex_param1	indicates if optional attribute param_1 is set
	double *param_1	the first trimming point in the para metric space of conic (OPTIONAL)
	long *ex_param2	indicates if optional attribute param_2 is set
	double *param_2	second trimming point in the para metric space of conic (OPTIONAL)
	long *sense	indicator of whether the trimmed curve direction coincides with that of the conic
	double *epsge	geometry resolution (conversion accuracy)

Output Parameters:	long *b_spline_curve_name	name of b_spline_curve
	*istat	return value: error code

Calls:	ccsibc, ccasbc, cvorax, cccobc, ccsibc, crossv, scalev, diffv,
Include Files:	-
Errors:	-
Author/Owner:	Per Evenson, Senter for Industriforskning P.O. Box 124, Blindern, N-0314 Oslo 3
Program Header:	long ccicbc (conic_name, pt_1, pt_2, ex_param1, param_1, ex_param2, param_2, sense, epsge, b_spline_curve_name, istat)

3.3.6.2.4 Module ccicoc

Name of File:	CCICOC.C
Function:	Converts a conic entity to a b_spline_curve

Input Parameters:	long *conic_name	name of the conic
	double *epsge	name of the conic

Output Parameters:	long *b_spline_curve_name	name of the b_spline_curve
	*istat	return value: error code

Calls:	cccobc
Include Files:	-
Errors:	-
Author/Owner:	Per Evenson, Senter for Industriforskning P.O. Box 124, Blindern, N-0314 Oslo 3
Program Header:	long ccicoc (conic_name, epsge, b_spline_curve_name, istat)

3.3.6.2.5 Module ccipac

Name of File:	CSICON.C
Function:	Converts a point on a circular curve from parametric to cartesian. Note! This routine is based on the FEGS software.

Input Parameters:	long *coord	name of the coordinate system
	double *radius	radius of circle
	long *ax2	name of referenced axis set
	double *param	the parameter of the point

Output Parameters:	double pnt_coord[3]	coordinates of the created cartesian point
	long *istat	return status

Calls:	-
Include Files:	-
Errors:	-
Author/Owner:	Per Evenson, Senter for Industriforskning P.O. Box 124, Blindern, N-0314 Oslo 3
Program Header:	long ccipac (coord,radius,ax2,param,pnt_coord,istat)

3.3.6.2.6 Module ccisic

Name of File:	CSISIC.C
Function:	Converts b_spline_curve data from ids to sisl format

Input Parameters:	SISLCurve *sisl_curve	name of b_spline_curve in sisl curve-struct

Output Parameters:	long *name	name of b_spline_curve (ids)
	long *istat	status indicator

Calls:	-
Include Files:	-
Errors:	

Author/Owner:	Per Evenson, Senter for Industriforskning
	P.O. Box 124, Blindern, N-0314 Oslo 3

Program Header: void ccisic (sisl_curve, name, istat)

3.3.6.2.7 Module ccisis

Name of File:	CSISIS.C	
Function:	Converts b_spline_surface data from ids to sisl format	
Input Parameters:	SISLSurf *sisl_surface	name of b_spline_surface in sisl curve-struct
Output Parameters:	long *name	name of b_spline_surface (ids)
	long *istat	status indicator
Calls:	-	
Include Files:	-	
Errors:	-	
Author/Owner:	Per Evenson, Senter for Industriforskning	
	P.O. Box 124, Blindern, N-0314 Oslo 3	

Program Header: void ccisis (sisl_surface, name, istat)

3.3.6.2.8 Module cciisc

Name of File:	CSIISC.C	
Function:	Converts b_spline_curve data from ids to sisl format	
Input Parameters:	long *name	name of b_spline_curve
	long *crat_flag	flag for conversion rational -> non-rational: 0 = no 1 = yes
	long *dim	dimension of b_spline_curve
Output Parameters:	SISLCurve **sisl_curve	b_spline_curve data in SISL curve-struct
	long *istat	status indicator
Calls:	-	
Include Files:	-	
Errors:	-	
Author/Owner:	Per Evenson, Senter for Industriforskning	
	P.O. Box 124, Blindern, N-0314 Oslo 3	

Program Header: void cciisc (name, crat_flag, dim, sisl_curve, istat)

3.3.6.2.9 Module cciiss

Name of File:	CSIISS.C
Function:	Converts b_spline_surface data from ids to sisl format

Input Parameters:	long *name	name of b_spline_surface
	long *crat_flag	flag for conversion rational -> non-rational: 0 -> no 1 -> yes
	long *dim	dimension of b_spline_surface
Output Parameters:	SISLCurve **sisl_surface	b_spline_surface data in SISL surface-struct
	long *istat	status indicator
Calls:	-	
Include Files:	-	
Errors:	-	
Author/Owner:	Per Evenson, Senter for Industriforskning P.O. Box 124, Blindern, N-0314 Oslo 3	
Program Header:	void cciiss (surface_name, crat_flag, dim, sisl_surface, istat)	

3.3.6.2.10 Module ccipbc

Name of File:	CSISIC.C
Function:	Gets b_spline_curve data from the APS-SS database and convert data to IDS format

Input Parameters:	long name	name of b_spline_surface
	long loc_cosy	name of a local_coordinate_system
	long deg	algebraic degree of the basis functions
	long up_ind_ctrl_pnts	upper index of array of control_points
	long n_ctrl_pnts	number of control_points
	long *ctrl_pnts	array with names of control_points
	long uniform	indicator of the type of knot set
	long ex_uikd	flag is set if there are knots_data (OPTIONAL)
	long up_ind_knots_data	upper index of array of knots_data
	long n_knot_mults	number of knot_multiplicities
	long *kn_mults	array with the multiplicities of knots
	long n_knots	number of knots
	double *knots	array with the knot values
	long n_weights	number of weights
	double *weights	array (of length n_weights) with weights
	long form_no	indicates the analytical type of surface
	long closed	closed-flag

	long self_int	indicates whether the b_spline self-intersects
	double *ca_pnts	array holding cartesian points
Output Parameters:	long *istat;	status indicator
Calls:	-	
Include Files:	-	
Errors:	-	
Author/Owner:	Per Evenson, Senter for Industriforskning P.O. Box 124, Blindern, N-0314 Oslo 3	
Program Header:	void ccipbc (name, loc_cosy, deg, up_ind_ctrl_pnts, n_ctrl_pnts, ctrl_pnts, uniform, ex_uikd, up_ind_knots_data, n_knot_mults, kn_mults, n_knots, knots, n_weights, weights, form_no, closed, self_int, ca_pnts, istat)	

3.3.6.2.11 Module ccipbs

Name of File:	CSISIS.C	
Function:	Gets b_spline_surface data from the APS-SS database and converts data to IDS format	
Input Parameters:	long name	name of b_spline_surface
	long loc_cosy	name of a local_coordinate_system
	long u_deg	algebraic degree of the basis functions in u
	long v_deg	algebraic degree of the basis functions in v
	long u_upper	upper index of array of control_points in u
	long v_upper	upper index of array of control_points in v
	long n_ctrl_pnts	number of control_points
	long *ctrl_pnts	array with names of control_points
	long u_uniform	indicator of the type of knot set in u
	long v_uniform	indicator of the type of knot set in v
	long ex_uiuk	flag is set if there are knots_data in u
	long up_ind_u_knot	upper index of array of knots_data in u
	long ex_uivk	flag is set if there are knots_data in v
	long up_ind_v_knot	upper index of array of knots_data in v
	long n_ukn_mults	number of knot_multiplicities in u
	long *u_kn_mults	array with the multiplicities of knots in u

long n_vkn_mults	number of knot_multiplicities in v
long *v_kn_mults	array with multiplicities of knots in v
long n_uknots	number of knots in u
double *u_knots	array with the knot values in u
long n_vknots	number of knots in v
double *v_knots	array with the knot values in v
long n_weights	number of weights
double *weights	array (of length n_weights) with weights
long u_closed	closed-flag in u
long v_closed	closed-flag in v
long form_no	indicates the analytical type of surface
double *ca_pnts	array holding cartesian points

Output Parameters: long *istat   status indicator

Calls: -

Include Files: -

Errors: -

Author/Owner: Per Evenson, Senter for Industriforskning
                          P.O. Box 124, Blindern, N-0314 Oslo 3

Program Header: void ccipbs (name, loc_cosy, u_deg, v_deg, u_upper, v_upper,
                          n_ctrl_pnts, ctrl_pnts, u_uniform, v_uniform, ex_uiuk,
                          up_ind_u_knot, ex_uivk, up_ind_v_knot, n_ukn_mults,
                          u_kn_mults, n_vkn_mults, v_kn_mults, n_uknots, u_knots,
                          n_vknots, v_knots, n_weights, weights, u_closed, v_closed,
                          form_no, ca_pnts, istat)

3.3.6.2.12 Module ccipoc

Name of File: CSIPNT.C

Function: Converts a point_on_curve to a cartesian_point

Input Parameters: long *pnt_on_curve_name   Entityname of point_on_curve

Output Parameters: long *capt_name   Identifier of the new cartesian_point
                          long *istat   return value: error code

Calls: ccpobc

Include Files: -

Errors: -

Author/Owner: Per Evenson, Senter for Industriforskning
                          P.O. Box 124, Blindern, N-0314 Oslo 3

Program Header: long ccipoc (pnt_on_curve_name,capt_name,istat)

3.3.6.2.13 Module ccipos

Name of File:	CSIPNT.C
Function:	Converts a point_on_surface to a cartesian_point
Input Parameters:	long *pnt_on_surface_name Entityname of point_on_surface
Output Parameters:	long *capt_name Identifier of the new cartesian_point
	long *istat return value: error code
Calls:	ccpobs
Include Files:	-
Errors:	-
Author/Owner:	Per Evenson, Senter for Industriforskning P.O. Box 124, Blindern, N-0314 Oslo 3
Program Header:	long ccipos (pnt_on_surf_name, capt_name,istat)

3.3.6.2.14 Module ccisbc

Name of File:	CCISBC.C	
Function:	Converts a b_spline_curve from IDS to sisl format	
Input Parameters:	long *name	name of b_spline_curve
	long *loc_cosy	name of a local_coordinate_system positioning the curve in space (OPTIONAL)
	long *deg	algebraic degree of the basis functions
	long *up_ind_ctrl_pnts	upper index of array of control_ points (as IPIM-definition)
	long *n_ctrl_pnts	number of control_points (not in IPIM) (= *up_ind_ctrl_pnts + 1)
	long ctrl_pnts[ ]	array (of length *n_ctrl_pnts) with the names of the control_points defining the b_spline_curve
	long *uniform	indicator of the type of knot set (OPTIONAL)
	long *ex_up_ind_knots_data	flag is set if knots_data array is filled with data (OPTIONAL)
	long *up_ind_knots_data	upper index of array of knots_data (as IPIM definition) (OPTIONAL)
	long *n_knot_mults	number of knot_multiplicities (not in IPIM; identical to *up_ind_knots_data)

	long kn_mults[ ]	array (of length *n_knot_mults) with the multiplicities of the knots (OPTIONAL)
	long *n_knots	number of knots defining the b_spline (not in IPIM; identical to *up_ind_knots_data) (OPTIONAL)
	double knots[ ]	array (of length *n_knots) with the knot values used for defining them b_spline basis functions (OPT.)
	long *n_weights	number of weights (not in IPIM; identical to *n_ctrl_pnts)
	double weights[ ]	array (of length *n_weights) with weights associated with the control_ points in the rational case (OPT.)
	long *form_no	indicates the analytical type of curve represented by the b_spline (OPT.)
	long *closed	indicates whether the b_spline is closed (OPTIONAL)
	long *self_int	indicates whether the b_spline self intersects (OPTIONAL)
	double ca_pnts[ ]	array holding cartesian points
Output Parameters:	SISLCurve **sisl_curve	b_spline_curve data in SISL curve-struct
	long *istat	status indicator
Calls:	-	
Include Files:	-	
Errors:	-	
Author/Owner:	Per Evenson, Senter for Industriforskning P.O. Box 124, Blindern, N-0314 Oslo 3	
Program Header:	void ccisbc (name, loc_cosy, deg, up_ind_ctrl_pnts, n_ctrl_pnts, ctrl_pnts, uniform, ex_up_ind_knots_data, up_ind_knots_data, n_knot_mults, kn_mults, n_knots, knots, n_weights, weights, form_no, closed, self_int, ca_pnts, sisl_curve, istat)	

3.3.6.2.15 Module compose_mult_knots

Name of File:	CCISBC.C	
Function:	Compose a knot array with multiple occurences based on the knot and mult arrays in the IDS database	
Input Parameters:	long *n_knots	size of arrays knot_v and mults
	double knot_v[ ]	single occurrence knot vector
	long mults[ ]	array of multiplicities

| Output Parameters: | double **et | array with multiple knots vector |
| | long *istat | status indicator |

Calls: -

Include Files: -

Errors: -

Author/Owner: Per Evenson, Senter for Industriforskning
                    P.O. Box 124, Blindern, N-0314 Oslo 3

Program Header: void compose_mult_knots (n_knots, knot_v, mults, et, istat)

3.3.6.2.16 Module default_mult_knots

Name of File: CCISBC.C

Function: Compose a default knot array with multiple occurences based on the uniform, degree and up_ind_ctrl_pnts information in the IDS database

Input Parameters:	long *uniform	indicator of the type of knot set
	long *deg	algebraic degree of the basis functions
	long *n_ctrl_pnts	number of control_points (not in IPIM) (= *up_ind_ctrl_pnts + 1)

| Output Parameters: | double **et | array with multiple knots vector |
| | long *istat | status indicator |

Calls: -

Include Files: -

Errors: -

Author/Owner: Per Evenson, Senter for Industriforskning
                    P.O. Box 124, Blindern, N-0314 Oslo 3

Program Header: void default_mult_knots (uniform, deg, n_ctrl_pnts, et, istat)

3.3.6.2.17 Module rat_descr

Name of File: CCISBC.C

Function: Decide if a description is rational

| Input Parameters: | long *n_ctrl_pnts | number of control_points (not in IPIM) (= *up_ind_ctrl_pnts + 1) |
| | double weights[ ] | array (of length *n_weights) with weights associated with the control_points in the rational case |

Output Parameters: -

Calls: -

Include Files: -

Errors: -

Author/Owner: Per Evenson, Senter for Industriforskning
                        P.O. Box 124, Blindern, N-0314 Oslo 3

Program Header: long rat_descr (n_ctrl_pnts, weights)

3.3.6.2.18 Module compose_rat_descr

Name of File: CCISBC.C

Function: Compose an array that contains the homogenous control points
                        of the rational description

Input Parameters: long *n_ctrl_pnts   number of control_points (not in
                                                IPIM) (= *up_ind_ctrl_pnts + 1)

                     double ca_pnts[ ]   array holding cartesian points
                     double weights[ ]   array (of length *n_weights) with
                                                weights associated with the control_
                                                points in the rational case
                     long dim   dimension of space

Output Parameters: double **rat_pnts   array to contain the homogenous
                                                control points of the rational
                                                description
                     long *istat   status indicator

Calls: -

Include Files: -

Errors: -

Author/Owner: Per Evenson, Senter for Industriforskning
                        P.O. Box 124, Blindern, N-0314 Oslo 3

Program Header: void compose_rat_descr (n_ctrl_pnts, ca_pnts, weights, dim,
                        rat_pnts, istat)

3.3.6.2.19 Module ccisbc2

Name of File: CSIIS2.C

Function: Convert b_spline_curve data from IDS- to SISL-format
                        Note! This version utilizing the newCurve2/NewSurf2
                        constructors, i.e. STEP-conformant representation of NURBS.

Input Parameters: long *name   name of b_spline_curve
                     long *loc_cosy   name of a local_coordinate_system
                                                positioning the curve in space (OPT)
                     long *deg   algebraic degree of the basis
                                                functions
                     long *up_ind_ctrl_pnts   upper index of array of control_
                                                points (as IPIM-definition)

long *n_ctrl_pnts	number of control_points (not in IPIM) (= *up_ind_ctrl_pnts + 1)
long ctrl_pnts[ ]	array (of length *n_ctrl_pnts) with the names of the control_points defining the b_spline_curve
long *uniform	indicator of the type of knot set (OPTIONAL)
long *ex_up_ind_knots_data	flag is set if knots_data array is filled with data (OPTIONAL)
long *up_ind_knots_data	upper index of array of knots_data (as IPIM-definition) (OPTIONAL)
long *n_knot_mults	number of knot_multiplicities (not in IPIM; identical to *up_ind_knots_data)
long kn_mults[ ]	array (of length *n_knot_mults) with the multiplicities of the knots (OPTIONAL)
long *n_knots	number of knots defining the b_spline (not in IPIM; identical to *up_ind_knots_data) (OPTIONAL)
double knots[ ]	array (of length *n_knots) with the knot values used for defining the b_spline basis functions (OPT.)
long *n_weights	number of weights (not in IPIM; identical to *n_ctrl_pnts)
double weights[ ]	array (of length *n_weights) with weights associated with the control_ points in the rational case (OPT.)
long *form_no	indicates the analytical type of curve represented by the b_spline (OPT.)
long *closed	indicates whether the b_spline is closed (OPTIONAL)
long *self_int	indicates whether the b_spline self intersects (OPTIONAL)
double ca_pnts[ ]	array holding cartesian points
Output Parameters: SISLCurve **sisl_curve	b_spline_curve data in SISL curve-struct
long *istat	status indicator
Calls:	compose_mult_knots, default_mult_knots, rat_descr
Include Files:	-
Errors:	-
Author/Owner:	Per Evenson, Senter for Industriforskning P.O. Box 124, Blindern, N-0314 Oslo 3

Program Header: void ccisbc2 (name, loc_cosy, deg, up_ind_ctrl_pnts,
                   n_ctrl_pnts, ctrl_pnts, uniform, ex_up_ind_knots_data,
                   up_ind_knots_data, n_knot_mults, kn_mults, n_knots, knots,
                   n_weights, weights, form_no, closed, self_int, ca_pnts,
                   sisl_curve, istat)

3.3.6.2.20 Module ccisbs

Name of File: CCISBS.C

Function: Convert a b_spline_surface from IDS to sisl format.

Input Parameters: long *name   name of b_spline_surface
                   long *loc_cosy   name of a local_coordinate_system
                                              positioning the surface in space
                                              (OPTIONAL)
                   long *u_deg   algebraic degree of the basis
                                              functions in u
                   long *v_deg   algebraic degree of the basis
                                              functions in v
                   long *u_upper   upper index of array of control_
                                              points in u (as IPIM definition)
                   long *v_upper   upper index of array of ontrol_points
                                              in v (as IPIM-definition)
                   long *n_ctrl_pnts   number of control_points (number
                                              of points in u times number of points
                                              in v: (u_upper+1) * (v_upper+1))
                                              (not in IPIM)
                   long ctrl_pnts[ ]   array (of length *n_ctrl_pnts) with
                                              the names of the control_points
                                              defining the b_spline_surface (in the
                                              same order as on the STEP-file, i.e.
                                              v lines of u CAPT-identifiers here in
                                              sequence) (differs from IPIM-
                                              definition)
                   long *u_uniform   indicator of the type of knot set in u
                                              (OPTIONAL)
                   long *v_uniform   indicator of the type of knot set in v
                                              (OPTIONAL)
                   long *ex_up_ind_u_knot   flag is set if there are knots_data in u
                                              (OPTIONAL)
                   long *up_ind_u_knot   upper index of array of knots_data
                                              in u (as IPIM-definition)
                                              (OPTIONAL)
                   long *ex_up_ind_v_knot   flag is set if there are knots_data in v
                                              (OPTIONAL)

long *up_ind_v_knot	upper index of array of knots_data in v (as IPIM-definition) (OPTIONAL)
long *n_ukn_mults	number of knot_multiplicities in u (not in IPIM; identical to *up_ind_u_knot)
long u_kn_mults[ ]	array (of length *n_ukn_mults) with multiplicities of the knots in u (OPTIONAL)
long *n_vkn_mults	number of knot_multiplicities in v (not in IPIM; identical to *up_ind_v_knot)
long v_kn_mults[ ]	array (of length *n_vkn_mults) with the multiplicities of the knots in v (OPTIONAL)
long *n_uknots	number of knots in u defining the b_spline_surface in u; (not in IPIM) (identical to *up_ind_u_knot) (OPTIONAL)
double u_knots[ ]	array (of length *up_ind_u_knot) with the knot values used for defining the b_spline basis functions in u (OPTIONAL)
long *n_vknots	number of knots in v defining the b_spline_surface in v; (not in IPIM) (identical to *up_ind_v_knot) (OPTIONAL)
double v_knots[ ]	array (of length *up_ind_v_knot) with the knot values used for defining the b_spline basis functions in v (OPTIONAL)
long *n_weights	number of weights (not in IPIM; identical to *n_ctrl_pnts)
long *n_weights	array (of length *n_weights) with weights associated with the control_points in the rational case (else = 1.0) (in the same order as on the STEP-file, i.e. v lines of u weights here in sequence) (OPT.) (differs from IPIM-definition)
long *u_closed	indicates whether the b_spline_ surface is closed in u (OPTIONAL)
long *v_closed	indicates whether the b_spline_ surface is closed in v (OPTIONAL)

	long *form_no	indicates the analytical type of curve represented by the b_spline_surface (OPTIONAL)
	double ca_pnts[ ]	array holding cartesian points
Output Parameters:	SISLSurf **sisl_surf	b_spline_surf data in SISL surf-struct
	long *istat	status indicator

Calls: compose_mult_knots, compose_rat_descr, default_mult_knots, rat_descr

Include Files: -

Errors: -

Author/Owner: Per Evenson, Senter for Industriforskning
                P.O. Box 124, Blindern, N-0314 Oslo 3

Program Header: void ccisbs (name, loc_cosy, u_deg, v_deg, u_upper, v_upper,
                n_ctrl_pnts, ctrl_pnts, u_uniform, v_uniform,
                ex_up_ind_u_knot, up_ind_u_knot, ex_up_ind_v_knot,
                up_ind_v_knot, n_ukn_mults, u_kn_mults, n_vkn_mults,
                v_kn_mults, n_uknots, u_knots, n_vknots, v_knots, n_weights,
                weights, u_closed, v_closed, form_no, ca_pnts, sisl_surf, istat)

3.3.6.2.21 Module ccisbs2

Name of File: CSIIS2.C

Function: To convert b_spline_surf data from ids to sisl format
                 Note! This version utilizing the newCurve2/NewSurf2
                 constructors, i.e. STEP-conformant representation of NURBS.

Input Parameters:	long *name	name of b_spline_surface
	long *loc_cosy	name of a local_coordinate_system positioning the surface in space (OPTIONAL)
	long *u_deg	algebraic degree of the basis functions in u
	long *v_deg	algebraic degree of the basis functions in v
	long *u_upper	upper index of array of control_ points in u (as IPIMdefinition)
	long *v_upper	upper index of array of control_ points in v (as IPIM definition)
	long *n_ctrl_pnts	number of control_points (number of points in u times number of points in v: (u_upper+1) * (v_upper+1)) (not in IPIM)

long ctrl_pnts[ ]	array (of length *n_ctrl_pnts) with the names of the control_points defining the b_spline_surface (in the same order as on the STEP-file, i.e. v lines of u CAPT-identifiers here in sequence) (differs from IPIM-defin.)
long *u_uniform	indicator of the type of knot set in u (OPTIONAL)
long *v_uniform	indicator of the type of knot set in v (OPTIONAL)
long *ex_up_ind_u_knot	flag is set if there are knots_data in u (OPTIONAL)
long *up_ind_u_knot	upper index of array of knots_data in u (as IPIM-definition) (OPTIONAL)
long *ex_up_ind_v_knot	flag is set if there are knots_data in v (OPTIONAL)
long *up_ind_v_knot	upper index of array of knots_data in v (as IPIM-definition) (OPTIONAL)
long *n_ukn_mults	number of knot_multiplicities in u (not in IPIM; identical to *up_ind_u_knot)
long u_kn_mults[ ]	array (of length *n_ukn_mults) with the multiplicities of the knots in u (OPTIONAL)
long *n_vkn_mults	number of knot_multiplicities in v (not in IPIM; identical to *up_ind_v_knot)
long v_kn_mults[ ]	array (of length *n_vkn_mults) with the multiplicities of the knots in v (OPTIONAL)
long *n_uknots	number of knots in u defining the b_spline_surface in u; (not in IPIM) (identical to *up_ind_u_knot) (OPTIONAL)
double u_knots[ ]	array (of length *up_ind_u_knot) with the knot values used for defining the b_spline basis functions in u (OPTIONAL)
long *n_vknots	number of knots in v defining the b_spline_surface in v; (not in IPIM) (identical to *up_ind_v_knot) (OPTIONAL)

	double v_knots[ ]	array (of length *up_ind_v_knot) with the knot values used for defining the b_spline basis functions in v (OPTIONAL)
	long *n_weights	number of weights (not in IPIM; identical to *n_ctrl_pnts)
	double weights[ ]	array (of length *n_weights) with weights associated with the control_ points in the rational case (else = 1.0) (in the same order as on the STEP-file, i.e. v lines of u weights here in sequence) (OPTIONAL) (differs from IPIM-definition)
	long *u_closed	indicates whether the b_spline_ surface is closed in u (OPTIONAL)
	long *v_closed	indicates whether the b_spline_ surface is closed in v (OPTIONAL)
	long *form_no	indicates the analytical type of curve represented by the b_spline_surface (OPTIONAL)
	double ca_pnts[ ]	array holding cartesian points
Output Parameters:	SISLSurf **sisl_surf	b_spline_surf data in SISL surf - struct
	long *istat	status indicator
Calls:	compose_mult_knots, default_mult_knots, rat_descr	
Include Files:	-	
Errors:	-	
Author/Owner:	Per Evenson, Senter for Industriforskning P.O. Box 124, Blindern, N-0314 Oslo 3	
Program Header:	void ccisbs2(name, loc_cosy, u_deg, v_deg, u_upper, v_upper, n_ctrl_pnts, ctrl_pnts, u_uniform, v_uniform, ex_up_ind_u_knot, up_ind_u_knot, ex_up_ind_v_knot, up_ind_v_knot, n_ukn_mults, u_kn_mults, n_vkn_mults, v_kn_mults, n_uknots, u_knots, n_vknots, v_knots, n_weights, weights, u_closed, v_closed, form_no, ca_pnts, sisl_surf, istat)	

3.3.6.2.22 Module ccpobc

Name of File:	CSIPNT.C
Function:	To convert a point_on_curve to a cartesian_point
Input Parameters:	long *curve_name	name of curve
	long *loc_cosy	name of a local_coordinate_system positioning the curve in space (OPT)

long *deg	algebraic degree of the basis functions
long *up_ind_ctrl_pnts	upper index of array of control_points (as IPIM-definition)
long *n_ctrl_pnts	number of control_points (not in IPIM) (= *up_ind_ctrl_pnts + 1)
long ctrl_pnts[ ]	array (of length *n_ctrl_pnts) with the names of the control_points defining the b_spline_curve
long *uniform	indicator of the type of knot set (OPTIONAL)
long *ex_up_ind_knots_data	flag is set if there are knots_data (OPTIONAL)
long *up_ind_knots_data	upper index of array of knots_data (as IPIM definition) (OPTIONAL)
long *n_knot_mults	number of knot_multiplicities (not in IPIM; identical to *up_ind_knots_data)
long kn_mults[ ]	array (of length *n_knot_mults) with the multiplicities of the knots (OPTTIONAL)
long *n_knots	number of knots defining the b_spline (not in IPIM; identical to *up_ind_knots_data) (OPTIONAL)
double knots[ ]	array (of length *n_knots) with the knot values used for defining the b_spline basis functions (OPT.)
long *n_weights	number of weights (not in IPIM; identical to *n_ctrl_pnts)
double weights[ ]	array (of length *n_weights) with weights associated with the control_points in the rational case
long *form_no	indicates the analytical type of curve represented by the b_spline (OPT.)
long *closed	indicates the analytical type of curve (OPTIONAL)
long *self_int	indicates whether the b_spline self intersects (OPTIONAL)
double *param	parameter value of the point location

Output Parameters:	double *x_coordinate	x_coordinate of cartesian_point
	double *y_coordinate	y_coordinate of cartesian_point
	double *z_coordinate	z_coordinate of cartesian_point
	long *istat	return value: error code

| Calls: | ccisbc |

Include Files:	-
Errors:	-
Author/Owner:	Per Evenson, Senter for Industriforskning
	P.O. Box 124, Blindern, N-0314 Oslo 3
Program Header:	long ccpobc(curve_name, loc_cosy, deg, up_ind_ctrl_pnts,
	n_ctrl_pnts, ctrl_pnts, uniform, ex_up_ind_knots_data,
	up_ind_knots_data, n_knot_mults, kn_mults, n_knots, knots,
	n_weights, weights, form_no, closed, self_int, param,
	x_coordinate, y_coordinate, z_coordinate, istat)

3.3.6.2.23 Module ccpobs

Name of File:	CSIPNT.C	
Function:	Convert a point_on_surface to a cartesian point on IDS-format	
Input Parameters:	long *surf_name	name of surface
	long *loc_cosy	name of a local_coordinate_system positioning the curve in space (OPT)
	long *u_deg	algebraic degree of the basis functions in u
	long *v_deg	algebraic degree of the basis functions in v
	long *u_upper	upper index of array of control_ points in u (as IPIM-definition)
	long *v_upper	upper index of array of control_ points in v (as IPIM-definition)
	long *n_ctrl_pnts	number of control_points (number of points in u times number of points in v: (u_upper+1) * (v_upper+1)) (not in IPIM)
	long ctrl_pnts[ ]	array (of length *n_ctrl_pnts) with the names of the control_points defining the b_spline_surface (in the same order as on the STEP-file, i.e. v lines of u CAPT-identifiers here in sequence) (differs from IPIM-defin.)
	long *u_uniform	indicator of the type of knot set in u (OPTIONAL)
	long *v_uniform	indicator of the type of knot set in v (OPTIONAL)
	long *ex_up_ind_u_knot	flag is set if there are knots_data in u (OPTIONAL)
	long *up_ind_u_knot	upper index of array of knots_data in u (as IPIM-definition) (OPT.)

long *ex_up_ind_v_knot	flag is set if there are knots_data in v (OPTIONAL)
long *up_ind_v_knot	upper index of array of knots_data in v (as IPIM-definition) (OPT.)
long *n_ukn_mults	number of knot_multiplicities in u (not in IPIM; identical to *up_ind_u_knot)
long u_kn_mults[ ]	array (of length *n_ukn_mults) with the multiplicities of the knots in u (OPTIONAL)
long *n_vkn_mults	number of knot_multiplicities in v (not in IPIM; identical to *up_ind_v_knot)
long v_kn_mults[ ]	array (of length *n_vkn_mults) with the multiplicities of the knots in v (OPTIONAL)
long *n_uknots	number of knots in u defining the b_spline_surface in u; (not in IPIM) (identical to *up_ind_u_knot) (OPT)
double u_knots[ ]	array (of length *up_ind_u_knot) with the knot values used for defining the b_spline basis functions in u (OPTIONAL)
long *n_vknots	number of knots in v defining the b_spline_surface in v; (not in IPIM) (identical to *up_ind_v_knot) (OPTIONAL)
double v_knots[ ]	array (of length *up_ind_v_knot) with the knot values used for efining the b_spline basis functions n v (OPTIONAL)
long *n_weights	number of weights (not in IPIM; dentical to *n_ctrl_pnts)
double weights[ ]	array (of length *n_weights) with eights associated with the control_points in the rational case else = 1.0) (in the same order as on he STEP-file, i.e. v lines of u eights here in sequence) (OPT.) dffers from IPIM-definition)
long *u_closed	indicates whether the b_spline_ urface is closed in u (OPTIONAL)
long *v_closed	indicates whether the b_spline_ urface is closed in v (OPTIONAL)

	long *form_no	indicates the analytical type of curve represented by the b_spline_surface OPTIONAL)
	double *param_1	first parameter value of point location
	double *param_2	second parameter value of point location
Output Parameters:	double *x_coordinate	x_coordinate of cartesian_point
	double *y_coordinate	y_coordinate of cartesian_point
	double *z_coordinate	z_coordinate of cartesian_point
	long *istat	return value: error code

Calls:	ccisbs
Include Files:	-
Errors:	-
Author/Owner:	Per Evenson, Senter for Industriforskning P.O. Box 124, Blindern, N-0314 Oslo 3
Program Header:	long ccpobs(surf_name, loc_cosy, u_deg, v_deg, u_upper, v_upper, n_ctrl_pnts, ctrl_pnts, u_uniform, v_uniform, ex_up_ind_u_knot, up_ind_u_knot, ex_up_ind_v_knot, up_ind_v_knot, n_ukn_mults, u_kn_mults, n_vkn_mults, v_kn_mults, n_uknots, u_knots, n_vknots, v_knots, n_weights, weights, u_closed, v_closed, form_no, param_1, param_2, x_coordinate, y_coordinate, z_coordinate, istat)

3.3.6.2.24 Module ccsdst

Name of File:	CCSDST.C
Function:	To calculate minimum, maximum and average distance between a polynomial b-spline curve and a polynomial b-spline surface.

Input Parameters:	surf	The surface object
	curv	The curve object
	epsge	Geometrical resolution.
	epsco	Computational resolution.
	maxstp	Maximal stepping length. Is neglected if maxstp<=aepsge If maxstp <= 0.0 the 3-D SISLbox of the surface is used for estimating max step length
Output Parameters:	mindst	Minimum distance between curve and surface
	averdst	Average distance between curve and surface

maxdst		Maximum distance between curve and surface
jstat		status messages: jstat = 0 : ok, jstat > 0 : warning , jstat < 0 : error

Calls: -

Include Files: -

Errors: -

Author/Owner: Per Evenson, Senter for Industriforskning
                    P.O. Box 124, Blindern, N-0314 Oslo 3

Program Header: void ccsdst (surf, curv, epsge, epsco, maxstp, mindst, averdst,
                    maxdst, jstat)

3.3.6.2.25 Module ccsibc

Name of File: CSISIC.C

Function: Converts a nonrational_b_spline_curve from SISL-format to
                    IDS-format. Cartesian points in array ca_pnts are stored as 3d-
                    points, i.e. 2d-curves get zero-values as z-coord.

Input Parameters: SISLCurve *sisl_curve   b_spline_curve data in SISL curve-
                                             struct

Output Parameters: long *name   name of b_spline_curve

                    long *loc_cosy   name of a local_coordinate_system
                                             positioning the curve in space (OPT)

                    long *deg   algebraic degree of the basis
                                             functions

                    long *up_ind_ctrl_pnts   upper index of array of control_
                                             points (as IPIM-definition)

                    long *n_ctrl_pnts   number of control_points (not in
                                             IPIM)

                    long *ctrl_pnts[ ]   array (of length *n_ctrl_pnts) with
                                             the names of the control_points
                                             defining the b_spline_curve

                    long *uniform   indicator of the type of knot set
                                             (OPTIONAL)

                    long *ex_up_ind_knots_data   flag is set if knots_data array
                                             is filled with data (OPTIONAL)

                    long *up_ind_knots_data upper index of array of knots_data
                                             (as IPIM-definition) (OPTIONAL)

                    long *n_knot_mults   number of knot_multiplicities (not
                                             in IPIM; identical to
                                             *up_ind_knots_data)

                    long *kn_mults[ ]   array (length *n_knot_mults) with
                                             the multiplicities of the knots (OPT.)

	long *n_knots	number of knots defining the b_spline (not in IPIM; identical to *up_ind_knots_data) (OPTIONAL)
	double *knots[ ]	array (of length *n_knots) with the knot values used for defining the b_spline basis functions
	long *n_weights	number of weights (not in IPIM; identical to *n_ctrl_pnts)
	double *weights[ ]	array (of length *n_weights) with weights associated with the control_ points in the rational case (OPT.)
	long *form_no	indicates the analytical type of curve represented by the b_spline (OPT.)
	long *closed	indicates whether the b_spline is closed (OPTIONAL)
	long *self_int	indicates whether the b_spline self intersects (OPTIONAL)
	double *ca_pnts[ ]	array holding cartesian points
	long *istat	status indicator

Calls:	-
Include Files:	-
Errors:	-
Author/Owner:	Per Evenson, Senter for Industriforskning P.O. Box 124, Blindern, N-0314 Oslo 3
Program Header:	void ccsibc (sisl_curve, name, loc_cosy, deg, up_ind_ctrl_pnts, n_ctrl_pnts, ctrl_pnts, uniform, ex_up_ind_knots_data, up_ind_knots_data, n_knot_mults, kn_mults, n_knots, knots, n_weights, weights, form_no, closed, self_int, ca_pnts, istat)

3.3.6.2.26 Module ccsibs

Name of File:	CSISIS.C
Function:	Convert a nonrational_b_spline_surface from SISL-format to IDS-format. Cartesion-ponts in array ca_pnts are stored as 3d-points, i.e. 2d-surfaces gets zero-values as z-coord.
Input Parameters:	SISLSurf *sisl_surf	b_spline_surf data in SISL surf-struct
Output Parameters:	long *name	name of b_spline_surface
	long *loc_cosy	name of a local_coordinate_system positioning the surface in space (OPTIONAL)
	long *u_deg	algebraic degree of the basis functions in u

long *v_deg	algebraic degree of the basis functions in v
long *u_upper	upper index of array of control_ points in u (as IPIM-definition)
long *v_upper	upper index of array of control_ points in v (as IPIM-definition)
long *n_ctrl_pnts	number of control_points (number of points in u times number of points in v; (u_upper+1) * (v_upper+1)) (not in IPIM)
long *ctrl_pnts[ ]	array (of length *n_ctrl_pnts) with the names of the control_points defining the b_spline_surface (in the same order as on the STEP-file, i.e. v lines of u CAPT-identifiers here in sequence) (differs from IPIM-defin.)
long *u_uniform	indicator of the type of knot set in u (OPTIONAL)
long *u_uniform	indicator of the type of knot set in v (OPTIONAL)
long *ex_up_ind_u_knot	flag is set if there are knots_data in u (OPTIONAL)
long *up_ind_u_knot	upper index of array of knots_data in u (as IPIM-definition) (OPT.)
long *up_ind_v_knot	upper index of array of knots_data in v (as IPIM-definition) (OPT.)
long *n_ukn_mults	number of knot_multiplicities in u (not in IPIM; identical to *up_ind_u_knot)
long *u_kn_mults[ ]	array (of length *n_ukn_mults) with the multiplicities of the knots in u (OPTIONAL)
long *n_vkn_mults	number of knot_multiplicities in v (not in IPIM; identical to *up_ind_v_knot)
long *v_kn_mults[ ]	array (of length *n_vkn_mults) with the multiplicities of the knots in v (OPTIONAL)
long *n_uknots	number of knots in u defining the b_spline_surface in u; (not in IPIM) (identical to *up_ind_u_knot) (OPT)
double *u_knots[ ]	array (of length *up_ind_u_knot) with the knot values used for defining the b_spline basis functions in u (OPTIONAL)

	double *v_knots[ ]	array (of length *up_ind_v_knot) with the knot values used for defining the b_spline basis functions in v (OPTIONAL)
	long *n_weights	number of weights (not in IPIM; identical to *n_ctrl_pnts)
	double *weights[ ]	array (of length *n_weights) with weights associated with the control_ points in the rational case (else = 1.0) (in the same order as on the STEP-file, i.e. v lines of u weights here in sequence) (OPTIONAL) (differs from IPIM-definition)
	long *u_closed	indicates whether the b_spline_ surface is closed in u (OPTIONAL)
	long *v_closed	indicates whether the b_spline_ surface is closed in v (OPTIONAL)
	long *form_no	indicates the analytical type of curve represented by the b_spline_surface (OPTIONAL)
	double *ca_pnts[ ]	array holding cartesian points
	long *istat	status indicator

Calls: -

Include Files: -

Errors: -

Author/Owner: Per Evenson, Senter for Industriforskning
                  P.O. Box 124, Blindern, N-0314 Oslo 3

Program Header: void ccsibs(sisl_surf, name, loc_cosy, u_deg, v_deg, u_upper,
                  v_upper, n_ctrl_pnts, ctrl_pnts, u_uniform, v_uniform,
                  ex_up_ind_u_knot, up_ind_u_knot, ex_up_ind_v_knot,
                  up_ind_v_knot, n_ukn_mults, u_kn_mults, n_vkn_mults,
                  v_kn_mults, n_uknots, u_knots, n_vknots, v_knots, n_weights,
                  weights, u_closed, v_closed, form_no, ca_pnts, istat)

3.3.6.2.27 Module cctcoc

Name of File: CSICON.C

Function: To convert parameters of a trimmed conic to nonrational
                  b_spline_curve parameters

Input Parameters: long *type   type of conic
                  long *coord   name of coordinate system
                  double pt1_coord[3]   coordinates of first trimming point

	double pt2_coord[3]	coordinates of second trimming point
	long *sense	indicator of whether the trimmed curve direction coincides with that of the conic
	long *ax2	name of axis2-placement
	double *co_par_1	1st parameter of conic
	double *co_par_2	2nd parameter of conic
	double *epsge	geometry resolution
Output Parameters:	long *deg	algebraic degree of the b_spline_curve
	long *up_ind_ctrl_pnts	upper index of array of control_ points (as IPIM-definition)
	long *n_ctrl_pnts	number of control_points (not in IPIM) (= *up_ind_ctrl_pnts + 1)
	long **ctrl_pnts	array (of length *n_ctrl_pnts) with the names of the control_points defining the b_spline_curve
	long *uniform	indicator of the type of knot set (OPTIONAL)
	long *ex_up_ind_knots_data	flag is set if knots_data array is filled with data (OPTIONAL)
	long *up_ind_knots_data	upper index of array of knots_data (as IPIM-definition) (OPTIONAL)
	long *n_knot_mults	number of knot_multiplicities (not in IPIM; identical to *up_ind_knots_data)
	long **kn_mults	array (of length *n_knot_mults) with the multiplicities of the knots (OPTIONAL)
	long *n_knots	number of knots defining the b_spline (not in IPIM; identical to *up_ind_knots_data) (OPTIONAL)
	double **knots	array (of length *n_knots) with the knot values used for defining the b_spline basis functions (OPT.)
	long *n_weights	number of weights (not in IPIM; identical to *n_ctrl_pnts)
	double **weights	array (of length *n_weights) with weights associated with the control_ points in the rational case (OPT.)
	long *form_no	indicates the analytical type of curve represented by the b_spline (OPT.)
	long *closed	indicates whether the b_spline is closed (OPTIONAL)

long *self_int	indicates whether the b_spline self intersects (OPTIONAL)
double **ca_pnts	array holding cartesian points
long *istat	status indicator

Calls: -

Include Files: -

Errors: -

Author/Owner: Per Evenson, Senter for Industriforskning
                  P.O. Box 124, Blindern, N-0314 Oslo 3

Program Header: long cctcoc (type, coord, pt1_coord, pt2_coord, sense, ax2,
                  co_par_1, co_par_2, epsge, deg, up_ind_ctrl_pnts, n_ctrl_pnts,
                  ctrl_pnts, uniform, ex_up_ind_knots_data, up_ind_knots_data,
                  n_knot_mults, kn_mults, n_knots, knots, n_weights, weights,
                  form_no, closed, self_int, ca_pnts,istat)

3.3.6.2.28 Module celpac

Name of File: CSICOC.C

Function: Converts a point on a elliptic curve from parametric to
                  cartesian.   Note! This routine is based on the FEGS software.

Input Parameters:
long *coord	name of the coordinate system
double *alpha	major axis parameter
double *beta	minor axis parameter
long *ax2	name of referenced axis set
double *param	the parameter of the point

Output Parameters:
double pnt_coord[3]	coordinates of the created cartesian
long *istat	return status

Calls: -

Include Files: -

Errors: -

Author/Owner: Per Evenson, Senter for Industriforskning
                  P.O. Box 124, Blindern, N-0314 Oslo 3

Program Header: long celpac (coord,alpha,beta,ax2,param,pnt_coord,istat)

3.3.6.2.29 Module csifun

Name of Routines:
init_name,	get_up_ind_knots,
fill_knots_mults,	compose_mult_knots,
default_mult_knots,	rat_descr,
compose_rat_descr,	compose_coef,
reverse,	itoa

Name of File:	CSIFUN.C
Function:	Necessary functions for SI tools (See Headers)
Calls:	-
Include Files:	-
Errors:	-
Author/Owner:	Per Evenson, Senter for Industriforskning P.O. Box 124, Blindern, N-0314 Oslo 3

Program Headers:

void init_name(val)Purpose: initialize start-value for b_spline name/control_point name-generation

long get_up_ind_knots (knot_v, size)
Purpose: Get upper index of knot_v with duplicated elements removed

void fill_knots_mults (et, et_size, up_ind_knots, knot_v, mults)
Purpose: Convert array with duplicated values into two arrays, one containing the values and the other the multiplicities of the values. All elements in array "mults" must be initialized to 0.

void compose_mult_knots (n_knots, knot_v, mults, et, istat)
Purpose: Compose a knot array with multiple occurences based on the knot and mult arrays in the IDS database

void default_mult_knots (uniform, deg, n_ctrl_pnts, et, istat)
Purpose: Compose a default knot array with multiple occurences based on the uniform, degree and up_ind_ ctrl_pnts information in the IDS database.

long rat_descr (n_ctrl_pnts, weights)
Purpose: Decide if a description is rational

void compose_rat_descr (n_ctrl_pnts, ca_pnts, weights, dim, rat_pnts, istat)
Purpose: Compose an array that contain the homogenous control points of the rational description

void compose_coef (n_ctrl_pnts, ca_pnts, dim, coef, istat)
Purpose: Compose an array that contain the control points of the non- rational description

void reverse(s) Purpose: reverse string s

void itoa (n,s) char s[];
int n;

3.3.6.2.30 Module ccilbc

Name of File:	CSILIN.C
Function:	To convert a trimmed line to a non-rational b_spline_curve

Input Parameters:

long *line_name	name of line
long *pt_1	name of first trimming point in cartesian space (OPTIONAL)
long *pt_2	name of second trimming point in cartesian space (OPTIONAL)
long *ex_param1	indicates if optional attribute param_1 is set
double *param_1	the first trimming point in the parametric space of line (OPT.)
long *ex_param2	indicates if optional attribute param_2 is set
double *param_2	the second trimming point in the parametric space of line (OPT.)
long *sense	indicator of whether the trimmed curve direction coincides with that of the line
double *epsge	geometry resolution (conversion accuracy)

Output Parameters:

long *b_spline_curve_name	name of b_spline_curve
*istat	return value: error code

Calls:	ccisbc
Include Files:	cdtget.h, cdtput.h, cdefin.h, enum.h, cidsut.h, lists.h, y_tab.h, cerror.h, - sisl.h, csifun.h, csisic.h, csilin.h
Errors:	-
Author/Owner:	Per Evenson, Senter for Industriforskning P.O. Box 124, Blindern, N-0314 Oslo 3
Program Headers:	long ccilbc (line_name, pt_1, pt_2, ex_param1, param_1, ex_param2, param_2, sense, epsge, b_spline_curve_name, istat)

3.3.6.2.31 Module cctlic

Name of File:	CSILIN.C
Function:	To convert parameters of a trimmed line to non-rational b_spline_curve parameters

Input Parameters:

long *type	type of curve
long *coord	name of coordinate system
double pt1_coord[3]	coordinates of first trimming point

	double pt2_coord[3]	coordinates of second trimming point
	long *sense	indicator of whether the trimmed curve direction coincides with that of the line
	long *capt	name of cartesian point
	long *dir	name of direction vector
	double *epsge	geometry resolution
Output Parameters:	long *deg	algebraic degree of the b_spline_curve
	long *up_ind_ctrl_pnts	upper index of array of control_ points (as IPIM-definition)
	long *n_ctrl_pnts	number of control_points (not in IPIM) (= *up_ind_ctrl_pnts + 1)
	long **ctrl_pnts	array (of length *n_ctrl_pnts) with the names of the control_points defining the b_spline_curve
	long *uniform	indicator of the type of knot set (OPTIONAL)
	long *ex_up_ind_knots_data	flag is set if knots_data array is filled with data (OPTIONAL)
	long *up_ind_knots_data	upper index of array of knots_data (as IPIM-definition) (OPTIONAL)
	long *n_knot_mults	number of knot_multiplicities (not in IPIM; identical to *up_ind_knots_data)
	long **kn_mults	array (of length *n_knot_mults) with the multiplicities of the knots (OPTIONAL)
	long *n_knots	number of knots defining the b_spline (not in IPIM; identical to *up_ind_knots_data) (OPTIONAL)
	double **knots	array (of length *n_knots) with the knot values used for defining the b_spline basis functions (OPT.)
	long *n_weights	number of weights (not in IPIM; identical to *n_ctrl_pnts)
	double **weights	array (of length *n_weights) with weights associated with the control points in the rational case (OPT.)
	long *form_no	indicates the analytical type of curve represented by the b_spline (OPT.)
	long *closed	indicates whether the b_spline is closed (OPTIONAL)

	long *self_int	indicates whether the b_spline self intersects (OPTIONAL)
	double **ca_pnts	array holding cartesian points
	long *istat	status indicator

Calls:	ccisbc

Include Files:	cdtget.h, cdtput.h, cdefin.h, enum.h, cidsut.h, lists.h, y_tab.h, cerror.h, - sisl.h, csifun.h, csisic.h, csilin.h

Errors:	-

Author/Owner:	Per Evenson, Senter for Industriforskning P.O. Box 124, Blindern, N-0314 Oslo 3

Program Headers:	long cctlic (type, coord, pt1_coord, pt2_coord, sense, capt, dir, epsge, deg, up_ind_ctrl_pnts, n_ctrl_pnts, ctrl_pnts, uniform, ex_up_ind_knots_data, up_ind_knots_data, n_knot_mults, kn_mults, n_knots, knots, n_weights, weights, form_no, closed, self_int, ca_pnts, istat)

3.3.6.2.32 Module clipac

Name of File:	CSILIN.C

Function:	Converts a point on a line from parametric to cartesian

Input Parameters:	long *coord	name of the coordinate system
	long *capt	name of referenced cartesian point
	long *dir	name of referenced direction vector
	double *param	the parameter of the point

Output Parameters:	double pnt_coord[3]	coordinates of the created cartesian point
	long *istat	return status

Calls:	ccisbc

Include Files:	cdtget.h, cdtput.h, cdefin.h, enum.h, cidsut.h, lists.h, y_tab.h, cerror.h, - sisl.h, csifun.h, csisic.h, csilin.h

Errors:	-

Author/Owner:	Per Evenson, Senter for Industriforskning P.O. Box 124, Blindern, N-0314 Oslo 3

Program Header:	long clipac (coord,capt,dir,param,pnt_coord,istat)

3.3.6.2.33 Module ccbbpc

Name of File:	CSIPAR.C

Function:	Convert a 3D b_spline_curve to 2D b_spline_curve in the parameter plane of a b_spline_surface (all on sisl format)

Input Parameters:	SISLSurf *psurf	The B-spline surface
	SISLCurve *pcurv	The input B-spline curve

double epsge	Maximal deviation allowed between true 2-D curve and the approximated 2-D curve.
Output Parameters: SISLCurve **rcpos	Pointer to approximated 2D curve
int *jstat	Return value: error code

Calls: -

Include Files: cerror.h
                  sislP.h, csifun.h, csipar.h

Errors: -

Author/Owner: Per Evenson, Senter for Industriforskning
                  P.O. Box 124, Blindern, N-0314 Oslo 3

Programs Header: void ccbbpc (psurf, pcurv, epsge, rcpos, jstat)

3.3.6.2.34 Module c2000

Name of File: CSIPAR.C

Function: Projecting a 3D curve into the parametric plane of a surface.
                  Generate an array of points in the parameter plane of the
                  surface.

Input Parameters: SISLCurve *pcurve	Pointer to the curve
SISLSurf *psurf	Pointer to the surface
double *aepsge	Geometry resolution
Output Parameters: int *nmbp	Number of points generated
double **epnts	Array of points in the surface corresponding to the curve
double *mxerr	Maximum distance between curve and surface
int *jstat	Return value: error code

Calls: -

Include Files: cerror.h,
                  sislP.h, csifun.h, csipar.h

Errors: -

Author/Owner: Per Evenson, Senter for Industriforskning
                  P.O. Box 124, Blindern, N-0314 Oslo 3

Programs Header: void c2000 (pcurve, psurf, aepsge, nmbp, epnts, mxerr, jstat)

3.3.6.2.35 Module s9eval

Name of File: CSIPAR.C

Function: Compute the position, first and second derivative of a curve
                  going through a given point of the surface when the 0-2'th
                  derivatives of the surface is given. The tangent of the wanted

curve is parallel to the projection of a given vector into the tangent plane of the surface.

Input Parameters:	eders	0-2'th derivatives of the surface. Dimension is 6*idim.
	enorms	Normal vector of the surface. Dimension is idim.
	etanc	Vector to be projected into the tangent plane of the surface. Dimension is idim.
	idim	Dimension of geometry space.
Output Parameters:	ederc	0-2'th derivative of the curve in the surface.
	jstat	status messages: jstat = 0 : Ok; jstat > 0 : Warning; jstat < 0 : Error

Calls:	-
Include Files:	cerror.h, sislP.h, csifun.h, csipar.h
Errors:	-
Author/Owner:	Per Evenson, Senter for Industriforskning P.O. Box 124, Blindern, N-0314 Oslo 3
Programs Header:	void s9eval (eders, enorms, etanc, ederc, idim, jstat)

3.3.6.2.36 Module csi221

Name of File:	CSI221.C
Function:	Converts STEP-models from FL3 to FL2
Input Parameters:	long fbsm_name	the name of the face_based_surface_model entity
	double *epsge	geometry resolution
Output Parameters:	long *g3ss_name	the name of the geometric_3d_surface_set entity
	long *stat	error code
Calls:	-
Include Files:	-
Errors:	-
Author/Owner:	Per Evenson, Senter for Industriforskning P.O. Box 124, Blindern, N-0314 Oslo 3
Program Header:	void csi221 (long fbsm_name, double *epsge, long *g3ss_name, long *stat)

3.3.6.2.37 Module csicfasf

Name of File:	CSI221.C
Function:	To convert a face entity to a b_spline_surface

Input Parameters: long *face_name   name of face
                   double *epsge   geometry resolution

Output Parameters: long *surface_name   name of surface
                   long *istat   return value: error code

Calls: -

Include Files: -

Errors: -

Author/Owner: Per Evenson, Senter for Industriforskning
                   P.O. Box 124, Blindern, N-0314 Oslo 3

Program Header: void csicfasf (long *face_name, double *epsge, long
                   *surface_name, long *istat);

3.3.6.2.38 Module csicvxlp

Name of File:	CSI221.C
Function:	To convert a vertex loop entity to a cartesian_point entity

Input Parameters: long *vertex_loop_name name of vertex_loop
                   double *epsge   geometry resolution

Output Parameters: long *capt_name   name of cartesian_point
                   long *istat   return value: error code

Calls: -

Include Files: -

Errors: -

Author/Owner: Per Evenson, Senter for Industriforskning
                   P.O. Box 124, Blindern, N-0314 Oslo 3

Program Header: long csicvxlp (vertex_loop_name, epsge, capt_name, istat)

3.3.6.2.39 Module csicedbc

Name of File:	CSI221.C
Function:	To convert a edge entity to a b_spline_curve entity

Input Parameters: long *edge_name   name of edge
                   long *edge_flag   sense of edge flag
                   long *surface_name   name of surface of current face
                   double *epsge   geometry resolution

Output Parameters: long *b_spline_curve_name   name of b_spline_curve
                    long *start_pt_name   name of start point of curve
                    long *end_pt_name   name of end point of curve
                    long *istat   return value: error code

Calls:	-
Include Files:	-
Errors:	-
Author/Owner:	Per Evenson, Senter for Industriforskning P.O. Box 124, Blindern, N-0314 Oslo 3
Program Header:	long csicedbc (edge_name, edge_flag, surface_name, epsge, b_spline_curve_name, start_pt_name, end_pt_name, istat)

3.3.6.2.40 Module csicvxcapt

| Name of File: | CSI221.C |
| Function: | To convert a vertex entity to a cartesian_point entity |

Input Parameters: long *vertex_name   name of vertex
                  double *epsge   geometry resolution

Output Parameters: long *point_name   name of point
                    long *istat   return value: error code

Calls:	-
Include Files:	-
Errors:	-
Author/Owner:	Per Evenson, Senter for Industriforskning P.O. Box 124, Blindern, N-0314 Oslo 3
Program Header:	long csicvxcapt (vertex_name, epsge, point_name, istat)

3.3.6.2.41 Module csi322

| Name of File: | CSI322.C |
| Function: | Converts STEP-models from FL2 to FL1 |

Input Parameters: long sbsm_name   the name of the shell_based_
                                              surface_model entity

Output Parameters: long *fbsm_name   the name of the face_based_
                                              surface_model entity
                    long *stat   error code

Calls:	-
Include Files:	-
Errors:	-

Author/Owner:	Per Evenson, Senter for Industriforskning
	P.O. Box 124, Blindern, N-0314 Oslo 3

Program Header: void csi322 (long sbsm_name, long *fbsm_name, long *stat)

3.3.6.2.42 Module csi_construct

Name of File:	CSI_CONSTRUCT.C
Function:	Special constructs
Purposes:	- Construct a SISLFace entity
	- Create and initialize a face (instance of SISLFace)
	- Add outer boundary to a face (instance of SISLFace)
	- Add inner loop to a face (instance of SISLFace)
	- Add inner boundary to a face (instance of SISLFace)
	- Create and initialize a SISLCurve instance
	- Create and initialize a SISLSurf instance

3.3.6.2.43 Module s1221

Name of File:	S1221.C
Function:	To compute the value and the first derivatives of the B-spline curve pointed to by curve, at the point with parameter value "parvalue". Evaluation starts from the right hand side.

Input Parameters:	Curve *pc1	Pointer to the curve for which position and derivatives are to be computed.
	int ider	The number (order) of derivatives to compute.
		<0: Error;
		=0: Compute position;
		=1: Compute position and derivative etc.
	double ax	The parameter value at which to compute position and derivatives.
Input/Output Parameters:	int *ileft	Pointer to the interval in the knot vector where ax is located. If et is the knot vector, the relation: et[leftknot] <= parvalue < et[leftknot+1] should hold. If (parvalue < et[in]) then leftknot should be "in-1". Here "in" is the number of B-spline coefficients. If leftknot does not have the right value when entering

		the routine, its value will be changed to the value satisfying the above condition.
Output Parameters:	double eder[ ]	Double array of dimension (der+1) * dim containing the position and derivative vectors. (dim is the number of components of each B-spline coefficient, i.e. the dimension of the Euclidean space in which the curve lies.) These vectors are stored in the following order: First the dim components of the position vector, then the dim components of the tangent vector, then the dim components of the second derivative vector, and so on. (The C declaration of derive as a two dimensional array would therefore be derive[dim][der+1].)
	int *jstat	Status messages >0 : warning =0 : ok <0 : error

Calls:	s1220, s6err, s6ratder
Include Files:	-
Errors:	-
Author/Owner:	Per Evenson, Senter for Industriforskning P.O. Box 124, Blindern, N-0314 Oslo 3
Program Header:	#define S1221

```
#define S1221
#ifndef lint
static char SISL_SccsId[] = "@(#)s1221.c      1.3 09/06/90";
#endif
#include "sislP.h"
#ifdef __STDC__
void s1221(Curve *pc1,int ider,double ax,int *ileft,double
eder[],int *jstat)
#else
void s1221(pc1,ider,ax,ileft,eder,jstat)
                         Curve *pc1;
                         int ider;
```

```
                                    double ax;
                                    int *ileft;
                                    double eder[];
                                    int *jstat;
             #endif
```

3.3.6.2.44 Module s1227

Name of File: S1227.C

Function: To compute the value and the first derivatives of the B-spline curve pointed to by curve, at the point with parameter value "parvalue. Evaluation from the left hand side.

Input Parameters: Curve *pc1   Pointer to the curve for which position and derivatives are to be computed.

                    int ider   The number of derivatives to compute.
                                          >0: Error.
                                          =0: Compute position.
                                          <0: Compute position and derivative etc.

                    double ax   The parameter value at which to compute position and derivatives.

Input/Output Parameters: int *ileft   Pointer to the interval in the knot vector where parvalue is located. If et is the knot vector, the relation: et[leftknot] <= parvalue <= et[leftknot+1]   should hold. If (parvalue < et[ik-1]) then leftknot should be "ik-1". Here "ik" is the order of the B-spline curve. If leftknot does not have the right value when entering the routine, its value will be changed to the value satisfying the above condition.

Output Parameters: double eder[ ]   Double array of dimension containing the position and derivative vectors.(dim is the number of components of each B-spline coefficient, i.e. the dimension of the Euclidean space in which the curve lies.) These vectors are stored in the following order: First the dim components of the

		position vector, then the dim components of the tangent vector, then the dim components of the second derivative vector, and so on. The C declaration of derive as a two dimensional array would therefore be: derive[dim][der+1].)
	int *jstat	Status messages >0 : warning =0 : ok <0: error

Calls:	s6err, s1219, s1220, s6knotmult
Include Files:	-
Errors:	-
Author/Owner:	Per Evenson, Senter for Industriforskning P.O. Box 124, Blindern, N-0314 Oslo 3

Program Header:
```
#define S1227
#ifndef lint
static char SISL_SccsId[] = "@(#)s1227.c      1.3 09/06/90";
#endif
#include "sislP.h"
#ifdef__STDC__
void s1227(Curve *pc1,int ider,double ax,int *ileft,double
eder[],int *jstat)
#else
void s1227(pc1,ider,ax,ileft,eder,jstat)
                      Curve *pc1;
                      int   ider;
                      double ax;
                      int   *ileft;
                      double eder[];
                      int   *jstat;
#endif
```

3.3.6.2.45 Module s1240

Name of File:	S1240.C
Function:	Calculates the length of a B-spline curve.The length calculated will not deviate more than (epsco/divided by the length calculated) from the real length of the curve.
Input Parameters:	SISLCurve *pcurve The B-spline curve double aepsco Computer resolution

Output Parameters:

double *clength	Length of the B-spline curve.
int *jstat	Status messages
	=1 : warning
	=0 : ok
	<0 : error

Calls:	s6dist, s6err,s1251
Include Files:	-
Errors:	-
Author/Owner:	Per Evenson, Senter for Industriforskning
	P.O. Box 124, Blindern, N-0314 Oslo 3

Program Header:

```
#define S1240
#ifndef lint
static char SISL_RCSID[] = "$Header:s1240.c,v 1.2 91/03/17
14:09:50 sisl Exp $";
#endif
#include "sislP.h"
#if defined(SISLNEEDPROTOTYPES)
void s1240(SISLCurve *pcurve,double aepsco,double
*clength,int *jstat)
#else
void s1240(pcurve,aepsco,clength,jstat)
                    SISLCurve *pcurve;
                    double aepsco;
                    double *clength;
                    int   *jstat;
#endif
```

3.3.6.2.46 Module s1302

Name of File:	S1302 C
Function:	To create a B-spline rotational surface by rotating a curve a given angle around the axis defined by point[ ] and axis[ ]. The maximal deviation allowed between the true rotational surface and the generated surface, is epsge.

Input Parameters:	Curve *pc	Pointer to the curve that is to be rotated.
	double aepsge	Maximal deviation allowed between the true rotational surface and the generated surface.
	double angle	The rotational angle. The angle is counter clockwise around axis. If the absolute value of the angle is greater

		than 2¶, then a rotational surface that is closed in the rotation direction is made.
	double ep[ ]	One point and the direction of the rotational axis.
Output Parameters:	Surf **rs	Pointer to the produced surface.
	int *jstat	Status messages >0 : warning =0 : ok <0 : error

Calls:	s6norm, s6scpr, s1301, s6rotax, s6mvec, s6err
Include Files:	-
Errors:	-
Author/Owner:	Per Evenson, Senter for Industriforskning P.O. Box 124, Blindern, N-0314 Oslo 3
Program Header:	#define S1302 #ifndef lint static char SISL_SccsId[] = "@(#)s1302.c 1.2 06/15/90"; #endif #include "sislP.h" #ifdef __STDC__ void s1302(Curve *pc,double aepsge,double angle,double ep[],double eaxis[], Surf **rs,int *jstat) #else void s1302(pc,aepsge,angle,ep,eaxis,rs,jstat) Curve *pc; double aepsge; double angle; double ep[]; double eaxis[]; Surf **rs; int *jstat; #endif

3.3.6.2.47 Module s1332

Name of File:	S1332.C
Function:	To create a swept B-spline surface by making the tensor-product of two B-spline curves.
Input Parameters:	Curve *pc1 Pointer to curve 1. Curve *pc2 Pointer to curve 2. double aepsge Maximal deviation allowed between the true swept

| | double ep[ ] | Point near the curve to be swept. If the point lies on curve 2, then curve 2 is swept along curve 1 with the point as contact point. If the point lies on curve 1,then curve 1 is swept along curve 2 with the point as contact point. If the point is not lying on any of the curves, then the surface will not interpolate any of the curves. |

Output Parameters:	Surf **rs	Pointer to the surface produced.
	int *jstat	Status messages
		>0 : warning
		=0 : ok
		<0 : error

Calls: s1707, s6err

Include Files: -

Errors: -

Author/Owner: Per Evenson, Senter for Industriforskning
                  P.O. Box 124, Blindern, N-0314 Oslo 3

Program Header:
```
#define S1332
#ifndef lint
static char SISL_SccsId[] = "@(#)s1332.c    1.2 06/15/90";
#endif
#include "sislP.h"
#ifdef __STDC__
void s1332(Curve *pc1,Curve *pc2,double aepsge,double
ep[],Surf **rs,int *jstat)
#else
void s1332(pc1,pc2,aepsge,ep,rs,jstat)
                        Curve *pc1;
                        Curve *pc2;
                        double aepsge;
                        double ep[];
                        Surf **rs;
                        int *jstat;
#endif
```

3.3.6.2.48 Module s1388

Name of File: S1388.C

Function: Convert a B-spline surface of order less than or equal to 4 in
                  both directions to a mesh of Coons patches with uniform

parameterisation. The function assumes that the B-spline surface is C.

Input Parameters:	Surf *ps1	Pointer to the surface that is to be converted
Output Parameters:	double *gcoons[ ]	Array containing the (sequence of) Coons patches. The total number of patches is numcoons1*numcoons2. The patches are stored in sequence with dim*16 doubles for each patch. For each corner of the patch we store in sequence: position, deriv ative, find area in second parameter direction, and twist. This array is allocated inside the routine and must be released by the calling routine.
	int *jnumb1	Number of Coons patches in first parameter direction.
	int *jnumb2	Number of Coons patches in second parameter direction.
	int *jdim	The dimension of the geometric space.
	int *jstat	Status messages >0 : Order too high, surface is interpolated. =0 : ok <0 : error

Calls:	s1424, s6err
Include Files:	-
Errors	:
Author/Owner:	Per Evenson, Senter for Industriforskning P.O. Box 124, Blindern, N-0314 Oslo 3
Program Header:	#define S1388

```
#define S1388
#ifndef lint
static char SISL_SccsId[] = "@(#)s1388.c     1.2 06/15/90";
#endif
#include "sislP.h"
#ifdef __STDC__
void s1388(Surf *ps1,double *gcoons[],int *jnumb1,int
*jnumb2,int *jdim,int *jstat)
#else
void s1388(ps1,gcoons,jnumb1,jnumb2,jdim,jstat)
                      Surf  *ps1;
                      double *gcoons[];
```

```
                                        int   *jnumb1;
                                        int   *jnumb2;
                                        int   *jdim;
                                        int   *jstat;
              #endif
```

3.3.6.2.49 Module s1389

Name of File:	S1389.C

Function: Convert a B-spline curve of order up to 4 to a sequence of cubic segments with uniform parameterisation.

Input Parameters: Curve *pc1   Pointer to the curve that is to be converted

Output Parameters: double *gcubic[ ]   Array containing the sequence of cubic segments. Each segment is represented by the start point, followed by the start tangent, end point and end tangent. Each segment needs 4*dim doubles for storage. This array is allocated inside the function and must be released by the calling function.

int *jnumb   Number of elements of length (4*dim) in the array cubic

int *jdim   The dimension of the geometric space.

int *jstat   Status messages
                       >0 : warning
                       =0 : ok
                       <0 : error

Calls: s1221, s1227, s62rr

Include Files: -

Errors: -

Author/Owner: Per Evenson, Senter for Industriforskning
                   P.O. Box 124, Blindern, N-0314 Oslo 3

Program Header: #define S1389
                   #ifndef lint
                   static char SISL_SccsId[] = "@(#)s1389.c   1.2 06/15/90";
                   #endif
                   #include "sislP.h"
                   #ifdef __STDC__
                   void s1389(Curve *pc1,double *gcubic[],int *jnumb,int

```
                    *jdim,int *jstat)
                    #else
                    void s1389(pc1,gcubic,jnumb,jdim,jstat)
                                        Curve  *pc1;
                                        double *gcubic[];
                                        int    *jnumb;
                                        int    *jdim;
                                        int    *jstat;
          #endif
```

3.3.6.2.50 Module s1421

Name of File:	S1421.C
Function:	Evaluate the surface pointed at by surf at the parameter value parvalue. Compute the derivatives and the normal if der>=1.

Input Parameters: SISLSurf *ps1   Pointer to the surface to evaluate.

int ider   Number (order) of derivatives to evaluate.

<0 : No derivatives evaluated.

=0 : Position evaluated.

>0 : Position and derivatives evaluated.

double epar[ ]   Parameter-value at which to evaluate. Dimension of parvalue is 2.

Input/Output Parameters: int *ilfs   Pointer to the interval in the knot vector in first parameter direction where parvalue [0] is found. The relation etl[leftknot1] <= parvalue[0] < etl[leftknot1+1] where etl is the knot vector should hold. leftknot1 should be set equal to zero at the first call to the routine. Do not change leftknot during a section of calls to s1421.

int *ilft   Corresponding to leftknot1 in the second parameter direction.

Output Parameters: double eder[ ]   Array where the derivatives of the surface in parvalue are placed. The sequence is position, first derivative in first parameter direction, first derivative in second parameter direction, (2,0) derivative, (1,1) derivative, (0,2) derivative, etc.

	Dimension of derive is dim*(1+2+...+(der+1)) = dim *(der+1)(der+2)/2.
double enorm[ ]	Normal of surface. Is evaluated if dimension is dim. The normal is not normalised.
int *jstat	Status messages
	=2 : Surface is degenerate at the point, normal has zero length.
	=1 : Surface is close to degenerate at the point. Angle between tangents is less than the angular tolerance.
	=0 : Ok.
	<0 : Error.

Calls:	s6err, s6crss, s1424
Include Files:	-
Errors	:
Author/Owner:	Per Evenson, Senter for Industriforskning P.O. Box 124, Blindern, N-0314 Oslo 3
Program Header:	

```
#define S1421
#ifndef lint
static char SISL_RCSID[] = "$Header: s1421.c,v 1.2 91/03/17
14:13:32 sisl Exp $";
#endif
#include "sislP.h"
#if defined(SISLNEEDPROTOTYPES)
void s1421(SISLSurf *ps1, int ider, double epar[], int *ilfs,
                        int *ilft, double eder[], double
                        enorm[], int *jstat)
#else
void s1421(ps1,ider,epar,ilfs,ilft,eder,enorm,jstat)
                        SISLSurf  *ps1;
                        int   ider;
                        double epar[];
                        int   *ilfs;
                        int   *ilft;
                        double eder[];
                        double enorm[];
                        int   *jstat;
#endif
```

3.3.6.2.51 Module s1424

Name of File: S1424.C

Function: Evaluate the surface pointed at by surf at the parameter value
                     (parvalue[0],parvalue[1]). Compute the der1*der2 first
                     derivatives. The derivatives that will be computed are $D_{i,j}$
                     with $i=0,1,...,der1$ and $j=0,1,...,der2$.

Input Parameters: Surf *ps1   Pointer to the surface to evaluate.

                     int ider1   Number (order) of derivatives to be
                                                  evaluated in first parameter
                                                  direction. The possible values are:
                                                  $0 <= der1 < surf->ik1$,
                                                  where surf->ik1 is the order of the
                                                  surface in first parameter direction.

                     int ider2   Number (order) of derivatives to be
                                                  evaluated in second parameter
                                                  direction. The possible values are:
                                                  $0 <= der1 < surf->ik2$,   where
                                                  surf->ik2 is the order of the surface
                                                  in second parameter direction.

                     double epar[ ]   Parameter-value at which to
                                                  evaluate. The dimension of parvalue
                                                  is 2.

Input/Output Parameters: int *ileft1   Pointer to the interval in the knot
                                                  vector in first parameter direction
                                                  where parvalue [0] is found.The
                                                  relation
                                                  etl[leftknot1] <= parvalue[0] <
                                                        etl[leftknot1+1]
                                                  where etl is the knot vector should
                                                  hold. leftknot1 should be set equal to
                                                  zero at the first call to the routine.
                                                  Do not change the value of leftknot1
                                                  between calls to the routine.

                     int *ileft2   Corresponding to leftknot1 in the
                                                  second parameter direction.

Output Parameters: double eder[ ]   Array of size d(der1+1)(der2+1)
                                                  where the position and the derivative
                                                  vectors of the surface in
                                                  (parvalue[0],parvalue[1]) is placed.
                                                  d=surf->dim is the number of
                                                  elements in each vector and is equal
                                                  to the geometrical dimension. The
                                                  vectors are stored in the following

		order: First the d components of the position vector, then the d compo nents of the D1,0 vector,and so on up to the d components of the Dder1,0 vector, then the d compo nents of the D1,1 vector etc.If derive is considered to be a three dimen sional array, then its declaration in C would be derive[d][der1+1][der2+1].
int	*jstat	Status messages =1: Warning. =0: Ok. <0 : Error.

Calls: s1220, s1219, s6err, s6sratder

Include Files: -

Errors: -

Author/Owner: Per Evenson, Senter for Industriforskning
                  P.O. Box 124, Blindern, N-0314 Oslo 3

Program Header: #define S1424
                  #ifndef lint
                  static char SISL_SccsId[] = "@(#)s1424.c   1.3 09/06/90";
                  #endif
                  #include "sislP.h"
                  #ifdef __STDC__
                  void s1424(Surf *ps1,int ider1,int ider2,double epar[], int
                  *ileft1, int *ileft2,double eder[],int *jstat)
                  #else
                  void s1424(ps1,ider1,ider2,epar,ileft1,ileft2,eder,jstat)
                              Surf *ps1;
                              int   ider1;
                              int   ider2;
                              double epar[];
                              int   *ileft1;
                              int   *ileft2;
                              double eder[];
                              int   *jstat;
                  #endif

3.3.6.2.52 Module s1602

Name of File: S1602.C

Function: To make a straight line represented as a B-spline curve
                  between two points.

Input Parameters:	double estapt[ ]	Start point of the straight line
	double endpt[ ]	End point of the straight line
	int ik	The order of the B-spline curve to be made.
	int idim	The dimension of the geometric space
	double astpar	Start value of the parameterisation of the curve
Output Parameters:	double *cendpar	Parameter value used at the end of the curve
	Curve **rc	Pointer to the found curve
	int *jstat	Status messages
		>0 : warning
		=0 : ok
		<0 : error

Calls:	s6dist, s6err
Include Files:	-
Errors:	-
Author/Owner:	Per Evenson, Senter for Industriforskning
	P.O. Box 124, Blindern, N-0314 Oslo 3

Program Header:
```
#define S1602
#ifndef lint
static char SISL_SccsId[] = "@(#)s1602.c     1.2 06/15/90";
#endif
#include "sislP.h"
#ifdef __STDC__
void s1602(double estapt[],double endpt[],int ik,int
idim,double astpar, double *cendpar,Curve **rc,int *jstat)
#else
void s1602(estapt,endpt,ik,idim,astpar,cendpar,rc,jstat)
                     double estapt[];
                     double endpt[];
                     int    ik;
                     int    idim;
                     double astpar;
                     double *cendpar;
                     Curve  **rc;
                     int    *jstat;
#endif
```

3.3.6.2.53 Module s1613

Name of File: S1613.C

Function:	To calculate a set of points on a B-spline curve. The straight lines between the points will not deviate more than epsge from the B-spline curve at any point.

Input Parameters:	SISLCurve *pc	The input B-spline curve.
	double aepsge	Geometry resolution, maximum distance allowed between the curve and the straight lines that are to be calculated.

Output Parameters:	double **gpoint	Calculated points, the array is allocated by this function.
	int *jnbpnt	Number of calculated points.
	int *jstat	Status messages
		>0 : warning
		=0 : ok
		<0 : error

Calls:	-

Include Files:	-

Errors:	-

Author/Owner:	Per Evenson, Senter for Industriforskning
	P.O. Box 124, Blindern, N-0314 Oslo 3

Program Header:

```
#define S1613
#ifndef lint
static char SISL_RCSID[] = "$Header: s1613.c,v 1.3 91/03/17
14:14:43 sisl Exp $";
#endif
#include "sislP.h"
#if defined(SISLNEEDPROTOTYPES)
void s1613(SISLCurve *pc,double aepsge,double **gpoint, int
*jnbpnt, int *jstat)
#else
void s1613(pc,aepsge,gpoint,jnbpnt,jstat)
                    SISLCurve *pc;
                    double aepsge;
                    double **gpoint;
                    int   *jnbpnt;
                    int   *jstat;
#endif
```

3.3.6.2.54 Module S1710

Name of File:	S1710.C

Function:	Subdivide a B-spline curve at a given parameter-value.

Input Parameters:	SISLCurve *pc1	SISLCurve to subdivide.
	double apar	Parameter value at which to subdivide.

Input/Output Parameters:	SISLCurve **rcnew1	First part of the subdivided curve.
	SISLCurve **rcnew2	Second part of the subdivided curve. If the parameter value is at the end of a B-spline curve. NULL pointers might be returned.
	int *jstat	Status messages >0 : warning =0 : ok <0 : error

Calls: s1700, s6err

Include Files: -

Errors: -

Author/Owner: Per Evenson, Senter for Industriforskning
                 P.O. Box 124, Blindern, N-0314 Oslo 3

Program Header: #define S1710
                 #ifndef lint
                 static char SISL_RCSID[] = "$Header: s1710.c,v 1.3 91/03/17
                 14:14:57 sisl Exp $";
                 #endif
                 #include "sislP.h"
                 #if defined(SISLNEEDPROTOTYPES)
                 void s1710(SISLCurve *pc1,double apar,SISLCurve
                 **rcnew1, SISLCurve **rcnew2,int *jstat)
                 #else
                 void s1710(pc1,apar,rcnew1,rcnew2,jstat)
                                    SISLCurve *pc1;
                                    double apar;
                                    SISLCurve **rcnew1;
                                    SISLCurve **rcnew2;
                                    int *jstat;
                 #endif

3.3.6.2.55 Module s1730

Name of File:	S1730.C
Function:	To convert a B-spline curve to a sequence of Bezier curves. The Bezier curves are stored as one B-spline curve with all knots of multiplicity newcurve- >ik (order of the curve).
Input Parameters: Curve *pc	SISLCurve to convert.

Output Parameters:	Curve **rcnew	The new B-spline curve containing all the Bezier curves.
	int *jstat	Status messages >0 : warning =0 : ok <0 : error

Calls:	s1701, s6err
Include Files:	-
Errors:	-
Author/Owner:	Per Evenson, Senter for Industriforskning P.O. Box 124, Blindern, N-0314 Oslo 3

Program Header:	#define S1730
	#ifndef lint
	static char SISL_SccsId[] = "@(#)s1730.c 1.2 06/15/90";
	#endif
	#include "sislP.h"
	#ifdef __STDC__
	void s1730(Curve *pc,Curve **rcnew,int *jstat)
	#else
	void s1730(pc,rcnew,jstat)
	                          Curve *pc;
	                          Curve **rcnew;
	                          int *jstat;
	#endif

3.3.6.2.56 Module s1731

Name of File:	S1731.C

Function:	To convert a B-spline surface to Bezier surfaces.The Bezier surfaces are stored in a B-spline surface with all knots having multiplisity equal to the order of the surface in the corresponding parameter direction.	
Input Parameters:	Surf *ps	Surface to convert.
Output Parameters:	Surf **rsnew	The new B-spline surface storing Bezier represented surfaces.
	int *jstat	Status messages >0 : warning =0 : ok <0 : error

Calls:	s1701, s6err
Include Files:	-
Errors:	-

Author/Owner: Per Evenson, Senter for Industriforskning
P.O. Box 124, Blindern, N-0314 Oslo 3

Program Header: define S1731
#ifndef lint
static char SISL_SccsId[] = "@(#)s1731.c   1.2 06/15/90";
#endif
#include "sislP.h"
#ifdef __STDC__
void s1731(Surf *ps,Surf **rsnew,int *jstat)
#else
void s1731(ps,rsnew,jstat)
                              Surf *ps;
                              Surf **rsnew;
                              int *jstat;

#endif

3.3.6.2.57 Module ccasbs

File: csia2s.c

Function: converts B-spline surface from APS-SS to SISL-format

Input Parameters:

int idim	dimension in which the curve lies
int ik1	order in 1. par. direction
int ik2	order in 2. par. direction
int in1	no. of vertices in 1. parameter direction
int in2	no. of vertices in 2. parameter direction
int ikp1	pointer to start of 1. knot vector in escr
int ikp2	pointer to start of 2. knot vector in escr
int ivp	pointer to start of vertices in escr
double escr[ ]	array holding vertices and knots
int kinxtfr	pointer to next free element in escr
int kimxscr	pointer to last free element in escr

Output Parameters:

SISLSurf **sisl_surf	b_spline_curve data in SISL surf-struct
long* istat	status indicator

Calls: SISLSurf *newSurf()   ...
                       newarray   ...

Include Files: <stdio.h>, "escr.h", "sisl.h", "ccmem.h", "csi_error.h", "IDS_error.h"

| Errors: | CNOERR | no error |
| | CMEERR | memory allocation fault |

Author/Owner: Per Evensen, Senter for Industriforskning,
P.O.Boks 124, Blindern, N-0314 Oslo

Last Modification: 26.6.92

3.3.6.2.58 Module ccsabc

| File: | csis2a.c |
| | |

Function: converts B-spline curve from SISL- to APS-SS format

| Input Parameters: | SISLCurve *sisl_curve | b_spline_curve data in SISL curve-struct |

Output Parameters: int *idim	dimension in which the curve lies
int *ik	order of curve
int *in	number of vertices
int *ipknt	pointer to start of knot vector in scrat
int *ipvert	pointer to start of vertices in scrat
float *scrat[ ]	array holding vertices and knots
long *istat	status indicator

| Calls: | malloc |

| Include Files: | <stdio.h>, "sisl.h", "cerror.h" |

Errors:	CNOERR	no error
	CMEERR	memory allocation fault
	CDTILL	SISLcurve illegally defined

Author/Owner: Per Evensen, Senter for Industriforskning,
P.O.Boks 124, Blindern, N-0314 Oslo

Last Modification: 26.6.92

3.3.6.2.59 Module ccsabs

| File: | csis2a.c |

Function: converts B-spline surface from SISL- to APS-SS format

| Input Parameters: | SISLSurf *sisl_surf | b_spline_surface data in SISL surf-struct |

Output Parameters: int *idim	dimension in which the surface lies
int *ik1	order in 1. par. direction
int *ik2	order in 2. par. direction
int *in1	no. of vertices in 1. paramater direction
int *in2	no. of vertices in 2. parameter direction

	int *ikp1	pointer to start of 1. knot vector in scrat
	int *ikp2	pointer to start of 2. knot vector in scrat
	int *ivp	pointer to start of vertices in scrat
	float *scrat[ ]	array holding vertices and knots
	long *istat	status indicator
Calls:	malloc	
Include Files:	<stdio.h>	
	"sisl.h"	
	"cerror.h"	
Errors:	CNOERR	no error
	CMEERR	memory allocation fault
	CDTILL	SISLsurf illegally defined
Author/Owner:	Per Evensen, Senter for Industriforskning,	
	P.O.Boks 124, Blindern, N-0314 Oslo	

Last Modification: 26.6.92

3.3.6.3 Conversion tools by GfS

Grouping of modules: GfS provides 3 groups of modules, which are character-ised here shortly:

1. B-spline conversions take b-spline curves and surfaces and produce entities appropriate for an FEM modeler. They calculate points on curves to discretise them and divide faces into patches to create a mesh of geometrical macros which may be divided further into finite elements.
2. Utility modules mainly support the use of Common Tools e.g. for Fortran developers. They provide keyword lists and entity types and change enumeration values to integer (or vice versa). Other utilities support the interactive use of Common Tools. An important part of this group deals with model conversions, e.g. from Wireframe to Surface, or with topology generation for geometric sets.
3. The Fortran-IDS is a complete database with the same functionality as described in chapter 4. It was created at a time when the CADEX-IDS (written in C) was not yet operational or did not yet cover the full range of entities. Its put, get, and attribute functions are 100% compatible. As all data are stored in Fortran arrays, this data structure allows an easy and quick dump of all data to a disk file. From this file the data structure can be read easily and quickly. This feature is important for processors that need a lot of interactive conversions. The user can safe the present status in order to interrupt and later resume his work. This

allows also to save an intermediate status which does not comply with any application protocol or a model which in a mixed way refers to several APs. A tool to copy from one IDS to the other is provided at the end of the data base section.

The GfS Common Tools software comprises 30 files, each belonging to one of the 3 groups:

File	Group	Size [kB]
bsmacu.f	1	6
bspocu.f	1	13
bsuti.f	1	3
mebssu.f	1	20
nurbs.f	1	17
pnombs.f	1	10
checker.f	2	2
creawf.f	2	4
crefac.f	2	3
crtopo.f	2	12
delent.c	2	2
eval.f	2	27
fenums.f	2	8
heatop.f	2	5
interr.f	2	17
parsers.c	2	4
sortop.f	2	3
statis.c	2	3
stypes.f	2	13
utbase.f	3	12
mbbase.f	3	18
ssbase.f	3	16
wfbase.f	3	8
csbase.f	3	11
cbbase.f	3	4
fcbase.f	3	10
pdbase.f	3	9
hdbase.f	3	10
dabase.f	3	20
idscfo.f	3	28

Most of the GfS modules are written in Fortran. Software design and module parameters allow a full compatibility with C-software. Tests showed no problems in linking and execution. In the following description all parameters are explained according to C syntax.

B-spline Modules

3.3.6.3.1 Module bsmacu

File:	bsmacu.f
Function:	computes 4 b-spline curves that are round a b-spline surface control by dialog, works on the IDS
Input Parameters:	long ia[ ] work space long lenwsp size of work space
Output Parameters:	-
Calls:	IDS get/put/list routines langbs in file bsmacu.f
Include Files:	impl.inc (implicit variable type declarations) , type.inc (entity types)
Errors:	-
Author/Owner:	Dr. Helmut J. Helpenstein, Gesellschaft fuer Strukturanalyse mbH, Pascalstr.17, D-52076 Aachen
Last Modification:	22.5.92

3.3.6.3.2 Module langbs

File:	bsmacu.f
Function:	computes and adds the distances between point i1 - i2 and i2 - i3. If these are control points, then the result gives an estimate of the length of the b-spline curve
Input Parameters:	long i1 name of beginning point of the curve long i2 name of an appropriate intermediate point of the curve long i3 name of end point of the curve
Output Parameters:	double bsl sum of distances (i1-i2) and (i2-i3)
Calls:	IDS get/put/list routines langbs in file bsmacu.f
Include Files:	impl.inc
Errors:	-
Author/Owner:	Dr. Helmut J. Helpenstein Gesellschaft fuer Strukturanalyse mbH, Pascalstr.17, D-52076 Aachen
Last Modification:	5.4.91

3.3.6.3.3 Module bspocu

File:	bspocu.f
Function:	conversion b-spline_curve to polyline with interactive control and branching

Input Parameters:

	long *nn	number of b-spline curves
	long li[ ]	array for list of b-spline curves
	long icapt[ ]	array for point names of generated polyline
	long *maxcpt	maximum numbers of points in a polyline
	long ipoi[ ]	array for points of a geometric set
	long *maxpoi	their maximum number
	long icurv[ ]	array for curves of a geometric set
	long *maxcur	their maximum number
	long isurf[ ]	array for surfaces of a geometric set
	long *maxsur	their maximum number
	long iscr[ ]	working space
	long *isize	its size

Output Parameters: -

Calls:	IDS: cglist, cglile, clipar, clfrst, clnext, clkill, cgculs, cmculs, cged, cmed, cgg3ss, cmg3ss
	In file bspocu.f: cobcpo, intbs
Include Files:	type.inc
Errors:	-
Author/Owner:	Dr. Helmut J. Helpenstein Gesellschaft fuer Strukturanalyse mbH, Pascalstr.17, D-52076 Aachen
Last Modification:	22.5.92

3.3.6.3.4 Module cobcpo

File:	bspocu.f
Function:	convert a b-spline curve to a polyline using the old APS routine

Input Parameters:

	long *icurv	name of the b-spline curve
	long *npoint	number of desired polyline points (if 0, the number is controlled by the tolerance)
	double *tol	tolerance: largest distance between b-spline curve andpolyline (not used if npoint > 0)
	long kont[ ]	array for names of control points

long *mxkont	maximum number of control points
long mult[ ]	array for multiplicities
long *mxmult	maximum number of multiplicities
double rknot[ ]	array for knot values
long *mxknot	maximum number of knot values
double weig[ ]	array for weights
long *mxweig	maximum number of weights
float eb[ ]	array for control point coordinates
long *lcoco	number of control points
float et[ ]	array for knots
long *leknot	length of knot vector
float escr[ ]	scratch array for SI routine P19503
long *imxscr	its size

Output Parameters:
long *ipoly	name of the generated polyline
long icapt[ ]	names of generated cartesian points
long *ncapt	number of generated points (length of icapt,if npoint>0 then ncapt=npoint)
long *istat	error status (0=ok)

Calls: In IDS: cglile, cgbscu, cgapfl, cgcapt, cfname, cpcapt, cpapfl,
                    cppocu
           In file si1.f: p19503

Include Files: -

Errors: -

Author/Owner: Dr. Helmut J. Helpenstein
                  Gesellschaft fuer Strukturanalyse mbH,
                  Pascalstr.17, D-52076 Aachen

Last Modification: 5.4.91

3.3.6.3.5 Module intbs

File: bspocu.f

Function: conversion b-spline curve to polyline

Input Parameters: long *isteu   control parameter:
                      1: not used here
                      2: npoint equidistant parameter
                          values are used
                      3: npoint is calculated from:
                          factor times no.of knots
                      4: npoint is input here (interactively)

long *npoint	number of points to be computed
double *factor	determines number of points from number of knots

	long *icurv	entity name of b-spline curve to be converted
	long kont[ ]	array for control point names
	long *mxkont	maximum number of control points
	long mult[ ]	array for multiplicities
	long *mxmult	maximum number of multiplicities
	double rknot[ ]	array for knot values
	long *mxknot	maximum number of knot values
	double weig[ ]	array for weights
	long *mxweig	maximum number of weights
	double xcont[ ]	array for control point coordinates
	long *lcoco	length of this array
	double uk[ ]	array for knots (after expanding multiplicities)
	long *leknot	number of knots; uk has this length; b has twice this length
Output Parameters:	long *ipoly	name of created polyline
	long icapt[ ]	array with names of created cartesian points of the polyline
	long *ncapt	number of created points
	long *istat	error parameter (0=ok)

Calls: In IDS: cglile, cgbscu, cgapfl, cgcapt, cfname, cpcapt, cpapfl, cppocu
                  In file bsuti.f: checkn ; in file pnombs.f: pobscu

Include Files: -

Errors: Error return from IDS routines
                  Overflow of arrays for b-spline curve entity

Author/Owner: Dr. Helmut J. Helpenstein
                  Gesellschaft fuer Strukturanalyse mbH,
                  Pascalstr.17, D-52076 Aachen

Last Modification: 22.5.92

3.3.6.3.6 Module checkn

File: bsuti.f

Function: checks, if parameter u is equal to a knot value; if yes, u is slightly changed;
                  checks, if parameter u is outside the knot range; if yes it is set to first/last knot.

Input Parameters:	double ukn[ ]	knots
	long *nk	number of knots
	double *u	parameter value to be checked

Output Parameters:	double *u	parameter value, eventually changed
Calls:	-	
Include Files:	-	
Errors:	-	
Author/Owner:	Dr. Helmut J. Helpenstein	
	Gesellschaft fuer Strukturanalyse mbH,	
	Pascalstr.17, D-52076 Aachen	

Last Modification: 25.10.91

3.3.6.3.7 Module conins

File:	bsuti.f
Function:	stores coordinates of all control points in array xyz

Input Parameters:	long list[ ]	names of control points
	long *nkont	number of control points
Output Parameters:	double xyz[ ]	array for coordinates (x1,y1,z1, x2,y2,z2, x3,...)
Calls:	cgcapt (IDS)	
Include Files:	-	
Errors:	-	
Author/Owner:	Dr. Helmut J. Helpenstein	
	Gesellschaft fuer Strukturanalyse mbH,	
	Pascalstr.17, D-52076 Aachen	

Last Modification: 25.10.91

3.3.6.3.8 Module knoins

File:	bsuti.f
Function:	stores all knots in array ukn according to their multiplicity

Input Parameters:	long *iex	existence flag
	long mult[ ]	multiplicities
	long *lmult	length of array mult
	double uknot[ ]	array of knot values returned from IDS
	long *uknot	length of array uknot
Output Parameters:	double ukn[ ]	knot array to be filed here
	long *nukn	resulting length of ukn
	double *u0	smallest knot value
	double *umax	greatest knot value
Calls:	-	

Include Files:	-
Errors:	-

Author/Owner:	Dr. Helmut J. Helpenstein
	Gesellschaft fuer Strukturanalyse mbH,
	Pascalstr.17, D-52076 Aachen

Last Modification: 25.10.91

3.3.6.3.9 Module mebssu

File:	mebssu.f
Function:	mesh b-spline surface

Input Parameters:	long *ns	number of b-spline surfaces
	long li[ ]	array for storing the list of b-spline surfaces
	long iwsp[ ]	working space
	long *lewsp	its size

Output Parameters: -

Calls:	IDS: cglist, cglile
	In file mebssu.f: bsflae

Include Files:	type.inc
Errors:	-

Author/Owner:	Dr. Helmut J. Helpenstein
	Gesellschaft fuer Strukturanalyse mbH,
	Pascalstr.17, D-52076 Aachen

Last Modification: 25.10.91

3.3.6.3.10 Module bsflae

File:	mebssu.f
Function:	control module for subdividing b-spline surfaces; patches are stored in a face_based_surface_model

Input Parameters:	long *isurf	name of b-spline surface
	long *ncont	number of control points
	long *nmultu	number of knot multiplicities in u
	long *nmultv	number of knot multiplicitiesin v
	long *nknotu	number of knot values in u
	long *nknotv	number of knot values in v
	long *nweigh	number of weights
	long *n	number of desired divisions in u
	long *m	number of desired divisions in v
	long *knz	number of desired additional points on each edge of the created faces

long ia[ ]	scratch array
long maxdim	size of scratch array (required size is computed in module)

Output Parameters: -

Calls: IDS: cgbssu
                    in file mebssu.f: fltopo, femwri
                    in file bsuti.f: knoins, conins, checkn

Include Files: -

Errors: -

Author/Owner: Dr. Helmut J. Helpenstein
                    Gesellschaft fuer Strukturanalyse mbH,
                    Pascalstr.17, D-52076 Aachen

Last Modification: 25.10.91

3.3.6.3.11 Module fltopo

File: mebssu.f

Function: create topology (face_based_surface_model) for subfaces of a
                    b-spline surface

Input Parameters: long *ibssu   name of the b-spline surface
                    double xyz[ ]   coordinates of points created by
                                         stepping the u,v parameters
                    long *npoi   number of points created
                    long *n   number of elements in u-direction
                    long *m   number of elements in v-direction
                    long *knz   number of intermediate points per
                                         edge
                    long kan[ ]   array for storing edge-related values
                                         (not yet used !)
                    long *maxkan   total number of edges
                    long *maxfla   maximum number of faces

Output Parameters: long ifla[ ]   face names for use in a
                                         connected_face_set
                    long *ifbsm   name of the generated
                                         face_based_surface_model

Calls: IDS: cgapfl, cpapfl, cfname, cdelet, cundel, cpcapt, cppocu,
                    cpvx, cped, cpedlo, cpfa, cpcfas, cpfbsm
                    In file mebssu.f: chkedl

Include Files: -

Errors: -

Author/Owner: Dr. Helmut J. Helpenstein
                  Gesellschaft fuer Strukturanalyse mbH,
                  Pascalstr.17, D-52076 Aachen

Last Modification: 2.5.91

3.3.6.3.12 Module chkedl

File: mebssu.f

Function: check the 4 edges of a loop, if one has zero length

Input Parameters: long loop[ ]   names of the 4 edges of the loop
                   long *ivier   number of edges (on input: 4)

Output Parameters: long *ivier   number of edges (on output: 4 or
                                            less)

                   long *istat   error return parameter of the IDS
                                            routines (0=no error)

Calls: cgvx, cgcapt, cged, cdelet, cmvx (all IDS)

Include Files: -

Errors: -

Author/Owner: Dr. Helmut J. Helpenstein
                   Gesellschaft fuer Strukturanalyse mbH,
                   Pascalstr.17, D-52076 Aachen

Last Modification: 2.5.91

3.3.6.3.13 Module femwri

File: mebssu.f

Function: writes directly an FEM file from a b-spline surface

Input Parameters: long *lfe   logical unit for FEM file output
                   double xyz[ ]   coordinates of generated points
                   long *npoi   number of generated points
                   long *n   number of elements in u-direction
                   long *m   number of elements in v-direction
                   long *knz   number intermediate points per edge

Output Parameters: -

Calls: chk4o3 (file mebssu.f)

Include Files: -

Errors: -

Author/Owner: Dr. Helmut J. Helpenstein
                   Gesellschaft fuer Strukturanalyse mbH,
                   Pascalstr.17, D-52076 Aachen

Last Modification: 2.5.91

3.3.6.3.14 Module chk4o3

File:	mebssu.f
Function:	checks if a given element/face has 3 or 4 vertices/sides

Input Parameters:	long ip[ ]	list of (up to 12) points of the face
	long *ivier	number of points (4, 8 or 12)
	long *knz	number of intermediate points per edge
	double xyz[ ]	coordinates of all points
	long *npoi	number of all points
Output Parameters:	long *ik	resulting number of points (in case of triangle: 3, 6 or 9)

Calls:	-
Include Files:	-
Errors:	-
Author/Owner:	Dr. Helmut J. Helpenstein
	Gesellschaft fuer Strukturanalyse mbH,
	Pascalstr.17, D-52076 Aachen
Last Modification:	2.5.91

3.3.6.3.15 Module prbssu

File:	nurbs.f
Function:	computes coordinates xyz of a point on a rational b-spline curve at parameter u

Input Parameters:	double xcont[ ]	coordinates of control points (their number: (iuup+1)*(ivup+1)
	long *iuup	upper index on control points in u-direction
	long *ivup	upper index on control points in v-direction
	double uk[ ]	u-knots, stored according to their multiplicity
	long *nuknu	upper index on u-knots (if a knot multiplicity was >1, this knot must appear several times in this list)
	double vk[ ]	v-knots, stored according to their multiplicity
	long *nvknv	upper index on v-knots (if a knot multiplicity was >1, this knot must appear several times in this list)
	double weight[ ]	weights (dimensions like control points)

	long *nw	number of weights (only the difference 0 or >0 matters and distinguishes polynomial case from rational case
	double bu[ ]	array for values of the basic functions in u (length: 2*(iuup+1))
	double bv[ ]	array for values of the basic functions in v (length: 2*(ivup+1))
	long *iudeg	degree of b-spline surface in u-direction
	long *ivdeg	degree of b-spline surface in v-direction
	double *u	parameter value u
	double *v	parameter value v
	double xyz[ ]	coordinate array with already calculated points
	long *npoi	number of points computed so far
Output Parameters:	double xyz[ ]	coordinate array for the calculated points (including the actual one)
	long *npoi	number of points computed so far (the last one isthe actual one)

Calls: -

Include Files: -

Errors: -

Author/Owner: Dr. Helmut J. Helpenstein
                    Gesellschaft fuer Strukturanalyse mbH,
                    Pascalstr.17, D-52076 Aachen

Last Modification: 22.5.92

3.3.6.3.16 Module ponurb

File:	nurbs.f
Function:	computes coordinates xyz of a point on a rational b-spline curve at parameter u

Input Parameters:	double xcont[ ]	coordinates of control points
	long *iupcon	upper index on control points (their number: iupcon+1)
	double weight[ ]	weights (their number is either 0 or iupcon+1)
	long *nw	number of weights (only the difference 0 or >0 matters and controls a switch in coordinates computation

double uk[ ]	knots, stored according to their multiplicity
long *iupkn	upper index on knots (and their number)
double b[ ]	array for values of the basic functions (length: 2*(iupkn+1))
long *ideg	degree of b-spline curve
double *u	parameter value

Output Parameters: double xyz[ ]   the three coordinates of the calculated point

Calls: -

Include Files: -

Errors: -

Author/Owner: Dr. Helmut J. Helpenstein
                    Gesellschaft fuer Strukturanalyse mbH,
                    Pascalstr.17, D-52076 Aachen

Last Modification: 22.5.92

3.3.6.3.17 Module dinurb

File: nurbs.f

Function: prepares a (rational) b-spline curve for display by dividing it into a number of straight lines

Input Parameters:

long *icurve	name of curve to be displayed (converted)
long kont[ ]	array for storing names of control points
long *maxkon	length of array kont (= maximum number of control points)
long mult[ ]	array for storing knot multiplicities
long *maxmul	maximum length of array mult
double rknot[ ]	array for storing knots (as received from IDS)
long *maxkno	maximum length of array rknot
double weight[ ]	array for storing weights (as received from IDS)
long *maxwei	maximum length of array weight
double eb[ ]	array for storing coordinates of control points
long *lcoco	number of control points; length of array eb = 3 * lcoco

	double et[ ]	array for storing expanded knots (stored according to their multiplicity)
	long *leknot	length of array et
	double escr[ ]	scratch area
	long *imxscr	length of scratch area

Output Parameters: - (calls repeatedly gralin with computed start and end coordinates of line segments)

Calls: in IDS: cgbscu, cgcapt
                    in bsuti.f: checkn
                    in nurns.f: ponurb
                    user-supplied: gralin

Include Files: -

Errors: -

Author/Owner: Dr. Helmut J. Helpenstein
                    Gesellschaft fuer Strukturanalyse mbH,
                    Pascalstr.17, D-52076 Aachen

Last Modification: 22.5.92

3.3.6.3.18 Module dinurs

File: nurbs.f

Function: prepares a (rational) b-spline surface for display by dividing it into a number of squares with straight edges

Input Parameters:	long *isurf	name of surface to be displayed (converted)
	long kont[ ]	array for storing names of control points
	long *nkont	length of array kont, number of control points
	long multu[ ]	array for storing knot multiplicities in u
	long *nmultu	length of array multu, number of u-multiplicities
	long multv[ ]	array for storing knot multiplicities in v
	long *nmultv	length of array multv, number of v-multiplicities
	double uknot[ ]	array for storing u-knots (as received from IDS)
	long *nknotu	length of array uknot

double vknot[ ]	array for storing v-knots (as received from IDS)
long *nknotv	length of array vknot
double weight[ ]	array for storing weights (as received from IDS, stored in same order as control points)
long *nw	length of array weight
double eb[ ]	array for storing coordinates of control points
long *lcoco	number of control points; length of array eb = 3 * lcoco
double uk[ ]	array for storing expanded u-knots (stored according to their multiplicity)
double vk[ ]	array for storing expanded v-knots (stored according to their multiplicity)
long *leknot	length of arrays uk and vk
double bu[ ]	array for computing u basic functions (length 2 * leknot)
double bv[ ]	array for computing v basic functions (length 2 * leknot)

Output Parameters: - (calls repeatedly gralin with computed start and end coordinates of line segments)

Calls: in IDS: cgbssu, cgcapt
                in bsuti.f: checkn
                in nurns.f: prbssu
                user-supplied: gralin

Include Files: -

Errors: -

Author/Owner: Dr. Helmut J. Helpenstein
                 Gesellschaft fuer Strukturanalyse mbH,
                 Pascalstr.17, D-52076 Aachen

Last Modification: 22.5.92

3.3.6.3.19 Module pobscu

File: pnombs.f

Function: computes coordinates xyz of a point on a b-spline curve at parameter u. For rational b-splines use routines in file nurbs.f.

Input Parameters: double xcont[ ]   coordinates of control points
                  long *iupcon   upper index on control points (their number: iupcon+1)

	double uk[ ]	knots, stored according to their multiplicity
	long *iupkn	upper index on knots (their number: iupkn+1)
	double b[ ]	array for values of the basic functions (length: 2*(iupkn+1))
	long *ideg	degree of b-spline curve
	double *u	parameter value
Output Parameters:	double xyz[ ]	the three coordinates of the calculated point

Calls: -

Include Files: -

Errors: -

Author/Owner: Dr. Helmut J. Helpenstein
                           Gesellschaft fuer Strukturanalyse mbH,
                           Pascalstr.17, D-52076 Aachen

Last Modification: 25.10.91

3.3.6.3.20 Module pobssu

File: pnombs.f

Function: compute a point on a b-spline surface given by control point
                          coordinates and a knot set

Input Parameters:	double xcont[ ]	coordinates of the control points (x1,y1,z1, x2,y2,z2, x3,...)
	long *iuup	upper index on control points in u-direction (point indices: 0,1,...,iuup)
	long *ivup	upper index on control points in v-direction
	double uknot[ ]	u-knots stored according to their multiplicity
	long *nuknu	upper index on u-knots (knot indices: 1,2,...,nuknu)
	double vknot[ ]	v-knots stored according to their multiplicity
	long *nvknu	upper index on v-knots (knot indices: 1,2,...,nvknu)
	double bu[ ]	array to store the basis functions in u (size 2*(nuknu+1))
	double bv[ ]	array to store the basis functions in v (size 2*(nvknu+1))
	long *iudeg	degree of the b-spline surface in u-direction

long *ivdeg	degree of the b-spline surface in v-direction
double *u	u parameter value for which the point must be computed
double *v	v parameter value for which the point must be computed
double xyz[ ]	coordinate array for the calculated points
long *npoi	number of points (including this new one)

Output Parameters: -

Calls: -

Include Files: -

Errors: -

Author/Owner: Dr. Helmut J. Helpenstein
                     Gesellschaft fuer Strukturanalyse mbH,
                     Pascalstr.17, D-52076 Aachen

Last Modification: 25.10.91

Utility Modules

3.3.6.3.21 Module stypes

File: stypes.f

Function: reads file y_tab.h and puts information into Fortran common
                     block /TYPES/, presets all itxxxx variables (including those
                     for Step file header)

Input Parameters: long *lu   logical unit to read the file y_tab.h
                     char filnam[ ]   tree name of file y_tab.h
                     long *nchar   number of characters in filnam

Output Parameters: long *istat   error parameter (0=ok)

Calls: tytest (in the same file)

Include Files: type.inc (common block /TYPES/ containing all entity names)

Errors: istat=1 : could not open file

Author/Owner: Dr. Helmut J. Helpenstein
                     Gesellschaft fuer Strukturanalyse mbH,
                     Pascalstr.17, D-52076 Aachen

Last Modification: 22.5.92

3.3.6.3.22 Module tytest

File:	stypes.f
Function:	counts how many itxxxx variables have been set by module stypes
Input Parameters:	-
Output Parameters:	number of variables in common /TYPES/ that have been set
Calls:	-
Include Files:	type.inc
Errors:	-
Author/Owner:	Dr. Helmut J. Helpenstein Gesellschaft fuer Strukturanalyse mbH, Pascalstr.17, D-52076 Aachen

Last Modification: 22.5.92

3.3.6.3.23 Module enumev

File:	fenums.f	
Function:	transforms from enumerations into integer values	
Input Parameters:	char kenn[ ]	enumeration class (e.g. 'uniform_type')
	char text[ ]	enumeration string (e.g. '.UNIFORM_KNOTS.') (Fortran declaration for both is 'character*(*)')
Output Parameters:	long *num	integer value of the enumeration
Calls:	-	
Include Files:	-	
Errors:	num=0 means: no integer transformation possible	
Author/Owner:	Dr. Helmut J. Helpenstein Gesellschaft fuer Strukturanalyse mbH, Pascalstr.17, D-52076 Aachen	

Last Modification: 28.4.92

3.3.6.3.24 Module enumco

File:	fenums.f	
Function:	transforms from integers into enumerations	
Input Parameters:	char kenn[ ]	enumeration class (e.g. 'curve_transition_code')
	long *num	integer value

Output Parameters: char text[ ]	enumeration string (e.g. '.CONT_SAME_GRADIENT.')
long *lt	number of characters in text

Calls:	-
Include Files:	-
Errors:	lt=0 means: no transformation possible
Author/Owner:	Dr. Helmut J. Helpenstein Gesellschaft fuer Strukturanalyse mbH, Pascalstr.17, D-52076 Aachen

Last Modification: 28.4.92

3.3.6.3.25 Module checker

File:	checker.f
Function:	"main" program for checker with interactive control
Input Parameters:	long *iakti dummy parameter (for future use)
Output Parameters:	-
Calls:	In file check.c (test tools): ckini, ckent, ckall, ckend In file heatop.f: namlan
Include Files:	-
Errors:	-
Author/Owner:	Dr. Helmut J. Helpenstein Gesellschaft fuer Strukturanalyse mbH, Pascalstr.17, D-52076 Aachen

Last Modification: 2.5.91

3.3.6.3.26 Module creawf

File:	creawf.f
Function:	creates wireframe models from any brep or ss models, topology models are converted to edge_based_wireframe_ model, geometric sets are converted to geometric_3d_curve_ set. On request all non-wireframe entities can be deleted.

Input Parameters:	long i1[ ]	array for: names of edge_loops, surfaces of a geometric_set, names of created connected_edge_sets
	long *n1	length of array i1
	long i2[ ]	array for: edges (or edge_logical_ structures) of an edge_loop, curves of a geometric set, edges of a connected_edge_set

long *n2	length of array i2
long i3[ ]	array for: points of a geometric set
long *n3	length of array i3

Output Parameters: -

Calls: In IDS: cgnumb, cglist, cgtype, client, cgapfl, clfrst, clnext, clkill, cgedlo, cgedls, cgg3ss, cfname, cpceds, cpebwm, cpg3cs, cpapfl, cpscpe, cdelet
In file sortop.f: sortop

Include Files: impl.inc (implicit variable type declarations),
type.inc (entity types)

Errors: -

Author/Owner: Dr. Helmut J. Helpenstein
Gesellschaft fuer Strukturanalyse mbH,
Pascalstr.17, D-52076 Aachen

Last Modification: 22.5.92

3.3.6.3.27 Module crefac

File: crefac.f

Function: creates edges from edge_loops and makes a face_based_ surface_model from a shell_based_wireframe_model; deletes old model on request.

Input Parameters:	long ia[ ]	working space
	long *lenwsp	length of array ia

Output Parameters: -

Calls: In IDS: get/put/list routines, cdelet
In file sortop.f: sortop

Include Files: impl.inc (implicit variable type declarations),
type.inc (entity types)

Errors: -

Author/Owner: Dr. Helmut J. Helpenstein
Gesellschaft fuer Strukturanalyse mbH,
Pascalstr.17, D-52076 Aachen

Last Modification: 22.5.92

3.3.6.3.28 Module crtopo

File: crtopo.f

Function: Topology generation for geometric sets: For bounded geometry (b-spline surfaces, b-spline curves, polylines, trimmed_curves) edges (and faces) are created and a new top model is created (edge_based_wireframe model or face_based_surface_model).

Input Parameters: long *isteu   control variable:
                                              1 = process only curves (and check for common vertices)
                                              2 = process only surfaces (and compute bounding edges)
                                              3 = process curves and surfaces (and check bounding edges)
                   long ia[ ]   working space
                   long *lenwsp   length of array ia

Output Parameters: -

Calls: In this file (crtopo.f): suchpt, suchvx, cufrsu (not for independent use)
                   In IDS: put/get/list/flag routines
                   In file sortop.f: sortop

Include Files: impl.inc (implicit variable type declarations),
                   type.inc (entity types)

Errors: -

Author/Owner: Dr. Helmut J. Helpenstein
                   Gesellschaft fuer Strukturanalyse mbH,
                   Pascalstr.17, D-52076 Aachen

Last Modification: 22.5.92

3.3.6.3.29 Module delent

File: delent.c

Function: deletes (and undeletes) entities interactively

Input Parameters: -

Output Parameters: -

Calls: In IDS: cdelet, cundel
                   In file delent.c: clean

Include Files: -

Errors: -

Author/Owner: Dr. Helmut J. Helpenstein
                   Gesellschaft fuer Strukturanalyse mbH,
                   Pascalstr.17, D-52076 Aachen

Last Modification: 22.5.92

3.3.6.3.30 Module clean

File:	delent.c
Function:	deletes all unused entities
Input Parameters:	-

Output Parameters: long *number   number of deleted entities
long *istat   status indicator from IDS

Calls:	In IDS (file ctable.c): clear_visited, mark_as_visited
Include Files:	-
Errors:	-
Author/Owner:	Dr. Helmut J. Helpenstein
	Gesellschaft fuer Strukturanalyse mbH,
	Pascalstr.17, D-52076 Aachen

Last Modification: 22.5.92

3.3.6.3.31 Module headit

File:	heatop.f
Function:	shows and allows editing of header information,
	lists top entities,
	is always used in interactive mode
Input Parameters:	-
Output Parameters:	-
Calls:	In IDS: cgwrld, cgtype, all routines in chdget.c and chdput.c
	in this file: showld, namlan
Include Files:	-
Errors:	-
Author/Owner:	Dr. Helmut J. Helpenstein
	Gesellschaft fuer Strukturanalyse mbH,
	Pascalstr.17, D-52076 Aachen

Last Modification: 22.5.92

3.3.6.3.32 Module showld

File:	heatop.f
Function:	shows all world top entities (type and name)

Input Parameters: long itops[ ]   list of world top entity names
long *ntops   number of world top entities

Output Parameters:	-
Calls:	In IDS: cgwrld, cgtype

Include Files:	type.inc
Errors:	-
Author/Owner:	Dr. Helmut J. Helpenstein Gesellschaft fuer Strukturanalyse mbH, Pascalstr.17, D-52076 Aachen

Last Modification: 25.10.91

3.3.6.3.33 Module namlan

File:	heatop.f
Function:	give length of a text (name, string); end of string is assumed, if no more non-blank characters are found or if \0 (C syntax) is encountered.

Input Parameters:	char nam[ ]	string to be examined
	long *lges	number of characters to be checked

Output Parameters:	long *lact	length of the string (without \0)

Calls:	-
Include Files:	-
Errors:	lact=0 means: no end of string found (string contains only blanks and no \0)
Author/Owner:	Dr. Helmut J. Helpenstein Gesellschaft fuer Strukturanalyse mbH, Pascalstr.17, D-52076 Aachen

Last Modification: 2.5.91

3.3.6.3.34 Module interr

File:	interr.f
Function:	interrogate contents of IDS interactively

Input Parameters:	long ia[ ]	integer array (needed for lists in entity parameters)
	double ra[ ]	real array (needed for lists in entity parameters)
	long *lenwsp	length of arrays ia and ra

Output Parameters: -

Calls:	All (!) IDS get routines, cgtype, cgscpe, cgapfl
Include Files:	impl.inc (implicit variable type declarations), type.inc (entity types)
Errors:	-

Author/Owner: Dr. Helmut J. Helpenstein
                     Gesellschaft fuer Strukturanalyse mbH,
                     Pascalstr.17, D-52076 Aachen

Last Modification: 22.5.92

3.3.6.3.35 Module parsers

File: parsers.c

Function: "main" program for parser with AP control

Input Parameters: long *debug   parser's verbose flag (0=off, 1=on)
                     long *write_log   parser's log-file flag (0=off, 1=on)
                     long *size   (not used, was previously provided
                                              for combined IDS init)
                     char *file   name of the STEP file to be read

Output Parameters: long *errors   number of errors occured during
                                              parsing

Calls: In the scanner/parser: init_parser, yystepparse,
                     yybrep_apparse, yysf_apparse, yywf_apparse, yycsg_apparse,
                     yycbr_apparse, fetch_parse_results

Include Files: parse.h   parse_results_struct and error types
                     enum.h   provides the AP names

Errors: syntax errors/warnings as encountered by the different parsers

Author/Owner: Dr. Helmut J. Helpenstein
                     Gesellschaft fuer Strukturanalyse mbH,
                     Pascalstr.17, D-52076 Aachen

Last Modification: 22.5.92

3.3.6.3.36 Module sortop

File: sortop.f

Function: generates representation entities for each top model

Input Parameters: -

Output Parameters: -

Calls: Only in IDS: get routines for top models, cgwrld, cgapfl
                                   put routines for representation entities

Include Files: -

Errors: -

Author/Owner: Dr. Helmut J. Helpenstein
                     Gesellschaft fuer Strukturanalyse mbH,
                     Pascalstr.17, D-52076 Aachen

Last Modification: 22.5.92

3.3.6.3.37 Module statis

File: statis.c

Function: "main" program for BMW statistics reporter

Input Parameters: -

Output Parameters: -

Calls: In file bmtool.c: headinf, get_ent
                     In IDS: cgname

Include Files: -

Errors: -

Author/Owner: Dr. Helmut J. Helpenstein
                     Gesellschaft fuer Strukturanalyse mbH,
                     Pascalstr.17, D-52076 Aachen

Last Modification: 25.10.91

3.3.6.3.38 Modules eval, heakfk, putdat

File: eval.f

Function: link from KfK parser to IDS (put routines)
                     can easily be adapted for any parser output stream
                     heakfk: header section, eval: data section, putdat: IDS calls

Input Parameters: -

Output Parameters: -

Calls: get_stream_of_tokens, all (!) IDS put routines

Include Files: -

Errors: -

Author/Owner: Dr. Helmut J. Helpenstein
                     Gesellschaft fuer Strukturanalyse mbH,
                     Pascalstr.17, D-52076 Aachen

Last Modification: 2.5.91

Fortran-IDS Modules

3.3.6.3.39 Module finids

File: utbase.f

Function: initialise Fortran-IDS
                     refer to introduction (3.3.6.3.1) for basic information on this
                     package

Input Parameters:	long *ment	maximum number of entities
	long *mtop	maximum number of world_top_entities
Output Parameters:	long *ment	number of possible entities (if istat<0)
	long *mtop	number of possible world_top_ entities (if istat<0)
	long *istat	error return parameter

Calls: stypes (file stypes.f)

Include Files: impl.inc (implicit variable typing)
                base.inc (The data base)

Errors: istat=0 : ok, >0 initialisation failed, <0 : initialisation with reduced number of entities

Author/Owner: Dr. Helmut J. Helpenstein
                Gesellschaft fuer Strukturanalyse mbH,
                Pascalstr.17, D-52076 Aachen

Last Modification: 2.5.91

3.3.6.3.40 List and Utility Modules

File: utbase.f

Function: Perform exactly the same functions as the respective IDS routines described in chapter 3.3.1/3.4.1. Names begin with f (instead of c), 2nd to 6th characters are identical:
fgwrld, fgtype, fliall, fglist, fgnumb, fgcuap, fpcuap, fgapfl, fpapfl, fgscpe, fpscpe, fgflag, fpflag, frflag, fexchi, fglile, fdelet, fundel

Input Parameters: see chapter 3.3.1

Output Parameters: see chapter 3.3.1

Calls: gibnum (in file dabase.f)

Include Files: impl.inc (implicit variable typing)
                base.inc (The data base)
                type.inc (entity types)

Errors: error return parameter istat: 0=ok,
                possible errors: entity not found

Author/Owner: Dr. Helmut J. Helpenstein
                Gesellschaft fuer Strukturanalyse mbH,
                Pascalstr.17, D-52076 Aachen

Last Modification: 2.5.91

3.3.6.3.41 Modules fpxxxx

File:	mbbase.f for BREP_AP entities
	sfbase.f for SS_AP entities not yet included in mbbase
	wfbase.f for WF_AP entities not yet included in mbbase and ssbase
	csbase.f for CSG_AP entities not yet included in a previous file
	cbbase.f for CBR_AP entities not yet included in a previous file
	fcbase.f for fac_brep and additional entities
	pdbase.f for product definition / representation entities
	hdbase.f for header entities
Function:	put routines for all entities
Input Parameters:	described in 3.3.1
Output Parameters:	described in 3.3.1
Calls:	stoent, stoint, storea, stocha, stonam, addtop (all in file dabase.f)
Include Files:	impl.inc (implicit variable typing)
	type.inc (entity types)
Errors:	2 = name already exists
	3 = table overflow (parameters, integers, reals, characters, names)
	4 = maximum number of entities exceeded
	8 = max. number of world_top_entities exceeded
Author/Owner:	Dr. Helmut J. Helpenstein
	Gesellschaft fuer Strukturanalyse mbH,
	Pascalstr.17, D-52076 Aachen
Last Modification:	22.5.92

3.3.6.3.42 Modules fgxxxx

File:	mbbase.f for BREP_AP entities
	ssbase.f for SS_AP entities not yet included in mbbase
	wfbase.f for WF_AP entities not yet included in mbbase and ssbase
	csbase.f for CSG_AP entities not yet included in a previous file
	cbbase.f for CBR_AP entities not yet included in a previous file
	fcbase.f for fac_brep and additional entities
	pdbase.f for product definition / representation entities
	hdbase.f for header entities

Function:	get routines for all entities
Input Parameters:	described in 3.3.1
Output Parameters:	described in 3.3.1
Calls:	retent, retint, retrea, retcha, retnam (all in file dabase.f)
Include Files:	impl.inc (implicit variable typing) type.inc (entity types)
Errors:	2 = wrong entity name 5 = entity does not exist 6 = wrong entity type 7 = entity is deleted
Author/Owner:	Dr. Helmut J. Helpenstein Gesellschaft fuer Strukturanalyse mbH, Pascalstr.17, D-52076 Aachen
Last Modification:	22.5.92

3.3.6.3.43 Module stoent

File:	dabase.f
Function:	store a new entity in the data base, fill a new entry in the lookup table
Input Parameters:	long *name name of entity to be stored long *ientyp type of entity (as in y_tab.h) long *nattri number of attributes
Output Parameters:	long *iadpar address of attributes long *istat error code
Calls:	namsor
Include Files:	impl.inc, base.inc
Errors:	3 = memory exceeded for attribute pointers 4 = memory exceeded for lookup table
Author/Owner:	Dr. Helmut J. Helpenstein Gesellschaft fuer Strukturanalyse mbH, Pascalstr.17, D-52076 Aachen
Last Modification:	5.7.91

3.3.6.3.44 Modules stoint, storea, stocha, stonam

File:	dabase.f
Function:	store an integer/real/character/name attribute (single value or array/string) in the data base
Input Parameters:	long *name name of entity to which the attribute belongs

	long *iattri	address in attribute table (computed by stoent, incremented here)
	long *number	number of values to be stored
	long/double/char/long val[ ]	array or single value to be stored

Output Parameters: long *istat   error code

Calls: -

Include Files: impl.inc, base.inc

Errors: 3 = memory exceeded for integer/real/character/name values

Author/Owner: Dr. Helmut J. Helpenstein
                    Gesellschaft fuer Strukturanalyse mbH,
                    Pascalstr.17, D-52076 Aachen

Last Modification: 5.7.91

3.3.6.3.45 Module addtop

File: dabase.f

Function: adds an entity name to the list of world_top_entities

Input Parameters: long *name   name of top entity to be added

Output Parameters: long *istat   error code

Calls: -

Include Files: impl.inc, base.inc

Errors: 8 = memory exceeded for world_top_entities

Author/Owner: Dr. Helmut J. Helpenstein
                    Gesellschaft fuer Strukturanalyse mbH,
                    Pascalstr.17, D-52076 Aachen

Last Modification: 5.7.91

3.3.6.3.46 Module retent

File: dabase.f

Function: retrieves an entity from the data base, gives address of attributes

Input Parameters: long *name   name of entity to be retrieved
                    long *ientyp   type of entity (as in y_tab.h), used for consistency check

Output Parameters: long *nattri   number of attributes
                    long *iadpar   address of attributes
                    long *istat   error code

Calls: gibnum

Include Files:	impl.inc, base.inc
Errors:	2 = wrong entity name (does not exist)
	6 = wrong entity type
	7 = entity is deleted
Author/Owner:	Dr. Helmut J. Helpenstein
	Gesellschaft fuer Strukturanalyse mbH,
	Pascalstr.17, D-52076 Aachen

Last Modification: 5.7.91

3.3.6.3.47 Modules retint, retrea, retcha, retnam

File:	dabase.f
Function:	gets an integer/real/character/name attribute (single value or array/string) from the data base

Input Parameters:	long *name	name of entity to which the attribute belongs
	long *iattri	address in attribute table (computed by retent, incremented here)

Output Parameters:	long *number	number of values returned
	long/double/char/long val[ ]	array or single value returned
	long *istat	error code

Calls:	-
Include Files:	impl.inc, base.inc
Errors:	6 = wrong attribute type
Author/Owner:	Dr. Helmut J. Helpenstein
	Gesellschaft fuer Strukturanalyse mbH,
	Pascalstr.17, D-52076 Aachen

Last Modification: 5.7.91

3.3.6.3.48 Module namsor

File:	dabase.f
Function:	puts a new entity name into the lookup table, updates pointer list
Input Parameters:	long *name	name of entity to be stored
Output Parameters:	long *istat	error code
Calls:	-
Include Files:	impl.inc, base.inc
Errors:	2 = entity already exists

Author/Owner: Dr. Helmut J. Helpenstein
                   Gesellschaft fuer Strukturanalyse mbH,
                   Pascalstr.17, D-52076 Aachen

Last Modification: 5.7.91

3.3.6.3.49 Module gibnum

File: dabase.f

Function: returns the address of an entity in the lookup table, uses binary
                   search

Input Parameters: long *name   name of requested entity
                   long *num   address of the entity in the lookup
                                           table

Output Parameters: long *istat   error code

Calls: -

Include Files: impl.inc, base.inc

Errors: 5 = entity not found/does not exist

Author/Owner: Dr. Helmut J. Helpenstein
                   Gesellschaft fuer Strukturanalyse mbH,
                   Pascalstr.17, D-52076 Aachen

Last Modification: 5.7.91

3.3.6.3.50 Module cstati

File: dabase.f

Function: prints statistics overview of data base
                   - version and size of data base
                   - number of parameters
                   - filling of tables
                   - maximum sizes

Input Parameters: -

Output Parameters: -

Calls: -

Include Files: impl.inc, base.inc

Errors: -

Author/Owner: Dr. Helmut J. Helpenstein
                   Gesellschaft fuer Strukturanalyse mbH,
                   Pascalstr.17, D-52076 Aachen

Last Modification: 5.7.91

3.3.6.3.51 Module cstore

File:	dabase.f
Function:	writes the whole data base to an unformatted file, can be read again by crcall
Input Parameters:	long *iu unit on which binary file was opened
Output Parameters:	long *istat error code
Calls:	arifil, arrfil, arcfil
Include Files:	impl.inc, base.inc
Errors:	1 = parameter error 2 = write error
Author/Owner:	Dr. Helmut J. Helpenstein Gesellschaft fuer Strukturanalyse mbH, Pascalstr.17, D-52076 Aachen

Last Modification: 5.7.91

3.3.6.3.52 Module crcall

File:	dabase.f
Function:	reads the whole data base from an unformatted file, must have been written by cstore
Input Parameters:	long *iu unit on which binary file was opened
Output Parameters:	long *istat error code
Calls:	arifil, arrfil, arcfil
Include Files:	impl.inc, base.inc
Errors:	1 = parameter error 2 = read error
Author/Owner:	Dr. Helmut J. Helpenstein Gesellschaft fuer Strukturanalyse mbH, Pascalstr.17, D-52076 Aachen

Last Modification: 5.7.91

3.3.6.3.53 Modules arifil, arrfil, arcfil

File:	dabase.f
Function:	reads from or writes to a binary file an integer/real/character array; uses blocked reading/writing to avoid operating system dependencies

Input Parameters:	long *lflag	flag: 1=read, 2=write
	long *iu	unit on which binary file was opened
	long/double/char array [ ]	array to be read/written
	long *length	length of the array
	long *lblock	number of values per block

Output Parameters:	long *istat	error code

Calls: -

Include Files: -

Errors:	1 = parameter error
	2 = read error

Author/Owner: Dr. Helmut J. Helpenstein
                   Gesellschaft fuer Strukturanalyse mbH,
                   Pascalstr.17, D-52076 Aachen

Last Modification: 5.7.91

3.3.6.3.54 Module wrdaba

File: dabase.f

Function: writes the whole data base to a formatted file for service purposes

Input Parameters:	long *iu	unit on which formatted file was opened

Output Parameters: -

Calls: -

Include Files: impl.inc, base.inc

Errors: -

Author/Owner: Dr. Helmut J. Helpenstein
                   Gesellschaft fuer Strukturanalyse mbH,
                   Pascalstr.17, D-52076 Aachen

Last Modification: 5.7.91

3.3.6.3.55 Module idscfo

File: idscfo.f

Function: copies C data base to Fortran data base or vice versa; allocates memory for called routines

Input Parameters:	long *isteu	1 = C -> Fortran
		2 = Fortran -> C

| long iwsp[ ] | working space (only one entity is stored at a time) |
| long *lenwsp | size of working space |

Output Parameters: -

Calls: ctofor, fortoc (both in file idscfo.f)

Include Files: - (see next entries)

Errors: Message: Memory allocation failed

Author/Owner: Dr. Helmut J. Helpenstein
                    Gesellschaft fuer Strukturanalyse mbH,
                    Pascalstr.17, D-52076 Aachen

Last Modification: 22.5.92

3.3.6.3.56 Module ctofor

File: idscfo.f

Function: call get routines of C-IDS, call put routines of Fortran-IDS

Input Parameters:	long li[ ]	array for list of all entities
	long *maxent	length of this list
	char text[ ]	array for string attributes
	long ia1[], ia2[], ia3[ ]	arrays for integer lists
	long *i1, *i2, *i3	lengths of these lists
	double ra1[], ra2[], ra3[ ]	arrays for real lists
	long *i4, *i5, *i6	lengths of these lists

Output Parameters: -

Calls: IDS files chdget.c, cdtget.c: all (!) IDS get routines
                    other IDS files: cliall, cgtype, cgscpe, cgapfl
                    hdbase.f, mb/ss/wf/csbase.f: all Fortran-IDS put routines
                    utbase.f: fpscpe, fpapfl

Include Files: impl.inc (implicit variable typing)
                    type.inc (entity types)

Errors: -

Author/Owner: Dr. Helmut J. Helpenstein
                    Gesellschaft fuer Strukturanalyse mbH,
                    Pascalstr.17, D-52076 Aachen

Last Modification: 22.5.92

3.3.6.3.57 Module fortoc

File: idscfo.f

Function: call get routines of Fortran-IDS, call put routines of C-IDS

Input Parameters:	long li[ ]	array for list of all entities
	long *maxent	length of this list
	char text[ ]	array for string attributes
	long ia1[], ia2[], ia3[ ]	arrays for integer lists
	long *i1, *i2, *i3	lengths of these lists
	double ra1[], ra2[], ra3[ ]	arrays for real lists
	long *i4, *i5, *i6	lengths of these lists

Output Parameters: -

Calls: hdbase.f, mb/ss/wf/csbase.f: all Fortran-IDS get routines
                utbase.f: fliall, fgtype, fgscpe, fgapfl
                IDS files chdput.c, cdtput.c: all (!) IDS put routines
                other IDS files: cpscpe, cpapfl

Include Files: impl.inc (implicit variable typing)
                type.inc (entity types)

Errors: -

Author/Owner: Dr. Helmut J. Helpenstein
                Gesellschaft fuer Strukturanalyse mbH,
                Pascalstr.17, D-52076 Aachen

Last Modification: 22.5.92

3.3.6.4 Conversion tools by FEGS Ltd

The input parameters of the modules described here are generically as follows:

TOP   The top level entity under which all entities of the required
                type will be converted. This enables either a particular instance
                of an entity to be converted as required for interactive
                conversion or all entities of the required type to be converted
                as in batch processing. This follows the pre-requisite of a top-
                down tree structure for the model.

CRITERION   The required criterion for conversion, omitted if only one.
                Most conversion tools will have several alternative ways to
                perform a conversion. For instance converting between
                elementary and b-spline surfaces may require the resulting b-
                spline to be of a particular order rational or non-rational, non-
                uniform or uniform, bezier etc. The door should be left open
                for new additions in the future.

PARAMETER 1   1st parameter of the conversion according to set criterion

PARAMETER 2   2nd parameter of the conversion according to set criterion

...   etc

The output parameters of the modules described here are generically as follows:

ISTAT   Return status. Conversion tools should as far as possible be fail safe when operating on a data model.

Routines description: The routines all make use of a set of vector functions and topological operator routines which have been developed for this purpose. These are all found in modules cvvect and cvtopo.

The routines are broken down as follows:

3.3.6.4.1 Module for b-spline to polyline conversions

File name: cvbspl.c
                        This file contains c conversion routines for b-spline curve to polyline conversions on the IDS.

Functions included are:

                        void cvbspl (top,criterion,parameter,gtol,istat)
                        void cvedbs (criterion,parameter,gtol,ed_name,num_par,istat)
                        void cvtrbs (criterion,parameter,gtol,trcu_name,num_par,istat)
                        void cvbspo (criterion,parameter,gtol,bscu_name,istat)

Function cvbspl

Function: Divides each b-spline curve into a polyline according to set criterion

Input Parameters: long top;   The top entity of the model to be transferred. 0 = all

                        long criterion;   The type of division required as follows:
                                                 0 = divide b-spline curve into 50 segments
                                                 1 = divide all b-spline curves into segments not longer than parameter units
                                                 2 = divide all b-spline curves into parameter segments

                        double parameter;   The distance of division
                        double gtol;   The geometric tolerance

Output Parameters: long istat;   The return status

Calls: In IDS: clipar, clfrst, clkill, clnext
                        In this file: cvedbs, cvtrbs, cvbspo

Include Files: cvbspl.h, cvvect.h, cdatas.h, ctable.h, lists.h, y_tab.h, cerror.h, cdefin.h, cdtget.h, cdtmod.h, cdtput.h, chdget.h, cidsut.h, lutils.h, enum.h

Errors: -

Author/Owner: C J Miles, FEGS Ltd

Last Modification: C J Miles 9th February 1992

Program Header: void cvbspl (top,criterion,parameter,gtol,istat)

Function cvedbs

Function:	Converts curves of type b-spline into curves of type polyline	
Input Parameters:	long criterion;	The type of division required as follows:
		0 = divide b-spline curve into 50 segments
		1 = divide all b-spline curves into segments not longer than parameter units
		2 = divide all b-spline curves into parameter segments
	double parameter;	The distance of division
	double gtol;	The geometric tolerance
	long ed_name;	Name of the edge
	long num_par;	Number of parents of spline
Output Parameters:	long istat;	The return status
Calls:	In IDS: cged, cgvx, cgculs, cglile, cgbscu, cgcapt, bsctop, bsctob, cfname, cpcapt, cmpocu In local library: bspocu	
Include Files:	cvbspl.h, cvvect.h, cdatas.h, ctable.h, lists.h, y_tab.h, cerror.h, cdefin.h, cdtget.h, cdtmod.h, cdtput.h, chdget.h, cidsut.h, lutils.h, enum.h	
Errors:	-	

Author/Owner: C J Miles, FEGS Ltd

Last Modification: C J Miles 9th February 1992

Program Header: void cvedbs (criterion,parameter,gtol,ed_name,num_par,istat)

Function cvtrbs

Function:	Converts a trimmed b-spline curve into a trimmed polyline with n cartesian points	
Input Parameters:	long criterion;	The type of division required as follows:
		0 = divide b-spline curve into 50 segments
		1 = divide all b-spline curves into segments not longer than parameter units

		2 = divide all b-spline curves into parameter segments
	double parameter;	The distance of division
	double gtol;	The geometric tolerance
	long trcu_name;	Name of the trimmed curve
	long num_par;	Number of parents of b-spline curve

Output Parameters:	long istat;	The return status

Calls:	In IDS: cgtrcu, cglile, cgbscu, cgcapt, bsctop, bsctob, cgcapt, bspocu, cfname, cpcapt, cmpocu

Include Files:	cvbspl.h, cvvect.h, cdatas.h, ctable.h, lists.h, y_tab.h, cerror.h, cdefin.h, cdtget.h, cdtmod.h, cdtput.h, chdget.h, cidsut.h, lutils.h, enum.h

Errors:	-

Author/Owner:	C J Miles, FEGS Ltd

Last Modification:	C J Miles 9th February 1992

Program Header:	void cvtrbs (criterion,parameter,gtol,trcu_name,num_par,istat)

Function cvbspo

Function:	Converts an un-trimmed b-spline curve into a un-trimmed polyline with n points

Input Parameters:	long criterion;	The type of division required as follows:
		0 = divide b-spline curve into 50 segments
		1 = divide all b-spline curves into segments not longer than parameter units
		2 = divide all b-spline curves into parameter segments
	double parameter;	the distance of division
	double gtol;	The geometric tolerance
	long bscu_name;	name of the b-spline_curve

Output Parameters:	long istat;	The return status

Calls:	In IDS: cglile, cgbscu, cgcapt, cfname, cpcapt, cmpocu, In local library: bspocu

Include Files:	cvbspl.h, cvvect.h, cdatas.h, ctable.h, lists.h, y_tab.h, cerror.h, cdefin.h, cdtget.h, cdtmod.h, cdtput.h, chdget.h, cidsut.h, lutils.h, enum.h

Errors:	-

Author/Owner:	C J Miles, FEGS Ltd

Last Modification: C J Miles 9th February 1992

Program Header: void cvbspo (criterion,parameter,gtol,bscu_name,istat)

3.3.6.4.2 Module for circle to circular arc conversions

File name: cvcici.c
                    This file contains functions for the conversion of a circle into a
                    number of circular arcs.

Routines included are:

                    void cvcici (top,maxang,istat)
                    void cvedci (maxang,ed_name,istat)
                    void cvtrci (maxang,trcu_name,istat)
                    void cvcirc (maxang, cu, cusense, start_pnt, end_pnt, ndiv,
                    angle, centre, vector1, binormal, istat)
                    void cipaca (coor, cu, param, pnt, istat)

Function cvcici.c

Function: Converts all circular arcs of angle greater than maxang into
                    arcs smaller than maxang and corrects all referencing entities.

Input Parameters: long top;   The top entity of the model to be
                                             transferred.   0 = all
                    double maxang;   The maximum angle a circular arc
                                             may subtend

Output Parameters: long istat;   The return status

Calls: In IDS: clfrst, clnext, clkill
                    In this file: cvedci, cvtrci

Include Files: cvcici.h, cvvect.h, cvtopo.h, cvorax.h, cdatas.h, ctable.h,
                    lists.h, y_tab.h, cerror.h, cdefin.h, cdtget.h, chdget.h, cidsut.h,
                    lutils.h, enum.h

Errors: -

Author/Owner: C J Miles, FEGS Ltd

Last Modification: C J Miles 14th April 1992

Program Header: void cvcici (top,maxang,istat)

Function cvedci

Function: Converts edges of type circle which span more than maxang
                    into smaller ones and adjusts the edge_loops or connected edge
                    set that reference them.

Input Parameters: double maxang;   The maximum angle a circular arc
                                             may subtend
                    long ed_name;   The name of the edge

Output Parameters: long istat;   The return status

Calls: In IDS: cged, cgvx, cgculs, cfname, cpcapt, cpvx, cpculs, cped, clipar, clfrst, clnext, clkill
                    In cvtopo.c: adedlo, adedces
                    In this file: cvcirc

Include Files: cvcici.h, cvvect.h, cvtopo.h, cvorax.h, cdatas.h, ctable.h, lists.h, y_tab.h, cerror.h, cdefin.h, cdtget.h, chdget.h, cidsut.h, lutils.h, enum.h

Errors: -

Author/Owner: C J Miles, FEGS Ltd

Last Modification: C J Miles 14th April 1992

Program Header: void cvedci (maxang,ed_name,istat)

Function cvtrci

Function: Converts trimmed curves of type circle which span more than maxang into smaller ones and adjusts the geometric 3d curve set that references them.

Input Parameters: double maxang;   The maximum angle a circular arc may subtend
                    long trcu_name;   The name of the trimmed curve

Output Parameters: long istat;   The return status

Calls: In IDS: cgtrcu, cipaca, cfname, cpcapt, cptrcu, clipar, clfrst, clnext, clkill
                    In this file: cvcirc
                    In cvtopo.c: adcug3cs, adcug3ss

Include Files: cvcici.h, cvvect.h, cvtopo.h, cvorax.h, cdatas.h, ctable.h, lists.h, y_tab.h, cerror.h, cdefin.h, cdtget.h, chdget.h, cidsut.h, lutils.h, enum.h

Errors: -

Author/Owner: C J Miles, FEGS Ltd

Last Modification: C J Miles 14th April 1992

Program Header: void cvtrci (maxang,trcu_name,istat)

Function cvcirc

Function: Calculates the number of divisions and angle required to split a STEP circular arc into circular arcs <= maxang

Input Parameters: double maxang;   The maximum angle for arcs
                    long cu;   The curve name
                    long cusense;   The curve sense

	long start_pnt;	The start point of the curve
	long end_pnt;	The end point of the curve
Output Parameters:	long ndiv;	The number of circular edges
	created	
	double angle;	The required division angle
	double centre[3];	The centre to the circle
	double vector1[3];	The vector from centre to point 1
	double binormal[3];	The vector binormal from the centre
	long istat;	The return status

Calls:	In IDS: cgcicu, cgcapt
	In cvorax.c: cvorax
	In cvvect.h: diffv, crossv, scalev, magv, dotv

| Include Files: | cvcici.h, cvvect.h, cvtopo.h, cvorax.h, cdatas.h, ctable.h, lists.h, y_tab.h, cerror.h, cdefin.h, cdtget.h, chdget.h, cidsut.h, lutils.h, enum.h |

| Errors: | - |

| Author/Owner: | C J Miles, FEGS Ltd |

| Last Modification: | C J Miles 14th April 1992 |

| Program Header: | void cvcirc (maxang, cu, cusense, start_pnt, end_pnt, ndiv, angle, centre, vector1, binormal, istat) |

Function cvpaca

| Function: | Converts a point on a circular curve from parametric to cartesian |

Input Parameters:	long coor;	The coordinate system
	long cu;	The curve name
	double param;	The parameter of the point

| Output Parameters: | long pnt; | The created cartesian point name |
| | long istat; | The return status |

Calls:	In IDS: cgcicu, cfname, cpcapt
	In cvorax.c: cvorax
	In cvvect.h: scalev

| Include Files: | cvcici.h, cvvect.h, cvtopo.h, cvorax.h, cdatas.h, ctable.h, lists.h, y_tab.h, cerror.h, cdefin.h, cdtget.h, chdget.h, cidsut.h, lutils.h, enum.h |

| Errors: | - |

| Author/Owner: | C J Miles, FEGS Ltd |

| Last Modification: | C J Miles 14th April 1992 |

| Program Header: | void cipaca (coor, cu, param, pnt, istat) |

3.3.6.4.3 Module for ellipse to elliptic arc conversions

File name: cvelel.c

This file contains conversion routines for conversion of an ellipse to a collection of elliptic arcs.

Routines included are:

    void cvelel (top,maxang,istat)
    void cvedel (maxang,ed_name,istat)
    void cvtrel (maxang,trcu_name,istat)
    void cvelip (maxang, cu, cusense, start_pnt, end_pnt, ndiv,
    starang, angle, centre, major, minor, istat)
    void elpaca (coor, cu, param, pnt, istat)

Function cvelel

Function:	Converts all elliptic arcs of angle greater than maxang into arcs smaller than maxang and corrects all referencing entities

Input Parameters:	long top;	The top entity of the model to be transferred. 0 = all
	double maxang;	The maximum angle a elliptic arc may subtend

Output Parameters:	long istat;	The return status
Calls:	In IDS: clfrst, cvedel, clnext, clkill, client	
	In this file: cvtrel	
Include Files:	cvelel.h, cvvect.h, cvtopo.h, cvorax.h, cdatas.h, ctable.h, lists.h, y_tab.h, cerror.h, cdefin.h, cdtget.h, cdtmod.h, chdget.h, cidsut.h, lutils.h, enum.h	
Errors:	-	
Author/Owner:	C J Miles, FEGS Ltd	
Last Modification:	C J Miles 26th April 1992	
Program Header:	void cvelel (top,maxang,istat)	

Function cvedel

Function:	Splits edges referencing curves of type ellipse which span more than maxang and adjusts the edge_loops or connected edge set that reference them.

Input Parameters:	double maxang;	The maximum angle an elliptic arc may subtend
	long ed_name;	The name of the edge

Output Parameters:	long istat;	The return status
Calls:	In IDS: cgvx, cgculs, cfname, cpcapt, cpvx, cpculs, cped, clipar, clfrst, clnext, clkill	

	In this file: cvelip
	In cvtopo.c: adedlo, adedces
Include Files:	cvelel.h, cvvect.h, cvtopo.h, cvorax.h, cdatas.h, ctable.h, lists.h, y_tab.h, cerror.h, cdefin.h, cdtget.h, cdtmod.h, chdget.h, cidsut.h, lutils.h, enum.h
Errors:	-
Author/Owner:	C J Miles, FEGS Ltd
Last Modification:	C J Miles 26th April 1992
Program Header:	void cvedel (maxang,ed_name,istat)

Function cvtrel

Function:	Converts trimmed curves of type ellipse which span more than maxang into smaller ones and adjusts the geometric 3d curve set that reference them
Input Parameters:	double maxang; The maximum angle a ellipse arc may subtend
	long trcu_name; The name of the trimmed curve
Output Parameters:	long istat; The return status
Calls:	In IDS: cgtrcu, elpaca, cfname, cpcapt, cptrcu, clipar, clfrst, clnext, clkill
	In cvsurf.c: cipaca
	In this file: cvelip
	. In cvtopo.c: adcug3cs, adcug3ss
Include Files:	cvelel.h, cvvect.h, cvtopo.h, cvorax.h, cdatas.h, ctable.h, lists.h, y_tab.h, cerror.h, cdefin.h, cdtget.h, cdtmod.h, chdget.h, cidsut.h, lutils.h, enum.h
Errors:	-
Author/Owner:	C J Miles, FEGS Ltd
Last Modification:	C J Miles 26th April 1992
Program Header:	void cvtrel (maxang,trcu_name,istat)

Function cvelip

Function:	Calculates the number of divisions and the angle required to split a STEP elliptic arc into elliptic arcs <= maxang
Input Parameters:	double maxang; The maximum angle for arcs
	long cu; The curve name
	long cusense; The curve sense
	long start_pnt; The start point of the curve
	long end_pnt; The end point of the curve

Output Parameters: long ndiv;   The number of circular edges
                                                     created
                   double starang;   The start angle director circle
                   double angle;   The required division angle
                   double centre[3];   The centre to the ellipse
                   double major[3];   The vector from centre to major axis
                                                     +ve
                   double minor[3];   The vector from centre to minor axis
                                                     +ve
                   long istat;   The return status

Calls: In IDS: cgelcu, cgcapt
                   In cvorax.c: cvorax
                   In cvvect.h: diffv, scalev, dotv, crossv, magv

Include Files: cvelel.h, cvvect.h, cvtopo.h, cvorax.h, cdatas.h, ctable.h,
                   lists.h, y_tab.h, cerror.h, cdefin.h, cdtget.h, cdtmod.h, chdget.h,
                   cidsut.h, lutils.h, enum.h

Errors: -

Author/Owner: C J Miles, FEGS Ltd

Last Modification: C J Miles 26th April 1992

Program Header: void cvelip (maxang, cu, cusense, start_pnt, end_pnt, ndiv,
                   starang, angle, centre, major, minor, istat)

Function elpaca

Function: Converts a point on an elliptic curve from parametric to
                   cartesian point.

Input Parameters: long coor;   The coordinate system
                   long cu;   The curve name
                   double param;   The parameter of the point

Output Parameters: long pnt;   The created cartesian point name
                   long istat;   The return status

Calls: In IDS: cgelcu, cfname, cpcapt
                   In cvorax.c: cvorax
                   In cvvect.h: scalev,

Include Files: cvelel.h, cvvect.h, cvtopo.h, cvorax.h, cdatas.h, ctable.h,
                   lists.h, y_tab.h, cerror.h, cdefin.h, cdtget.h, cdtmod.h, chdget.h,
                   cidsut.h, lutils.h, enum.h

Errors: -

Author/Owner: C J Miles, FEGS Ltd

Last Modification: C J Miles 26th April 1992

Program Header: void elpaca (coor, cu, param, pnt, istat)

3.3.6.4.4 Module for axis2_placements correction

File name:	cvorax.c
Function:	Get the normalised axis directions for an axis2_placement the axis placement may be an ill defined set according to STEP
Input Parameters:	long ax2; The axis2_placement
Output Parameters:	double centre[3]; The centre of the zaxis
	double xaxis[3]; The x-axis_vector
	double yaxis[3]; The y-axis vector
	double zaxis[3]; The z-axis vector
	long istat; The return status
Calls:	In IDS: cdax2, cgcapt, cgdi
	In cvvect.c: vnorm, crossv, dotv, scalev, diffv, crossv,
Include Files:	cvvect.h, cvorax.h, cdatas.h, ctable.h, lists.h, y_tab.h, cerror.h, cdefin.h, cdtget.h, chdget.h, cidsut.h, lutils.h, enum.h
Errors:	-
Author/Owner:	C J Miles, FEGS Ltd
Last Modification:	C J Miles 14th April 1992
Program Header:	void cvorax (ax2, centre, xaxis, yaxis, zaxis, istat)

3.3.6.4.5 Module for polyline to polyline conversions

File name:	cvplpl.c

This file contains conversion routines for converting a polyline into several polylines.

Routines included are:

```
void cvplpl (top,criterion,parameter,gtol,istat)
void cvedpl (criterion,parameter,gtol,ed_name,istat)
void cvplcu (criterion,parameter,gtol,plcu_name,istat)
void cvtrpl (criterion, parameter, gtol, trcu_name, plcu_name,
istat)
```

Function cvplpl

Function:	Divides a polyline into n new lopylines according to set criteria
Input Parameters:	long top; The top entity of the model to be converted. 0 = all
	long criterion; The type of division required as follows:
	0 = divide all closed polyline curves in four

		1 = divide all polyline curves into segments not longer than parameter units
		2 = divide all polyline curves into parameter segments
	double parameter;	The parameter
	double gtol;	The geometric tolerance

Output Parameters:	long istat;	The return status

Calls:	In IDS: client, clfrst, cvedpl, clnext, clkill, clipar, clnext
	In cvelel.c: cvtrpl
	In cvplpl.c: cvplcu

Include Files:	cvplpl.h, cvvect.h, cvtopo.h, cdatas.h, ctable.h, lists.h, y_tab.h, cerror.h, cdefin.h, cdtget.h, chdget.h, cidsut.h, lutils.h, enum.h

Errors:	-

Author/Owner:	C J Miles, FEGS Ltd

Last Modification:	C J Miles 25th April 1992

Program Header:	void cvplpl (top,criterion,parameter,gtol,istat)

Function cvedpl

Function:	Converts edges of type polyline into several edges with polylines in and adjusts the edge loops or connected edge sets that reference them

Input Parameters:	long criterion;	The type of division required as follows:
		0 = divide all closed polyline curves in four
		1 = divide all polyline curves into segments not longer than parameter units
		2 = divide all polyline curves into parameter segments
	double parameter;	The parameter
	double gtol;	The geometric tolerance
	long ed_name;	The name of the edge

Output Parameters:	long istat;	The return status

Calls:	In IDS: cged, cgvx, cgculs, cglile, cgpocu, cgcapt, cfname, cpcapt, cpvx, cpculs, cped, clipar, clfrst, clnext, clkill
	In cvtopo.c: adedlo, adedces

Include Files:	cvplpl.h, cvvect.h, cvtopo.h, cdatas.h, ctable.h, lists.h, y_tab.h, cerror.h, cdefin.h, cdtget.h, chdget.h, cidsut.h, lutils.h, enum.h

Errors: -

Author/Owner: C J Miles, FEGS Ltd

Last Modification: C J Miles 25th April 1992

Program Header: void cvedpl (criterion,parameter,gtol,ed_name,istat)

Function cvplcu

Function: Splits a polyline into n trimmed polylines and adjusts the
                       geometric 3d curve set that references it

Input Parameters: long criterion;   The type of division required as
                                                    follows:
                                                    0 = divide all closed polyline curves
                                                          in four
                                                    1 = divide all polyline curves into
                                                          segments not longer than
                                                          parameter units
                                                    2 = divide all polyline curves into
                                                          parameter segments

                       double parameter;   The parameter
                       double gtol;   The geometric tolerance
                       long plcu_name;   The name of the poly_line

Output Parameters: long istat;   The return status

Calls: In IDS: cglile, cgpocu, cgcapt, cfname, cpcapt, cptrcu, clipar,
                       clfrst, clnext, clkill
                       In cvvect.c: vdist
                       In cvtopo.c: adcug3cs, adcug3ss

Include Files: cvplpl.h, cvvect.h, cvtopo.h, cdatas.h, ctable.h, lists.h, y_tab.h,
                       cerror.h, cdefin.h, cdtget.h, chdget.h, cidsut.h, lutils.h, enum.h

Errors: -

Author/Owner: C J Miles, FEGS Ltd

Last Modification: C J Miles 25th April 1992

Program Header: void cvplcu (criterion,parameter,gtol,plcu_name,istat)

Function cvtrpl

Function: Converts trimmed curves of type polyline into several
                       trimmed_curves of type polylines and adjusts the geometric 3d
                       curve sets that reference them

Input Parameters: long criterion;   The type of division required as
                                                    follows:

		0 = divide all closed polyline curves in four
		1 = divide all polyline curves into segments not longer than parameter units
		2 = divide all polyline curves into parameter segments
	double parameter;	The parameter
	double gtol;	The geometric tolerance
	long trcu_name;	The name of the trimmed_curve
	long plcu_name;	The name of the polyline_curve
Output Parameters:	long istat;	The return status

Calls:	In IDS: cgtrcu, cglile, cgpocu, cgcapt, cfname, cpcapt, clipar, clfrst, clnext, clkill
	In cvtopo.c: adcug3cs, adcug3ss
	In cvvect.c: vdist

Include Files:	cvplpl.h, cvvect.h, cvtopo.h, cdatas.h, ctable.h, lists.h, y_tab.h, cerror.h, cdefin.h, cdtget.h, chdget.h, cidsut.h, lutils.h, enum.h

Errors:	-
Author/Owner:	C J Miles, FEGS Ltd
Last Modification:	C J Miles 25th April 1992
Program Header:	void cvtrpl (criterion,parameter,gtol,trcu_name,plcu_name,istat)

3.3.6.4.6 Module for elementary surface type sphere to b-spline surface conversions

File name:	cvspbs.c

This file contains conversions for converting an elementary surface of type sphere into a b-spline surface.

Functions included are:

> void cvspbs (top,istat)
> void cvsphere (sphere_name,istat)

Function cvspbs

Function:	Converts all spheres below the top entity in the IDS to b-spline surfaces
Input Parameters:	long top;	The top entity of the model to be transferred. 0 = all
Output Parameters:	long istat;	The return status

Calls:	In IDS: clfrst, clnext, clkill
	In this file: cvsphere

Include Files:	cvspbs.h, cvvect.h, cvorax.h, cdatas.h, ctable.h, lists.h, y_tab.h, cerror.h, cdefin.h, cdtget.h, cdtmod.h, chdget.h, cidsut.h, lutils.h, enum.h

Errors:	-
Author/Owner:	C J Miles, FEGS Ltd
Last Modification:	C J Miles 10th April 1992
Program Header:	void cvspbs (top,istat)

Function cvsphere

Function:	Converts a sphere to a b-spline surface
Input Parameters:	long sphere_name; The IDS sphere_name
Output Parameters:	long istat; The return status
Calls:	In IDS: cgspsu, cfname, cpcapt, cmbssu
Include Files:	cvspbs.h, cvvect.h, cvorax.h, cdatas.h, ctable.h, lists.h, y_tab.h, cerror.h, cdefin.h, cdtget.h, cdtmod.h, chdget.h, cidsut.h, lutils.h, enum.h
Errors:	-
Author/Owner:	C J Miles, FEGS Ltd
Last Modification:	C J Miles 10th April 1992
Program Header:	void cvsphere (sphere_name,istat)

3.3.6.4.7 Module for b-spline surface to point set conversions

File name:	cvsurf.c

This file contains conversion functions for conversion of a b-spline surface to a point set. The resulting point set is used as te basis for a Coon's blend surface definition.

Functions included are:
    void cvsurf (top,criterion,parameter1,parameter2,gtol,istat)
    void cvbsps
    (criterion,parameter1,parameter2,gtol,bssu_name,istat)

Function cvsurf

Function:	Divides b-spline surface into patches according to set criterion.
Input Parameters:	long top; The top entity of the model to be transferred. 0 = all

| | long criterion; | The type of conversion required as follows:
		0 = divide b-spline surface into a patch of n x m
	double parameter1;	The division param. in direction 1
	double parameter2;	The division parameter in direct. 2
	double gtol;	The geometric tolerance

Output Parameters: long istat;   The return status

Calls: In IDS: client, clfrst, clnext, clkill
                In this file: cvbsps

Include Files: cvsurf.h, cdatas.h, ctable.h, lists.h, y_tab.h, cerror.h, cdefin.h, cdtget.h, chdget.h, cidsut.h, lutils.h, enum.h

Errors: -

Author/Owner: C J Miles, FEGS Ltd

Last Modification: C J Miles 2 Dec 1991

Program Header: void cvsurf (top,criterion,parameter1,parameter2,gtol,istat)

Function cvbsps

Function: Converts an untrimmed b-spline surface into a grid of n points, which can be dumped into a native file format.

Input Parameters: long criterion;   The type of conversion required as follows:
                                          0 = divide b-spline surface into a patch of n x m
                  double parameter1;   the division parameter in direction 1
                  double parameter2;   the division parameter in direction 2
                  double gtol;   the geometric tolerance
                  long bssu_name;   the name of the b-spline_surface

Output Parameters: long istat;   The return status

Calls: In IDS: cglile, cgbssu, cgcapt, bsposu

Include Files: cvsurf.h, cdatas.h, ctable.h, lists.h, y_tab.h, cerror.h, cdefin.h, cdtget.h, chdget.h, cidsut.h, lutils.h, enum.h

Errors: -

Author/Owner: C J Miles, FEGS Ltd

Last Modification: C J Miles 2 Dec 1991

Program Header: void cvbsps (criterion, parameter1, parameter2, gtol, bssu_name, istat)

3.3.6.4.8 Module for elementary surface type torus to b-spline surface conversions

File name: cvtobs.c

This file contains C conversion routines for elementary torus to b-spline surface.

Functions included are:

        void cvtobs (top,istat)

        void cvtorus (torus_name,istat)

Function cvtobs

Function:	Converts all torus below the top entity in the IDS to b-spline surfaces
Input Parameters:	long top; The top entity of the model to be searched. 0 = all
Output Parameters:	long istat; The return status
Calls:	IN IDS: client, clfrst, clnext, clkill In this file: cvtorus
Include Files:	cvtobs.h, cvorax.h, cvvect.h, cdatas.h, ctable.h, lists.h, y_tab.h, cerror.h, cdefin.h, cdtget.h, cdtmod.h, chdget.h, cidsut.h, lutils.h, enum.h
Errors:	-
Author/Owner:	C J Miles, FEGS Ltd
Last Modification:	C J Miles 10 April 1992
Program Header:	void cvtobs (top,istat)

Function cvtorus

Function:	Converts a torus to a b-spline surface
Input Parameters:	long torus_name; The torus name
Output Parameters:	long istat; The return status
Calls:	In IDS: cgtosu, cfname, cpcapt, cmbssu In cvorax.c: cvorax
Include Files:	cvtobs.h, cvorax.h, cvvect.h, cdatas.h, ctable.h, lists.h, y_tab.h, cerror.h, cdefin.h, cdtget.h, cdtmod.h, chdget.h, cidsut.h, lutils.h, enum.h
Errors:	-
Author/Owner:	C J Miles, FEGS Ltd
Last Modification:	C J Miles 10 April 1992
Program Header:	void cvtorus (torus_name,istat)

3.3.6.4.9 Module for topological operators

File name: cvtopo.c

This file contains topological operators for the IDS.

Functions included are:

        void adedlo (elo_name, seg_name, nedges, edges, istat)
        void adedces (ceds_name, seg_name, nedges, edges, istat)
        void adcug3cs (g3cs_name, trcu_name, ncurves, trcu, istat)
        void adcug3ss (g3ss_name, trcu_name, ncurves, trcu, istat)

Function adedlo

Function:	Replace an edge with several in a loop in the correct order

Input Parameters:	long elo_name;	the edge loop name
	long seg_name;	the edge to be replaced
	long nedges;	the number of edges replacing the edge
	long edges;	the replacement edge names

Output Parameters: long istat;   the return status

Calls: In IDS: cglile, cgedlo, cgedls, cfname, cpedls, cmedlo

Include Files: cvtopo.h, cdatas.h, ctable.h, lists.h, y_tab.h, cerror.h, cdefin.h, cdtget.h, chdget.h, cidsut.h, lutils.h, enum.h

Errors: -

Author/Owner: C J Miles, FEGS Ltd

Last Modification: C J Miles 14th April 1992

Program Header: void adedlo (elo_name, seg_name, nedges, edges, istat)

Function adedces

Function:	Replace an edge with several edges in a connected_edge_set

Input Parameters:	long ceds_name;	the connected edge set
	long seg_name;	the edge to be replaced
	long nedges;	the number of edge replacing the original one
	long edges;	the replacement edges

Output Parameters: long istat;   The return status

Calls: In IDS: cglile, cgceds, cmceds

Include Files: cvtopo.h, cdatas.h, ctable.h, lists.h, y_tab.h, cerror.h, cdefin.h, cdtget.h, chdget.h, cidsut.h, lutils.h, enum.h

Errors: -

Author/Owner: C J Miles, FEGS Ltd

Last Modification: C J Miles 14th April 1992

Program Header: void adedces (ceds_name, seg_name, nedges, edges, istat)

Function adcug3cs

Function: Replace a trimmed curve with several trimmed curves in a
                   geometric 3d curve_set.

Input Parameters: long g3cs_name;   the geometric 3d curve set
                   long trcu_name;   the trimmed curve to be replaced
                   long ncurves;   the number of curves which replace
                                            the original
                   long trcu;   The list of the replacement trimmed
                                            curves

Output Parameters: long istat;   The return status

Calls: In IDS: cglile,cgg3cs, cmg3cs

Include Files: cvtopo.h, cdatas.h, ctable.h, lists.h, y_tab.h, cerror.h, cdefin.h,
                   cdtget.h, chdget.h, cidsut.h, lutils.h, enum.h

Errors: -

Author/Owner: C J Miles, FEGS Ltd

Last Modification: C J Miles 14th April 1992

Program Header: void adcug3cs (g3cs_name, trcu_name, ncurves, trcu, istat)

Function adcug3ss

Function: Replace a trimmed curve with several trimmed curves in a
                   geometric 3d surface_set.

Input Parameters: long g3ss_name;   the geometric 3d surface set
                   long trcu_name;   the trimmed curve to be replaced
                   long ncurves;   the number of curves which replace
                                            the original
                   long trcu;   The list of the replacement trimmed
                                            curves

Output Parameters: long istat;   The return status

Calls: In IDS: cglile,cgg3ss, cmg3ss

Include Files: cvtopo.h, cdatas.h, ctable.h, lists.h, y_tab.h, cerror.h, cdefin.h,
                   cdtget.h, chdget.h, cidsut.h, lutils.h, enum.h

Errors: -

Author/Owner: C J Miles, FEGS Ltd

Last Modification: C J Miles 14th April 1992

Program Header: void adcug3ss (g3ss_name, trcu_name, ncurves, trcu, istat)

3.3.6.4.10 Module for vector operations

File name: cvvect.c

This file contains a set of vector operators

Functions included are:

                void crossv (v1,v2,v3)
                void diffv (v1,v2,idim,v3)
                void sumv (v1,v2,idim,v3)
                double vdist (v1,v2,idim)
                double dotv (v1,v2,idim)
                void scalev (s1,v1,idim,v2)
                double magv (v1,idim)
                void vnorm (v1,idim,v2)

Function crossv

Function:	Calculates the cross product of two vectors of dimension 3

Input Parameters: double v1[ ];   the first vector
                    double v2[ ];   the second vector

Output parameters: double v3[ ];   the cross product

Calls: -

Include Files: cvvect.h, cdatas.h, ctable.h, lists.h, y_tab.h, cerror.h, cdefin.h, cdtget.h, chdget.h, cidsut.h, lutils.h, enum.h

Errors: -

Author/Owner: C J Miles, FEGS Ltd

Last Modification: C J Miles 14th April 1992

Program Header: void crossv (v1,v2,v3)

Function diffv

Function: Calculates the difference between two vectors

Input Parameters: double v1[ ];   the first vector
                    double v2[ ];   the second vector
                    long idim;   dimension is 2 or 3

Output Parameters: double v3[ ];   the difference vector

Calls: -

Include Files: cvvect.h, cdatas.h, ctable.h, lists.h, y_tab.h, cerror.h, cdefin.h, cdtget.h, chdget.h, cidsut.h, lutils.h, enum.h

Errors: -

Author/Owner: C J Miles, FEGS Ltd

Last Modification: C J Miles 14th April 1992

Program Header: void diffv (v1,v2,idim,v3)

Function sumv

Function: Calculates the sum of two vectors

Input Parameters: double v1[ ];   the first vector
                    double v2[ ];   the second vector
                    long idim;   dimension is 2 or 3

Output Parameters: double v3[ ];   the sum vector

Calls: -

Include Files: cvvect.h, cdatas.h, ctable.h, lists.h, y_tab.h, cerror.h, cdefin.h,
                    cdtget.h, chdget.h, cidsut.h, lutils.h, enum.h

Errors: -

Author/Owner: C J Miles, FEGS Ltd

Last Modification: C J Miles 14th April 1992

Program Header: void sumv (v1,v2,idim,v3)

Function vdist

Function: Calculates the distance between two vectors

Input Parameters: double v1[ ];   the first vector
                    double v2[ ];   the second vector
                    long   idim;   dimension is 2 or 3

Output Parameters: -

Calls: -

Include Files: cvvect.h, cdatas.h, ctable.h, lists.h, y_tab.h, cerror.h, cdefin.h,
                    cdtget.h, chdget.h, cidsut.h, lutils.h, enum.h

Errors: -

Author/Owner: C J Miles, FEGS Ltd

Last Modification: C J Miles 14th April 1992

Program Header: double vdist (v1,v2,idim)

Function dotv

Function: Calculates the dot product of two vectors

Input Parameters: double v1[ ];   the first vector
                    double v2[ ];   the second vector
                    long idim;   dimension is 2 or 3

Output Parameters: -

Calls: -

Include Files: cvvect.h, cdatas.h, ctable.h, lists.h, y_tab.h, cerror.h, cdefin.h, cdtget.h, chdget.h, cidsut.h, lutils.h, enum.h

Errors: -

Author/Owner: C J Miles, FEGS Ltd

Last Modification: C J Miles 14th April 1992

Program Header: double dotv (v1,v2,idim)

Function scalev

Function: Scales a vector

Input Parameters: double s1;   the scale
                   double v1[ ];   the first vector
                   long idim;   dimension is 2 or 3

Output Parameters: double v2[ ];   the scaled vector

Calls: -

Include Files: cvvect.h, cdatas.h, ctable.h, lists.h, y_tab.h, cerror.h, cdefin.h, cdtget.h, chdget.h, cidsut.h, lutils.h, enum.h

Errors: -

Author/Owner: C J Miles, FEGS Ltd

Last Modification: C J Miles 14th April 1992

Program Header: void scalev (s1,v1,idim,v2)

Function magv

Function: Find the magnitude of a vector

Input Parameters: double v1[ ];   the first vector
                   long idim;   dimension is 2 or 3

Output Parameters: -

Calls: -

Include Files: cvvect.h, cdatas.h, ctable.h, lists.h, y_tab.h, cerror.h, cdefin.h, cdtget.h, chdget.h, cidsut.h, lutils.h, enum.h

Errors: -

Author/Owner: C J Miles, FEGS Ltd

Last Modification: C J Miles 14th April 1992

Program Header: double magv (v1,idim)

Function vnorm

Function:	Normalises a vector	
Input Parameters:	double v1[ ]; long idim;	the first vector dimension is 2 or 3
Output Parameters:	double v2[ ];	the normalised vector
Calls:	-	
Include Files:	cvvect.h, cdatas.h, ctable.h, lists.h, y_tab.h, cerror.h, cdefin.h, cdtget.h, chdget.h, cidsut.h, lutils.h, enum.h	
Errors:	-	
Author/Owner:	C J Miles, FEGS Ltd	
Last Modification:	C J Miles 14th April 1992	
Program Header:	void vnorm (v1,idim,v2)	

3.4 User's Guide

3.4.1 IDS

3.4.1.1 Main functionality of the IDS

Before you do anything with the IDS you have to initialise it. This is done by the routine 'cinids' described above.

Afterwards you can store data in the IDS. This is done with the Scanner/Parser when you are in the postprocessing direction or from your CAD system when you run a preprocessor.

When the IDS is filled you can carry out a lot of actions: You can invoke other CADEX common tools like the IDS Data Checker, generate statistics and structure viewer files, do some conversions if necessary or call the Formatter. Or you can start the system dependent part of your postprocessor's back-end and fill the data structure in your CAD system.

When you intend to leave the IDS and want to free the memory which has been used by the IDS, you have to call the routine 'cfrmem'.

3.4.1.2 File organisation of the IDS

The IDS package consists of the following files:

```
cdatas.h     - the IDS data structure
cdefin.h     - some #defines of IDS tokens
cdtget.c     - data section get routines
cdtget.h     - function prototypes of routines of file "cdtget.c"
```

```
cdtmod.c      - data section modify routines
cdtmod.h      - function prototypes of routines of file "cdtmod.c"
cdtput.c      - data section put routines
cdtput.h      - function prototypes of routines of file "cdtput.c"
cptget.c      - product information get routines
cptget.h      - function prototypes of routines of file "cptget.c"
cptmod.c      - product information modify routines
cptmod.h      - function prototypes of routines of file "cptmod.c"
cptput.c      - product information put routines
cptput.h      - function prototypes of routines of file "cptput.c"
cerror.h      - #defines of IDS error tokens
chdget.c      - header section get routines
chdget.h      - function prototypes of routines of file "chdget.c"
chdput.c      - header section put routines
chdput.h      - function prototypes of routines of file "chdput.c"
cidsut.c      - IDS utility functions
cidsut.h      - function prototypes of routines of file "cidsut.c"
cinit.c       - IDS initialisation routine
cinit.h       - function prototype of routine of file "cinit.c"
cscope.c      - scope handling routines
cscope.h      - function prototypes of routines of file "cscope.c"
ctable.c      - internally used IDS routines (lookup table, ...)
ctable.h      - function prototypes of routines of file "ctable.c"
ids_XY.com    - description of new IDS version X.Y (release notes)
ids_get.man   - programmers' manual: headers of all IDS get routines
ids_mod.man   - programmers' manual: headers of all IDSmodify routines
ids_put.man   - programmers' manual: headers of all IDS put routines
ids_rest.man  - programmers' manual: headers of all other IDS access
                routines
ids_lut.man   - programmers' manual: headers of all internally used
                IDS routines
```

3.4.1.3 Dependencies from other parts of CADEX Common Tools

There are also some files from the Scanner/Parser package which are important for the IDS.

The file "**y_tab.h**" contains the integer values which represent the entity types. This is done by a set of '#define TOK_...' statements which have been automatically generated from '%token' statements in the file "yacc.hh" by the UNIX tool yacc. Whenever in the IDS an entity type is of interest, this file "y_tab.h" has to be included.

In the file "**enum.h**" there are some #define statements which are used to identify Application Protocols. This however is used only internally in the routine 'cpapfl' to check if the AP information of a specified entity to be stored in the IDS is among the values which are allowed. If the user explicitly wants to modify the

AP information of a specified entity or wants to interpret the integer number stored with the entity he has to know about these entries in "enum.h".

In the file file "enum.h" there are also some #define statements which describe the integer values representing enumeration classes and the alternatives within each class. The latter are stored in the IDS data structure. For interpretation or modification of these attribute values the user has to know about these entries in "enum.h".

In file **"lutils.c"** there are two arrays which might be of interest for the user of the IDS:

The first array 'step_keyword_table' in "lutils.c" describes a relationship between integer keyword token numbers (TOK_... from the file "y_tab.h") and the corresponding keyword strings. The user however should never use this array explicitly, but should use the function 'lookup_token_keyword' which returns the keyword string corresponding to a given token number (this function is at present used within the CADEX common tools formatter and the IDS checker).

The 2nd array in "lutils.c" is called 'step_enum_table', which holds the integers representing enumeration classes, the integers representing the alternatives within each class (those are defined in file "enum.h") and the corresponding strings to be written in the STEP physical file. The user however should use the function 'lookup_token_enum' to retrieve these strings from given entity attributes.

3.4.2 Scanner and parser

3.4.2.1 The procedural interface of the scanner/parser

The "how to" or user interface to the S/P is described in this section. As mentioned above, the S/P has been designed to be callable from either "C" or FORTRAN, the two languages in use in CADEX. The interface itself is very simple, several routines are all that is needed to access the entire functionality of the S/P software. Note, however, that the S/P makes extensive use of IDS so-called "put" routines to install into memory the entities parsed in the STEP file. In the second part of this section an example main file is described to demonstrate how the S/P can be called by a calling program. To integrate the Scanner/Parser into a STEP postprocessor, only the routines described here need to be called.

First the IDS must be initialized so that it will accept the STEP entities that the S/P will parse. The routine is called "cinids()", but refer to IDS documentation for a list of parameters and their meaning. The rest of the routines to be called are part of the S/P and are as follows:

Routine	Functionality
init_parser()	Initializes the S/P software.
yystepparse()	Invokes the STEP physical header file parser.
fetch_parse_results()	Retrieves results of last parsing session.

reset_parser()	Resets variables in between parse sessions.
yybrep_apparse()	Invokes the BREP AP parser.
yyss_apparse()	Invokes the SS AP parser.
yywf_apparse()	Invokes the WF AP parser.
yycsg_apparse()	Invokes the CSG AP parser.
yycbr_apparse()	Invokes the Compound Brep AP parser.

The routines are defined with the following parameters:

```
long init_parser(long write_log, char *fname, long debug)
```

The variable "write_log" is a flag telling the system whether or not to write a log file (NULL means do not); "fname" is the name of the file (with path) to be parsed; and "debug" a flag that indicates whether a trace of the parse tree should be output to STDOUT (NULL means do not). The routine returns NULL if everything is ok, non-NULL if no ok.

```
long yystepparse()
```

The routine takes no parameters, and returns NULL if OK (file parsed to EOF), and non-NULL if not OK. It should be called only once, as the first parser to operate on a file. This parsing routines reads the STEP header file section, calls the appropriate IDS "put" routines to store the header information, and then begin to read the product definition entities which are necessary for a valid STEP file conform with the current CADEX APs. Since the Common Toolkit release 6.0 these product definition entities are accepted by the Scanner/Parser and can be stored into the IDS. One of these entities must be APPLICATION_PROTOCOL which defines the used AP for all following shape, geometrical and topological entities. A change of an AP is only allowed after a new APPLICATION_ PROTOCOL entity is found in the file.

After APPLICATION_PROTOCOL is found in the file the actual parser returns to the calling software to let the user know which AP specific parser should be called next.

```
void fetch_parse_results(long *error, long *use_ap,
                         long *error_count, long*warning_count)
```

This routine retrieves the variables that if interpreted, tell in detail the results of the most recent parsing session. If "error" returns as NULL, no errors were found (error codes can be found in the file "parse.h"). The "use_ap" flag tells which AP attribute or EOF was detected at the end of the last parser session (see "parse.h" for list). "error_count" is a count of all the errors found, and "warning_count" a count of all warnings (for example, syntactical mistakes in the header section are not fatal errors).

Looking into the file "parse.h", the flags returned by fetch_parse_results() have the following meaning:

Return parameter "use_ap_flag":

This flag defines the domain for the AP_ENUM_CLASS defined in the file sp/enum.hh. In the APPLICATION_PROTOCOL entity the attribute "iso_10303_ part_number" is evaluated to get the right CADEX AP flag (204 BREP, 205 SF,

206 WF; since the CSG and the CBR AP have no valid ISO number assigned at the moment 250 was used for CSG and 299 for the CBR AP).

BREP_AP	means that the last call to the parser read the next APPLICATION_PROTOCOL entity and it said that a BREP AP based model will be next in the file.
SF_AP	means the same but for the SF AP.
WF_AP	means the same but for the WF AP.
CSG_AP	means the same but for the CSG AP.
CBR_AP	means the same but for the Compound Brep AP.
UNKNOWN_AP	means that no valid AP number was found in the APPLICATION_PROTOCOL entity or no such entity was found in the STEP file

Return parameter "error_flag":

CDX_SYNTAX_ERROR means that the parser found syntax errors in the file from which it could not parse the file successfully. It may still continue after encountering such an error, but it is letting you know that such an error (for example, a CARTESIAN_POINT with a missing attribute) was found.

CDX_SYNTAX_WARNING means a less important syntax error was found. Presently only syntax errors in the header section are recorded as warnings.

CDX_IDS_ERROR means that one or more IDS routines returned an error message. CDX_END_FILE means that the end of the file was found.

```
void reset_parser()
```

The routine takes no parameters and returns no value; it resets internal variable only. It should be called in between each parser call.

```
long yybrep_apparse()
```

The routine takes no parameters, and returns NULL if OK (file parsed to EOF), and non-NULL if not OK. It should be called when the next section of the file (indicated by the APPLICATION_PROTOCOL entity whose value is returned by the "fetch_parser_results") will contain BREP AP information.

```
long yysf_apparse()
```

The routine takes no parameters, and returns NULL if OK (file parsed to EOF), and non-NULL if not OK. It should be called when the next section of the file (indicated by the APPLICATION_PROTOCOL entity whose value is returned by the "fetch_parser_results") will contain SF AP information.

```
long yywf_apparse()
```

The routine takes no parameters, and returns NULL if OK (file parsed to EOF), and non-NULL if not OK. It should be called when the next section of the file (indicated by the APPLICATION_PROTOCOL entity whose value is returned by the "fetch_parser_results") will contain WF AP information.

```
long yycsg_apparse()
```

The routine takes no parameters, and returns NULL if OK (file parsed to EOF), and non-NULL if not OK. It should be called when the next section of the file (indicated by the APPLICATION_PROTOCOL entity whose value is returned by the "fetch_parser_results") will contain CSG AP information.

```
long yycbr_apparse()
```

The routine takes no parameters, and returns NULL if OK (file parsed to EOF), and non-NULL if not OK. It should be called when the next section of the file (indicated by the APPLICATION_PROTOCOL entity whose value is returned by the "fetch_parser_results") will contain CBR AP information.

Let's briefly summarize then the routines to be called, their order and additional routines from other parts of the CADEX toolkit necessary to use the Scanner Parser in a STEP post processor.

First call the IDS initialization routine cinids(), followed by the S/P initialization routine "init_parser()". Then the first parser function, yysteppparse(), followed by the routine fetch_parser_results(). If no errors were detected, the processing can continue. But first call the reset_parser() routine to reset internal variables. Then call one of the "yy<brep?sf?wf?csg?cbr>_apparse()" functions, depending on the value of the use_ap flag returned by the last call to fetch_parser_results(). Then call fetch_parser_results() again and depending on what it returns, call the parser it tells you to, and so on, until the use_ap flag returns with the value EOF, which means the parsing job is done.

At this point, the S/P's job is complete, and the users back-end processor takes over by operating on the IDS that the S/P built up.

3.4.2.2 A simple example for the use of the S/P in a post processor

As an aid to the user, a "main.c" file is included that demonstrates how the S/P can be called by a calling program. The executable "step_pp" is called with the following syntax:

```
step_pp [-l] [-v] [-s size] filename
```

-l - write a log file of parser sessions with name "filename.l"

-v - verbose options that records all state transitions that took place during parsing (output to standard out).

-s - tells the parser/scanner to initialize the IDS with a lookup table that contains "size" entries.

The routine "main()", as an example to the user, initializes the IDS, initializes the S/P, and begins to parse the specified file with the parser "yysteppparse" to read and insert the header information and the first APPLICATION_PROTOCOL entity (this STEP file structure tells which AP style of STEP data will be coming next in the file). Five APs (the BREP, the Surface or SF, the Wire frame or WF, the Constructive Solid Geometry or CSG and the Compound BREP or CBR) are supported. Depending on what is returned, the APs specific parsers mentioned in above are called. Each time one of the parsers returns, its passes back a flag specifying the next parser to be used, if any to be called.

After the last parser has returned with the value "EOF", the user has finished using the S/P, or front-end, of the processor, and can now run his system specific back-end to transfer what now sits in the IDS into his native environment. Obviously if the S/P is to be used as a tool, the present example "main()" must be modified to build the interface to a specific back-end. In the example case the CADEX toolkit formatter is called to write a file of the entities read in before by the Scanner / Parser.

3.4.3 Formatter

3.4.3.1 Characteristics of the formatter

The Cadex File Formatter is a functional module that takes the data stored in the Intermediate Data Structure (IDS) and writes them into a text file, formatted according to the Step specifications.

Since IDS format is assumed to be in one-to-one relation with Step specifications, the basic work of the Formatter is to correctly write the IDS data, leaving to other tools any function of type conversion and syntactic and semantic check.

In summary, the characteristics of the Formatter are:

* it creates Step files formatted according to the specifications of ISO 10303-21, "Clear text encoding of the exchange structure";
* the shape entities defined in the Cadex Boundary Representation (including Facetted BRep), Sculptured Surface, Wire Frame, and Compound Boundary Representation Application Protocols (APs) are recognized and written into the output Step file;
* the output Step file contains only backward referenced entities (i.e., an entity appears in the file as defining another entity only when it has already been defined);
* the case of multiple APs is handled;
* optionally, the entity identifiers are increasingly ordered in the output file;
* optionally, in case of multiple top entities, each sub-structure is written to a different output file;
* optionally, in case of multiple top entities, a single sub-structure is selected and written to the output file;
* the layout of the output file can be variously customized;
* the interface functions are clearly identified and declared in a specific include file, so that a main program can use the Formatter by simply including such a file and then calling the needed function;
* it is fully integrated with all other Cadex software.

3.4.3.2 Integrating the formatter with IDS and scanner/parser

In order to correctly integrate the Formatter with IDS and scanner/parser, the file cform.c must be compiled and linked together with all IDS and Scanner/Parser modules. During these operations, the file cform.h and a number of *.h files of IDS and SP must be reachable by the compiler.

The Formatter allows several configurable parameters to be set, each parameter having a default value which can be overridden by an explicit redefinition.

Two main possibilities are supported for a runtime configuration of the Formatter:

- by means of a configuration file;

- by (repeatedly) calling a function which sets a specified new value for a specified parameter.

Therefore, in order to call the Formatter, the main program must contain a call to the following functions.

```
1.   long cfcdef(long *)
```

(setting the default values for all parameters) and if one or more default values must be changed:

```
2a.  long cfcfil(char *, long *)
```

(reading the configuration file whose name is specified as the first parameter)

or:

```
2b.  long cfcpar(char *, char *, long *)
```

(setting the parameter whose name is specified as the first parameter to the value specified as the second parameter)

and finally:

```
3.   long cfmt(long *)
```

(calling the Formatter main algorithm)

All these functions return as their value a status/error variable; a pointer to that variable is also returned as the last argument of each function.

Example 1:

```
/* other lines of code */
{
long *istat = 0;                    /* status/error variable */
char conffile[30] = "config.dat"    /* configuration file */
cfcdef(istat);
if(!*istat)  cfcfil(conffile, istat);
if(!*istat)  cfmt(istat);
}
```

Example 2:

```
/* other lines of code */
{
long *istat = 0;
```

```
cfcdef(istat);
/* note that the parameter value must be passed as a     */
/* string see the next section for a discussion on       */
/* parameter namesand values                            . */
if(!*istat)   cfcpar("OUT_DEV", "stdout", istat);
if(!*istat)   cfcpar("CHAR_INDENT", "5", istat);
if(!*istat)   cfcpar("INCR_ID", "1", istat);
if(!*istat)   cfmt(istat);
}
```

Moreover, it is defined a "standard configuration behavior", corresponding to the configuration by means of a configuration file, obtained by just one call to the function:

```
long cform(char *, long *)
```

whose first parameter is the name of such a configuration file. A call to cform() is equivalent to the list of calls shown in the Example 1.

Example 3:

```
/* other lines of code */
{
long *istat = 0;                     /* status/error variable */
char conffile[30] = "config.dat"     /* configuration file */
cform(conffile, istat);
}
```

Note: This modularization allows to variously customize the user's environment; for instance it is possible to configure the Formatter at the beginning of a main program, and then call the function cfmt() each time the IDS contents are modified.

3.4.3.3 Configuring the formatter

The Formatter can be configured with respect to several parameters, all having a default value. The user can override a default value by writing in the configuration file a line with the format:

```
keyword_name    value
```

or by calling the function:

```
cfcpar(keyword_name, value, error_code);
```

All keyword_names must be written in upper case.

If a configuration file is used, the lines can appear in the file in any order; otherwise, the function cfcpar() can be repeatedly called.

In the following, the currently recognized parameters, their allowed values, their default value, and their semantics are listed.

Name:	BASE_ID
Range:	An integer in [0,1000]
Default:	1

Semantics:	Base of the increasing sequence of the entity names (only when INCR_ID is on)
Name:	CHAR_CONT
Range:	An integer in [0,5]
Default:	1
Semantics:	Number of characters added at the begin of the wrapped-around lines (only when pretty-printing)
Name:	CHAR_INDENT
Range:	An integer in [0,10]
Default:	2
Semantics:	Number of characters of indentation for scoped entities (only when pretty printing)
Name:	FLOAT_FORMAT
Range:	Any correct C format string for real variables
Default:	"%15.8e"
Semantics:	The C format string for the real variables (use the character "g" for removing trailing zeroes; e.g. %g)
Name:	INCR_ID
Range:	A positive integer
Default:	0
Semantics:	Flag for writing increasingly ordered entities in the output file
Name:	INT_FORMAT
Range:	Any correct C format string for integer variables
Default:	"%ld"
Semantics:	The C format string for the integer variables
Name:	LINE_LENGTH
Range:	An integer in [40,1000]
Default:	72
Semantics:	Max length of the line (only when pretty printing)
Name:	MULTI_FILE
Range:	A positive integer
Default:	0
Semantics:	Flag for creating multiple files: in the case of more than one top entity, if the field has a non-zero value each sub-structure (the top entity and its defining entities) is written on a different file (these files are called <OUT_DEV><n>, for n=1,...)
Name:	OUT_DEV
Range:	Any correct filename, or "stdout"
Default:	"stepfile"
Semantics:	Output stream: either "stdout" for the screen, or a name for the output file

Name:	PRETTY_PRINT
Range:	An integer in [0,4]
Default:	3
Semantics:	Print format

when = 0 (unformatted output),   the file does not contain any formatting character (therefore it appears as a single physical line);

when = 1 (fixed record length),   the behavior is similar to the unformatted output, but NewLine characters are added after LINE_LENGTH characters, so that all physical lines have the same length (this format is not currently supported by the Scanner/Parser);

when = 2 (basic printing),   all logical lines appear as physical lines (therefore a NewLine character is added at the end of each line defining an entity);

when = 3 (basic pretty printing),   the pretty printing is activated (all logical lines appears as physical lines and several additional parameters (LINE_LENGTH, CHAR_INDENT, CHAR_CONT) control the format of the output file);

when = 4 (enhanced pretty printing),   the behavior is similar to the basic pretty printing, but the file is generated in an even more "human readable" format: a blank is added after all commas, and a comment is added to all "ENDSCOPE" strings, indicating the name of the scoped entity

Name:	SINGLE_TOP
Range:	either 0 or an entity name (an integer in [1,999999999])
Default:	0
Semantics:	In the case more than one top entity is stored in IDS, for writing a specified one:

when = 0, all top entities are written;

when > 0, only the top entity whose name is specified is written

Name:	STEP_ID
Range:	An integer in [1,100]
Default:	1
Semantics:	Step of the increasing sequence of the entity names (only when INCR_ID is on)

3.4.3.4 Example of a configuration file

OUT_DEV	outfile	/* name of the output file */
PRETTY_PRINT	4	/* pretty printing fully activated */
LINE_LENGTH	80	/* the maximum length of the lines is set to 80 characters */

CHAR_INDENT	5	/* entities in scope are 5 chars indented with respect to scoped entities */
CHAR_CONT	3	/* 3 chars are added at the beginning of each wrapped around line */
FLOAT_FORMAT	%f	/* format for float variables is set */
INT_FORMAT	%ld	/* format for int variables is set */
MULTI_FILE	1	/* each top entity is written on a different file */
INCR_ID	1	/* entity renumbering is activated */
BASE_ID	100	/* the first entity has ID 100 */
STEP_ID	3	/* the increment of the renumbering is set to 3 */

3.4.4 Checker

The IDS data checker is dependent on the IDS and some means of populating it, such as the scanner/parser or a pre processor. Use of the checker is therefore a matter of writing a program with appropriate calls to the interface functions described in section 3.3.4. The following sections show an example syntactically correct STEP file and the results of running this through the checker.

3.4.4.1 Example STEP file

This file is according to the 1991 STEP version. Changes of the STEP DIS which occurred during the last year of the CADEX project have had effects on both the appearance of the file and on the Common Toolkit software. However the principles (sections, entity types, keywords, syntax) have not been changed significantly.

```
STEP;
HEADER;
FILE_NAME('block_open.step', 'Thu Jun 27 13:09:43 BST 1991',
   ('Ni, X.',
    'CADDETC, 171 Woodhouse Lane, Leeds LS2 3AR, UK',
    'Phone: +44 0532 334453, Fax: +44 0532 445270'),
   ('R&D Section', 'CADDETC'),
   'STEP PreVersion 1.0', 'preprocessor is a text editor',
   'originating system = NONAME_BREP');
FILE_DESCRIPTION(('primitive solid: open block', 'file'),
   'Brep level 2.0');
/* IMP_LEVEL('Brep level 1.0'); */
```

```
MAXIMUM_SIGNIFICANT_DIGIT(6);
CLASSIFICATION('test');
VECTOR_TOLERANCE(0.01);
DISTANCE_TOLERANCE(0.02);
ENDSEC;
DATA;
USE_AP(.BREP.);

    /* geometry data */
#1=CARTESIAN_POINT(-1.00000E+00,  2.00000E+00,  3.00000E+00);
#2=CARTESIAN_POINT( 1.00000E+00,  2.00000E+00,  3.00000E+00);
#3=CARTESIAN_POINT(-1.00000E+00, -2.00000E+00,  3.00000E+00);
#4=CARTESIAN_POINT( 1.00000E+00, -2.00000E+00,  3.00000E+00);
#5=CARTESIAN_POINT(-1.00000E+00,  2.00000E+00, -3.00000E+00);
#6=CARTESIAN_POINT( 1.00000E+00,  2.00000E+00, -3.00000E+00);
#7=CARTESIAN_POINT(-1.00000E+00, -2.00000E+00, -3.00000E+00);
#8=CARTESIAN_POINT( 1.00000E+00, -2.00000E+00, -3.00000E+00);
#101=DIRECTION(-2.00000E+00,  0.00000E+00,  0.00000E+00);
#102=DIRECTION(-2.00000E+00,  0.00000E+00,  0.00000E+00);
#103=DIRECTION( 0.00000E+00, -4.00000E+00,  0.00000E+00);
#104=DIRECTION( 0.00000E+00, -4.00000E+00,  0.00000E+00);
#105=DIRECTION(-2.00000E+00,  0.00000E+00,  0.00000E+00);
#106=DIRECTION(-2.00000E+00,  0.00000E+00,  0.00000E+00);
#107=DIRECTION( 0.00000E+00, -4.00000E+00,  0.00000E+00);
#108=DIRECTION( 0.00000E+00, -4.00000E+00,  0.00000E+00);
#109=DIRECTION( 0.00000E+00,  0.00000E+00, -6.00000E+00);
#110=DIRECTION( 0.00000E+00,  0.00000E+00, -6.00000E+00);
#111=DIRECTION( 0.00000E+00,  0.00000E+00, -6.00000E+00);
#112=DIRECTION( 0.00000E+00,  0.00000E+00,  6.00000E+00);
#121=DIRECTION(1.0,0.0,0.0);
#122=DIRECTION(-1.0,0.0,0.0);
#123=DIRECTION(0.0,1.0,0.0);
#124=DIRECTION(0.0,-1.0,0.0);
#125=DIRECTION(0.0,0.0,1.0);
#126=DIRECTION(0.0,0.0,-1.0);
#201=LINE($, #1, #101);
#202=LINE($, #3, #102);
#203=LINE($, #4, #103);
#204=LINE($, #3, #104);
#205=LINE($, #5, #105);
#206=LINE($, #7, #106);
#207=LINE($, #6, #107);
#208=LINE($, #5, #108);
#209=LINE($, #2, #109);
```

```
#210=LINE($, #1, #110);
#211=LINE($, #8, #111);
#212=LINE($, #7, #112);
#301=AXIS2_PLACEMENT(#3, #125, #125);
#302=AXIS2_PLACEMENT(#1, #123, #123);
#303=AXIS2_PLACEMENT(#7, #124, #124);
#304=AXIS2_PLACEMENT(#8, #121, #121);
#305=AXIS2_PLACEMENT(#3, #122, #122);
#306=AXIS2_PLACEMENT(#5, #126, #126);
#401=PLANE($,#301);
#402=PLANE($,#302);
#403=PLANE($,#303);
#404=PLANE($,#304);
#405=PLANE($,#305);
#406=PLANE($,#306);
    /* topology data */
#501=VERTEX(#1);
#502=VERTEX(#2);
#503=VERTEX(#3);
#504=VERTEX(#4);
#505=VERTEX(#5);
#506=VERTEX(#6);
#507=VERTEX(#7);
#508=VERTEX(#8);
#601 = CURVE_LOGICAL_STRUCTURE(#201, .T.);
#602 = CURVE_LOGICAL_STRUCTURE(#202, .T.);
#603 = CURVE_LOGICAL_STRUCTURE(#203, .T.);
#604 = CURVE_LOGICAL_STRUCTURE(#204, .T.);
#605 = CURVE_LOGICAL_STRUCTURE(#205, .T.);
#606 = CURVE_LOGICAL_STRUCTURE(#206, .T.);
#607 = CURVE_LOGICAL_STRUCTURE(#207, .T.);
#608 = CURVE_LOGICAL_STRUCTURE(#208, .T.);
#609 = CURVE_LOGICAL_STRUCTURE(#209, .T.);
#610 = CURVE_LOGICAL_STRUCTURE(#210, .T.);
#611 = CURVE_LOGICAL_STRUCTURE(#211, .T.);
#612 = CURVE_LOGICAL_STRUCTURE(#212, .T.);
#701=EDGE(#501, #502, #601);
#702=EDGE(#503, #504, #602);
#703=EDGE(#504, #502, #603);
#704=EDGE(#503, #501, #604);
#705=EDGE(#505, #506, #605);
#706=EDGE(#507, #508, #606);
#707=EDGE(#506, #508, #607);
#708=EDGE(#505, #507, #608);
```

```
#709=EDGE(#502, #506, #609);
#710=EDGE(#501, #505, #610);
#711=EDGE(#508, #504, #611);
#712=EDGE(#507, #503, #612);
#801=EDGE_LOGICAL_STRUCTURE(#701,.T.);
#802=EDGE_LOGICAL_STRUCTURE(#702,.T.);
#803=EDGE_LOGICAL_STRUCTURE(#703,.T.);
#804=EDGE_LOGICAL_STRUCTURE(#704,.T.);
#805=EDGE_LOGICAL_STRUCTURE(#705,.T.);
#806=EDGE_LOGICAL_STRUCTURE(#706,.T.);
#807=EDGE_LOGICAL_STRUCTURE(#707,.T.);
#808=EDGE_LOGICAL_STRUCTURE(#708,.T.);
#809=EDGE_LOGICAL_STRUCTURE(#709,.T.);
#810=EDGE_LOGICAL_STRUCTURE(#710,.T.);
#811=EDGE_LOGICAL_STRUCTURE(#711,.T.);
#812=EDGE_LOGICAL_STRUCTURE(#712,.T.);
#901=EDGE_LOGICAL_STRUCTURE(#701,.F.);
#902=EDGE_LOGICAL_STRUCTURE(#702,.F.);
#903=EDGE_LOGICAL_STRUCTURE(#703,.F.);
#904=EDGE_LOGICAL_STRUCTURE(#704,.F.);
#905=EDGE_LOGICAL_STRUCTURE(#705,.F.);
#906=EDGE_LOGICAL_STRUCTURE(#706,.F.);
#907=EDGE_LOGICAL_STRUCTURE(#707,.F.);
#908=EDGE_LOGICAL_STRUCTURE(#708,.F.);
#909=EDGE_LOGICAL_STRUCTURE(#709,.F.);
#910=EDGE_LOGICAL_STRUCTURE(#710,.F.);
#911=EDGE_LOGICAL_STRUCTURE(#711,.F.);
#912=EDGE_LOGICAL_STRUCTURE(#712,.F.);
#1001=EDGE_LOOP((#904, #802, #803, #901));
#1002=EDGE_LOOP((#910, #801, #809, #905));
#1003=EDGE_LOOP((#912, #806, #811, #902));
#1004=EDGE_LOOP((#911, #907, #909, #903));
#1005=EDGE_LOOP((#812, #804, #810, #808));
#1006=EDGE_LOOP((#908, #805, #807, #906));
#1101=LOOP_LOGICAL_STRUCTURE(#1001,.T.);
#1102=LOOP_LOGICAL_STRUCTURE(#1002,.T.);
#1103=LOOP_LOGICAL_STRUCTURE(#1003,.T.);
#1104=LOOP_LOGICAL_STRUCTURE(#1004,.T.);
#1105=LOOP_LOGICAL_STRUCTURE(#1005,.T.);
#1106=LOOP_LOGICAL_STRUCTURE(#1006,.T.);
#1201=SURFACE_LOGICAL_STRUCTURE(#401, .T.);
#1202=SURFACE_LOGICAL_STRUCTURE(#402, .T.);
#1203=SURFACE_LOGICAL_STRUCTURE(#403, .T.);
```

```
#1204=SURFACE_LOGICAL_STRUCTURE(#404, .T.);
#1205=SURFACE_LOGICAL_STRUCTURE(#405, .T.);
#1206=SURFACE_LOGICAL_STRUCTURE(#406, .T.);
#1301=FACE((#1101), #1201);
#1302=FACE((#1102), #1202);
#1303=FACE((#1103), #1203);
#1304=FACE((#1104), #1204);
#1305=FACE((#1105), #1205);
#1306=FACE((#1106), #1206);
#1401=FACE_LOGICAL_STRUCTURE(#1301,.T.);
#1402=FACE_LOGICAL_STRUCTURE(#1302,.T.);
#1403=FACE_LOGICAL_STRUCTURE(#1303,.T.);
#1404=FACE_LOGICAL_STRUCTURE(#1304,.T.);
#1405=FACE_LOGICAL_STRUCTURE(#1305,.T.);
#1406=FACE_LOGICAL_STRUCTURE(#1306,.T.);
#1501=CLOSED_SHELL((#1401, #1403, #1404, #1405, #1406));
#1502=SHELL_LOGICAL_STRUCTURE(#1501,.T.);
#1503=MANIFOLD_SOLID_BREP(#1502,());
END_AP(.BREP.);
ENDSEC;
ENDSTEP;
```

3.4.4.2 Error file as output of the data checker

```
error 101: the following entities are defined but not used by top
entity --
          #123, #302, #402, #801, #809
          #905, #910, #1002, #1102, #1202
          #1302, #1402
error 402: entity "#1501", where rule "topology constraints" violates
--
          edge entity "#701" should be referenced exactly twice.
error 402: entity "#1501", where rule "topology constraints" violates
--
          edge entity "#705" should be referenced exactly twice.
error 402: entity "#1501", where rule "topology constraints" violates
--
          edge entity "#709" should be referenced exactly twice.
error 402: entity "#1501", where rule "topology constraints" violates
--
          edge entity "#710" should be referenced exactly twice.
Total 5 error(s) found.
```

3.4.5 Statistics tools

3.4.5.1 General user information

Compiling and calling the STEP statistics programs. A few lines have been changed or added to the file "Makefile", originally provided by HP - Andrew Kutter and Peter J. Schild -, in order to compile and link the source code for the scanner/parser, the IDS and one or more modules of the statistics software. The executable file is either written to a separate file (e.g. "sstat") or is included in the processor together with all other functions.

Error handling. Two types of errors have been identified:

If an error in Scanner/Parser or IDS routines occurres, the user can decide, wether the program should be interrupted. Alternatively, if the program is not interrupted, the result of the program may be invalid or may contain "Null" values.

If any internal function returns an error, a message is displayed on the screen and the program is terminated.

The possible errors are listed in the header files "sstat.h", "scomp.h" and "sstruc.h".

3.4.5.2 STEP statistics viewer

The input. For input only a STEP filename is required. The existence of the STEP file is checked later on. This input is prompted by the program as shown in the listing below.

```
*********************************************************************
**                                                                 **
**      S T E P - F I L E - S T A T I S T I C - P R O G R A M      **
**                                                                 **
**                      SSTAT Version 1.0                          **
**                                                                 **
*********************************************************************
** This Program produces a statistic report of the contents        **
** of the STEP-Inputfile.                                          **
** This report will be stored with the STEP-Filename and the       **
** extension <.sta> in the default-directory.                      **
*********************************************************************
**                                                                 **
**        **************************************************        **
**        * STEP - Filename :>                          *           **
**        **************************************************        **
**                                                                 **
*********************************************************************
```

Example STEP file 1

```
STEP;
HEADER;
FILE_NAME('Test-file','2.5.90  10:03',('Dr. Helmut J. Helpenstein',
'Tel. +49 2408 6011'),('GfS','Pascalstr. 17','D-52076 Aachen'),
'May 90','Hel 1.2','GfS-FEM->PROLOG');
FILE_DESCRIPTION (('test file for curve_on_surface'),'apss_v1.1.1');
CLASSIFICATION ('');
MAXIMUM_SIGNIFICANT_DIGIT ( 9);
ENDSEC;
DATA;
USE_AP (.SS.);
#4 = FACE_LOGICAL_STRUCTURE (#7,.T.);
#10 = EDGE_LOGICAL_STRUCTURE (#17,.T.);
#11 = EDGE_LOGICAL_STRUCTURE (#16,.T.);
#12 = EDGE_LOGICAL_STRUCTURE (#15,.T.);
#20 = CARTESIAN_POINT (0.,0.,0.);
#15 = VERTEX (#20);
#19 = CARTESIAN_POINT (0.,0.,0.);
#16 = VERTEX (#19);
/* #18 =  */
#17 = CURVE_LOGICAL_STRUCTURE (#18,.T.);
#14 = EDGE (#15,#16,#17);
#13 = EDGE_LOGICAL_STRUCTURE (#14,.T.);
#9 = EDGE_LOOP ((#10,#11,#12,#13));
#7 = LOOP_LOGICAL_STRUCTURE (#9,.T.);
/* #8 =  */
#6 = FACE ((#7),#8);
#5 = FACE_LOGICAL_STRUCTURE (#6,.T.);
#3 = OPEN_SHELL ((#4,#5));
#2 = SHELL_LOGICAL_STRUCTURE (#3,.T.);
#1 = SHELL_BASED_SURFACE_MODEL ((#2));
END_AP (.SS.);
ENDSEC;
ENDSTEP;
```

Output of the STEP statistic viewer

```
Header-Informations about the STEP-File : cus.nf
********************************************************************
STEP-NAME              : Test-file
CREATION TIME          : 2.5.90  10:03
AUTHOR                 : Dr. Helmut J. Helpenstein
                         Tel.+49 2408 6011
ORGANISATION           : GfS  Pascalstr. 17  D-52076 Aachen
```

```
STEP-VERSION                    : May 90
PREPROCESSOR VERSION            : Hel 1.2
ORGINATING-SYSTEM-NAME          : GfS-FEM->PROLOG
STEP-FILE-DESCRIPTION           : test file for curve_on_surface
PREPROCESSOR-IMPLEMENT-LEVEL    : apss_v1.1.1
SECURITY-CLASSIFICATION         :
MAXIMUM SIGNIFICANT DIGITS      : 9
VECTOR TOLERANCE                : 0.000000
DISTANCE TOLERANCE              : 0.000000


*********************************************************************
*  Statistic-Report                                                 *
*********************************************************************


    Entity-Name :                        Entity-Type:  Occurences:
    ---------------------------------------------------------------
    SHELL_BASED_SURFACE_MODEL            333             1
    SHELL_LOGICAL_STRUCTURE              335             1
    OPEN_SHELL                           317             1
    FACE_LOGICAL_STRUCTURE               299             2
    FACE                                 297             1
    LOOP_LOGICAL_STRUCTURE               313             1
    EDGE_LOOP                            289             1
    EDGE_LOGICAL_STRUCTURE               288             4
    EDGE                                 286             1
    VERTEX                               352             2
    CURVE_LOGICAL_STRUCTURE              278             1
    CARTESIAN_POINT                      266             2

                                            Total: 18
```

3.4.5.3 STEP file comparator

The input. After starting the program the same input mask as with the statistic viewer is displayed. The user is prompted to identify both STEP files. The name of the output file is identical to the filename of the first input file and the extension is ".scp".

Example STEP file 2. The comparator needs two input files. The first is assumed to be the same as for the statistics viewer. The second is given here:

```
STEP;
HEADER;
FILE_NAME('Test-file', '3.5.90  15:13', ('Dr. Helmut J. Helpenstein',
'Tel. +49 2408 6011'), ('GfS', 'Pascalstr. 17', 'D-52076 Aachen'),
'IDS 5.3', 'May 90', 'IDS->Formatter by Italcad');
```

```
FILE_DESCRIPTION(('wireframe model with circular arcs'),
'apwf_v1.3');
CLASSIFICATION('');
MAXIMUM_SIGNIFICANT_DIGIT(8);
ENDSEC;
DATA;
USE_AP(.WF.);
#1 = CARTESIAN_POINT(0.0000000, 0.0000000, 0.0000000);
#2 = VERTEX(#1);
#3 = CARTESIAN_POINT(10.000000, 0.0000000, 0.0000000);
#4 = VERTEX(#3);
#5 = DIRECTION(1.0000000, 0.0000000, 0.0000000);
#6 = LINE($, #1, #5);
#7 = CURVE_LOGICAL_STRUCTURE(#6, .T.);
#8 = EDGE(#2, #4, #7);
#9 = CARTESIAN_POINT(0.0000000, 10.000000, 0.0000000);
#10 = VERTEX(#9);
#11 = DIRECTION(0.0000000, 0.0000000, 1.0000000);
#12 = AXIS2_PLACEMENT(#1, #11, #5);
#13 = CIRCLE($, 10.000000, #12);
#14 = CURVE_LOGICAL_STRUCTURE(#13, .T.);
#15 = EDGE(#4, #10, #14);
#16 = DIRECTION(0.0000000, 1.0000000, 0.0000000);
#17 = LINE($, #1, #16);
#18 = CURVE_LOGICAL_STRUCTURE(#17, .T.);
#19 = EDGE(#2, #10, #18);
#20 = CONNECTED_EDGE_SET((#8, #15, #19));
#21 = CARTESIAN_POINT(10.000000, 0.0000000, 8.0000000);
#22 = VERTEX(#21);
#23 = LINE($, #3, #11);
#24 = CURVE_LOGICAL_STRUCTURE(#23, .T.);
#25 = EDGE(#4, #22, #24);
#26 = CARTESIAN_POINT(0.0000000, 0.0000000, 8.0000000);
#27 = VERTEX(#26);
#28 = LINE($, #26, #5);
#29 = CURVE_LOGICAL_STRUCTURE(#28, .T.);
#30 = EDGE(#27, #22, #29);
#31 = LINE($, #1, #11);
#32 = CURVE_LOGICAL_STRUCTURE(#31, .T.);
#33 = EDGE(#2, #27, #32);
#34 = CONNECTED_EDGE_SET((#8, #25, #30, #33));
#35 = CARTESIAN_POINT(0.0000000, 10.000000, 8.0000000);
#36 = VERTEX(#35);
#37 = LINE($, #26, #16);
```

```
#38 = CURVE_LOGICAL_STRUCTURE(#37, .T.);
#39 = EDGE(#27, #36, #38);
#40 = LINE($, #9, #11);
#41 = CURVE_LOGICAL_STRUCTURE(#40, .T.);
#42 = EDGE(#10, #36, #41);
#43 = CONNECTED_EDGE_SET((#19, #33, #39, #42));
#44 = AXIS2_PLACEMENT(#26, #11, #5);
#45 = CIRCLE($, 10.000000, #44);
#46 = CURVE_LOGICAL_STRUCTURE(#45, .T.);
#47 = EDGE(#22, #36, #46);
#48 = CONNECTED_EDGE_SET((#15, #42, #47, #25));
#49 = CONNECTED_EDGE_SET((#30, #47, #39));
#50 = EDGE_BASED_WIREFRAME_MODEL((#20, #34, #43, #48, #49));
END_AP(.WF.);
ENDSEC;
ENDSTEP;
```

Output of the STEP file comparator

Header-Informations about the STEP-File 1: cus.nf
****

STEP-NAME	: Test-file
CREATION TIME	: 2.5.90 10:03
AUTHOR	: Dr. Helmut J. Helpenstein
	Tel.+49 2408 6011
ORGANISATION	: GfS Pascalstr. 17 D-52076 Aachen
STEP-VERSION	: May 90
PREPROCESSOR VERSION	: Hel 1.2
ORGINATING-SYSTEM-NAME	: GfS-FEM->PROLOG
STEP-FILE-DESCRIPTION	: test file for curve_on_surface
PREPROCESSOR-IMPLEMENT-LEVEL	: apss_v1.1.1
SECURITY-CLASSIFICATION	:
MAXIMUM SIGNIFICANT DIGITS	: 9
VECTOR TOLERANCE	: 0.000000
DISTANCE TOLERANCE	: 0.000000

Header-Informations about the STEP-File 2: wbogen.nf
****

STEP-NAME	: Test-file
CREATION TIME	: 3.5.90 15:13
AUTHOR	: Dr. Helmut J. Helpenstein
	Tel.+49 2408 6011
ORGANISATION	: GfS Pascalstr. 17 D-52076 Aachen
STEP-VERSION	: IDS 5.3
PREPROCESSOR VERSION	: May 90

```
ORGINATING-SYSTEM-NAME         : IDS->Formatter by Italcad
STEP-FILE-DESCRIPTION          : wireframe model with circular arcs
PREPROCESSOR-IMPLEMENT-LEVEL   : apwf_v1.3
SECURITY-CLASSIFICATION        :
MAXIMUM SIGNIFICANT DIGITS     : 8
VECTOR TOLERANCE               : 0.000000
DISTANCE TOLERANCE             : 0.000000
```

```
****************************************************************
*  Statistic - Comparation between :  File 1      and      File 2  *
****************************************************************
```

Occurences :
============

Token-Name	Entity-Typ	File 1	File 2
SHELL_BASED_SURFACE_MODEL	333	1	0
SHELL_LOGICAL_STRUCTURE	335	1	0
OPEN_SHELL	317	1	0
FACE_LOGICAL_STRUCTURE	299	2	0
FACE	297	1	0
LOOP_LOGICAL_STRUCTURE	313	1	0
EDGE_LOOP	289	1	0
EDGE_LOGICAL_STRUCTURE	288	4	0
EDGE	286	1	9
VERTEX	352	2	6
CURVE_LOGICAL_STRUCTURE	278	1	9
CARTESIAN_POINT	266	2	6
DIRECTION	284	0	3
LINE	311	0	7
AXIS2_PLACEMENT	259	0	2
CIRCLE	267	0	2
CONNECTED_EDGE_SET	274	0	5
EDGE_BASED_WIREFRAME_MODEL	287	0	1
	Total :	18	50

3.4.5.4 STEP structure viewer

The input. For input only a STEP filename is required. The existence of the STEP file is checked later on. This input is prompted by the program as shown in the listing below.

```
*************************************************************************
**                                                                     **
**          S T E P - F I L E - S T R U C T U R E - V I E W E R        **
**                                                                     **
**                        SSTRUC Version 1.0                           **
**                                                                     **
*************************************************************************
** This Program produces an overview about the structural              **
** contents of the STEP-Inputfile.                                     **
** The structure report will be saved with the STEP-Filename           **
** and the extension <.stc> in the default-directory                   **
*************************************************************************
*                                                                       *
*        ***************************************************            *
*        * STEP - Filename :>                            *              *
*        ***************************************************            *
*************************************************************************
```

Explanation of the STEP structure viewer output. The result of the structure analysis as listed in the example below is written into an output file as well as being displayed on the screen. The name of the output file is the same as the input file name. The extension of the output file is ".stc".

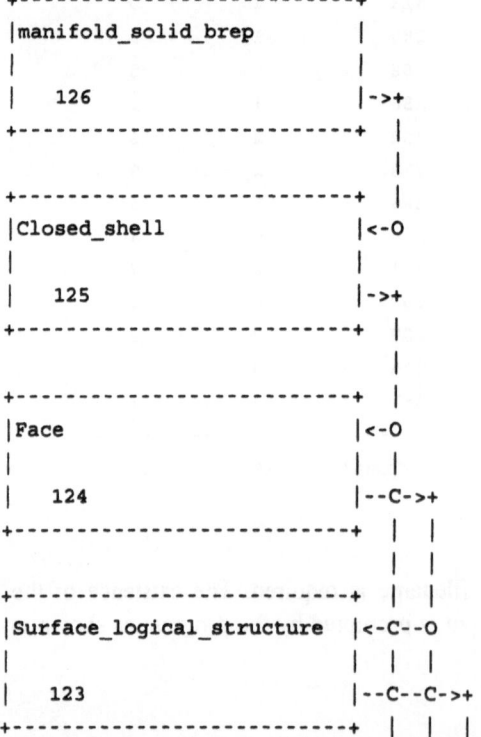

```
+---------------------------+
|manifold_solid_brep        |
|                           |
|    126                    |->+
+---------------------------+  |
                               |
+---------------------------+  |
|Closed_shell               |<-O
|                           |
|    125                    |->+
+---------------------------+  |
                               |
+---------------------------+  |
|Face                       |<-O
|                           |  |
|    124                    |--C->+
+---------------------------+  |  |
                               |  |
+---------------------------+  |  |
|Surface_logical_structure  |<-C--O
|                           |  |  |
|    123                    |--C--C->+
+---------------------------+  |  |  |
```

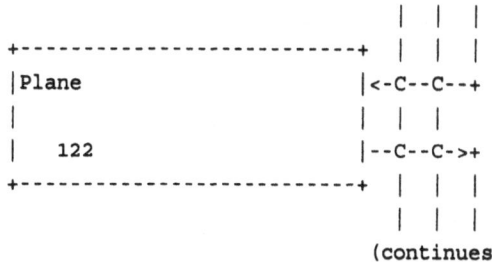

```
                                      |   |   |
      +-------------------------+     |   |   |
      |Plane                    |<-C--C--+
      |                         |     |   |   |
      |     122                 |--C--C->+
      +-------------------------+     |   |   |
                                      |   |   |
                            (continues)
```

In the output file, each entity contained in the STEP input file is represented by a box. The box contains the entity type and name. The first box represents the top entity of the model.

The connecting pointers between the entities are shown on the right hand side of the entity boxes. The following pointer notation is used:

">" or "<"	- Direction of the pointer
"+"	- Corner
"C"	- Bridge over a crossing pointer
"O"	- Pointer node

It is not always possible to show all references on one page (flag), because the number of columns on the screen and the printer is limited. The references which cannot be displayed on one page (flag), are represented on an extra page (flag).

Complete output of the STEP structure viewer, For the example STEP file 1 (listed in the section on the statistics viewer) here the output of the STEP structure viewer is shown:

```
**********************************************************************
Header-Informations about the STEP-File : cus.nf
**********************************************************************
STEP-NAME                      : Test-file
CREATION TIME                  : 2.5.90   10:03
AUTHOR                         : Dr. Helmut J. Helpenstein
                                 Tel.+49 2408 6011
ORGANISATION                   : GfS  Pascalstr. 17  D-52076 Aachen
STEP-VERSION                   : May 90
PREPROCESSOR VERSION           : Hel 1.2
ORGINATING-SYSTEM-NAME         : GfS-FEM->PROLOG
STEP-FILE-DESCRIPTION          : test file for curve_on_surface
PREPROCESSOR-IMPLEMENT-LEVEL   : apss_v1.1.1
SECURITY-CLASSIFICATION        :
MAXIMUM SIGNIFICANT DIGITS     : 9
VECTOR TOLERANCE               : 0.000000
DISTANCE TOLERANCE             : 0.000000
```

Structure of the STEP file cus.nf; 1. of 1 flags

```
+--------------------------------+
| Name: 1                        |
|                                |
| SHELL_BASED_SURFACE_MODEL      |->+
+--------------------------------+  |
                                     |
                                     |
+--------------------------------+  |
| Name: 2                        |<-O
|                                |
| SHELL_LOGICAL_STRUCTURE        |->+
+--------------------------------+  |
                                     |
                                     |
+--------------------------------+  |
| Name: 3                        |<-O
|                                |
| OPEN_SHELL                     |->+
+--------------------------------+  |
                                     |
+--------------------------------+  |
| Name: 4                        |<-O
|                                |  |
| FACE_LOGICAL_STRUCTURE         |--C->+
+--------------------------------+  |  |
                                    | C |
+--------------------------------+  | C |
| Name: 7                        |<-C--O--------O
|                                |  | |          |
| LOOP_LOGICAL_STRUCTURE         |--C->+          |
+--------------------------------+  | |           |
                                    | |  |         |
+--------------------------------+  | |  |         |
| Name: 9                        |<-C--O          |
|                                |  | |            |
| EDGE_LOOP                      |--C->+           |
+--------------------------------+  | |            |
                                    | |  |          |
+--------------------------------+  | |  |          |
| Name: 10                       |<-C--O           |
|                                |  | |  |          |
| EDGE_LOGICAL_STRUCTURE         |--C--C->+         |
+--------------------------------+  | |  |          |
                                    | |  |  |       |
```

```
                                     |  |  |  |   |
+------------------------------+  |  |  |   |
| Name: 17                     |<-C--C--O--O   |
|                              |  |  |  |   |  |
| CURVE_LOGICAL_STRUCTURE      |--C--C->+  |  |
+------------------------------+  |  |  |  |  |
                                  |  |  |  |  |
+------------------------------+  |  |  |  |  |
| Name: 18                     |<-C--C--O   |  |
|                              |  |  |  |   |  |
| (null)                       |  |  |   |  |
+------------------------------+  |  |   |  |
                                  |  |   |  |
+------------------------------+  |  |   |  |
| Name: 11                     |<-C--O   |  |
|                              |  |  |   |  |
| EDGE_LOGICAL_STRUCTURE       |--C--C->+  |  |
+------------------------------+  |  |  |  |  |
                                  |  |  |  |  |
+------------------------------+  |  |  |  |  |
| Name: 16                     |<-C--C--O--O   |
|                              |  |  |  |   |  |
| VERTEX                       |--C--C->+  |  |
+------------------------------+  |  |  |  |  |
                                  |  |  |  |  |
+------------------------------+  |  |  |  |  |
| Name: 19                     |<-C--C--O  |  |
|                              |  |  |  |   |  |
| CARTESIAN_POINT              |  |  |   |  |
+------------------------------+  |  |   |  |
                                  |  |   |  |
+------------------------------+  |  |   |  |
| Name: 12                     |<-C--O   |  |
|                              |  |  |   |  |
| EDGE_LOGICAL_STRUCTURE       |--C--C->+  |  |
+------------------------------+  |  |  |  |  |
                                  |  |  |  |  |
+------------------------------+  |  |  |  |  |
| Name: 15                     |<-C--C--O--O   |
|                              |  |  |  |   |  |
| VERTEX                       |--C--C->+  |  |
+------------------------------+  |  |  |  |  |
                                  |  |  |  |  |
```

```
                                      |  |  |  |  |
    +----------------------------------+  |  |  |  |  |
    | Name: 20                         |<-C--C--O  |  |
    |                                  |  |  |     |  |
    | CARTESIAN_POINT                  |  |  |     |  |
    +----------------------------------+  |  |     |  |
                                      |  |     |  |
                                      |  |     |  |
                                      |  |     |  |
    +----------------------------------+  |  |     |  |
    | Name: 13                         |<-C--O     |  |
    |                                  |  |        |  |
    | EDGE_LOGICAL_STRUCTURE           |--C->+     |  |
    +----------------------------------+  |  |     |  |
                                      |  |     |  |
                                      |  |     |  |
                                      |  |     |  |
    +----------------------------------+  |  |     |  |
    | Name: 14                         |<-C--O     |  |
    |                                  |  |        |  |
    | EDGE                             |--C------->+  |
    +----------------------------------+  |           |
                                      |           |
                                      |           |
    +----------------------------------+  |           |
    | Name: 5                          |<-O           |
    |                                  |              |
    | FACE_LOGICAL_STRUCTURE           |->+           |
    +----------------------------------+  |           |
                                      |           |
                                      |           |
    +----------------------------------+  |           |
    | Name: 6                          |<-O           |
    |                                  |              |
    | FACE                             |------------->+
    +----------------------------------+              |
                                                  |
                                                  |
    +----------------------------------+              |
    | Name: 8                          |<-------------O
    |                                  |
    | (null)                           |
    +----------------------------------+
```

3.4.6 User's guide of SI tools

General remark. More information concerning SISL is available in the SISL Reference Manual, version 2.1, SI, 5. April 1991. Identifier versions of Conversion Tools are dependent on the IDS. Parameter versions are independent of the IDS; the parameter list is - where relevant - conformant to STEP. Basic tools provide required functionality without satisfying any formats or the requirement of being callable from both C and FORTRAN.

3.4.6.1 File CSIA2S.C - APS to SISL conversion

This file contains routines especially designed for the conversion of B-spline entities from the APS-SS format to the SISL-format. Among these routines the following ones are considered to be independent conversion tools.

CCASBC

Category: basic; geometry conversion; C

Converts a B-spline curve from the APS-SS format to the SISL-format. APS-SS is a format that is used in some of the CADEX-systems. SISL is the SI Spline Library. Many of the Conversion Tools are based on the SISL-format. Input to this routine is a set of parameters that describes a B-spline curve according to the APS-SS format. Output is a pointer to a SISL curve-structure. Only one curve can be converted at a time. The input curve will not be modified.

There are tools converting SISL-curves to IDS conformant B-spline curves: ccisic() (identifier-version); ccsibc() (parameter-version).

CCASBS

Category: basic; geometry conversion; C

Converts a B-spline surface from the APS-SS format to the SISL-format. APS-SS is a format that is used in some of the CADEX-systems. SISL is the SI Spline Library. Many of the Conversion Tools are based on the SISL-format. Input to this routine is a set of parameters that describes a B-spline surface according to the APS-SS format. Output is a pointer to a SISL surface-structure. Only one surface can be converted at a time. The input surface will not be modified.

There are tools converting SISL-surfaces to IDS conformant B-spline surfaces: ccisis() (identifier-version); ccsibs() (parameter-version).

3.4.6.2 File CSIS2A.C - SISL to APS conversion

This file contains routines especially designed for the conversion of B-spline entities from the SISL-format to the APS-SS format. Among these routines the following ones are considered to be independent conversion tools.

CCSABC

Category: basic; geometry conversion; C

Converts a B-spline curve from the SISL-format to the APS-SS format. APS-SS is a format that is used in some of the CADEX-systems. SISL is the SI Spline Library. Many of the Conversion Tools are based on the SISL format. Input to this routine is a pointer to a SISL curve-structure. Output is a set of parameters that describes a B-spline curve according to the APS-SS format. Only one curve can be converted at a time. The input curve will not be modified.

There are tools converting IDS conformant B-spline curves to SISL-curves: cciisc() (identifier-version); ccisbc() (parameter-version).

CCSABS

Category: basic; geometry conversion; C

Converts a B-spline surface from the SISL-format to the APS-SS format. APS-SS is a format that is used in some of the CADEX-systems. SISL is the SI Spline Library. Many of the Conversion Tools are based on the SISL format. Input to this routine is a pointer to a SISL surface-structure. Output is a set of parameters that describes a B-spline surface according to the APS-SS format. Only one surface can be converted at a time. The input surface will not be modified.

There are tools converting IDS conformant B-spline surfaces to SISL-surfaces: cciiss() (identifier-version); ccisbs() (parameter-version).

3.4.6.3 File CSI221.C - removing topology

This file contains routines especially designed for the conversion of FL2-models to FL1-models. Among these routines the following ones are considered to be independent conversion tools.

CSI221

Category: identifier; topology conversion; C

Converts a surface model that is conformant to Functional Level 2 of the latest CADEX Surface AP to the corresponding Functional Level 1. Input and output to this routine are IDS-identifiers to entities of types face_based_surface_model respectively geometric_3d_ surface_set. Only one model can be converted at a time. The resulting model will be a pure geometric model without any references to topology. An absolute geometry resolution value (recommended: 0.1) is required input. A local, relative value might be calculated that overrules the given value. The input model will not be modified or deleted. The output model references copies of the geometrical instances that are used by the input model. Thus, if the input model is deleted, the output model will remain complete. There is no parameter version of this tool.

Restrictions: This version does handle FL2 models with references to untrimmed B-spline geometry only; analytic and trimmed geometry are not supported.

CSICFASF

Category: identifier; topology conversion; C

Converts an entity instance of type face to an instance of type B-spline surface. Input to this routine is an IDS-identifier to an instance of type face; output is an IDS-identifier to an instance of type B-spline surface. An absolute geometry resolution value (recommended: 0.1) is required input. A local, relative value might be calculated that overrules the given value. The face instance will not be modified or deleted. The B-spline surface instance is a new IDS instance and has no reference to the input instance. Thus, if the face instance is deleted, the B-spline surface instance will remain complete. Only one instance can be converted at a time. There is no parameter version of this tool.

Restrictions: This version does handle face instances with references to untrimmed B-spline geometry only; analytic and trimmed geometry are not supported.

CSICVXLP

Category: identifier; topology conversion; C

Converts an entity instance of type vertex loop to an instance of type cartesian point. Input to this routine is an IDS-identifier to an instance of type vertex loop; output is an IDS-identifier to an instance of type cartesian point. An absolute geometry resolution value (recommended: 0.1) is required input. A local, relative value might be calculated that overrules the given value. The input instance will not be modified or deleted. The output instance references a copy of the cartesian point that is used by the vertex loop. Thus, if the vertex loop is deleted, the cartesian point that is output of this routine will not be touched. Only one instance can be converted at a time. There is no parameter version of this tool.

CSICEDBC

Category: identifier; topology conversion; C

Converts an entity of type edge to an entity of type B-spline curve and to cartesian points at start and end. Input to this routine is an IDS-identifier to an instance of type edge; output is an IDS-identifier to an instance of type B-spline curve and two IDS-identifiers to instances of type cartesian point. The direction of the curve is correspondant to the direction of the edge and its sense-flag. An absolute geometry resolution value (recommended: 0.1) is required input. A local, relative value might be calculated that overrules the given value. The edge instance will not be modified or deleted. The B-spline curve instance is a new IDS instance and has no reference to the input instance. Thus, if the edge instance is deleted, the B-spline curve instance will remain complete. Only one instance can be converted at a time. There is no parameter version of this tool.

Restrictions: This version does handle edge instances with references to untrimmed B-spline geometry only; analytic and trimmed geometry are not supported.

CSICVXCAPT

Category: identifier; topology conversion; C

Converts an entity of type vertex to an entity of type cartesian point. Input to this routine is an IDS-identifier to an instance of type vertex; output is an IDS-identifier to an instance of type cartesian point. An absolute geometry resolution value (recommended: 0.1) is required input. A local, relative value might be calculated that overrules the given value. The input instance will not be modified or deleted. The output instance references a copy of the cartesian point that is used by the vertex. Thus, if the vertex is deleted, the cartesian point that is output of this routine will not be touched. Only one instance can be converted at a time. There is no parameter version of this tool.

3.4.6.4 File CSI322.C - shell-based to face-based conversion

This file contains routines especially designed for the conversion of FL3-models to FL2-models. Among these routines the following ones are considered to be independent conversion tools.

CSI322

Category: identifier; topology conversion; C

Converts a surface model that is conformant to Functional Level 3 of the latest CADEX Surface AP to the corresponding Functional Level 2. Input and output to this routine are IDS-identifiers to entities of types shell_based_surface_model respectively face_based_ surface_model. The input model will not be modified or deleted. The output model references from the faces and down the tree the same geometrical and topological instances that are used by the input model. Thus, if the input model is deleted, the output model will also be destroyed. There are no restrictions on the use of geometry for the input model. Only one model can be converted at a time. There is no parameter version of this tool.

3.4.6.5 File CSICON.C - conics to b-spline curves

This file contains routines especially designed for the conversion of both trimmed and untrimmed conics to B-spline curves. Among these routines the following ones are considered to be independent conversion tools.

CCICBC

Category: identifier; geometry conversion; C

Converts a trimmed conic curve of type circle, ellipse, hyperbola or parabola to a B-spline curve. Input to this routine is an IDS-identifier to an instance of type conic (circle, ellipse, hyperbola or parabola); output is an IDS-identifier to an instance of type B-spline curve. Trimming information must be provided by either identifying IDS-entities of type cartesian points that are located on the conic or by specifying the parameter values of start- and end-point relative to the conic. If no trimming information is present, the routine will return an error message. Routines ccicoc() (identifier-version) and cccobc() (parameter-version) handle untrimmed conics, i.e. circles and ellipses. A sense flag indicates whether the directions of the trimmed curve and the conic coincide. An absolute geometry resolution value (recommended: 0.1) is required input as well. A local, relative value might be calculated that overrules the given value. Only one instance can be converted at a time. The input instance will not be modified or deleted. Input and output instances will be totally independent from each other. Thus, if the input instance is deleted, the output instance will not be touched. There is a parameter version of this tool called cctcoc().

CCTCOC

Category: parameter; geometry conversion; C

Converts a trimmed conic curve of type circle, ellipse, hyperbola or parabola to a B-spline curve. This routine is the corresponding parameter version to the identifier tool ccicbc(). Input to this routine is a set of parameters describing a STEP-conformant conic instance, i.e. a circle, ellipse, hyperbola or parabola. The type of the conic must be given as well as its coordinate system. The coordinate system is described by three cartesian points, its origo, a point on the x-axis and a point on the z-axis. Each conic is defined by one or two parameters; these are

- for a circle   : radius

- for an ellipse : semi-axis 1 and 2

- for a hyperbola: semi-axis and semi-imaginary-axis

- for a parabola : focal distance.

For more explanations consult ISO 10303-205 version 1. May 1992, the underlying specification for this tool, i.e. the Surface Application Protocol. A sense flag indicates whether the directions of the trimmed curve and the conic coincide. An absolute geometry resolution value (recommended: 0.1) is required input. A local, relative value might be calculated that overrules the given value. Output from this routine is a set of parameters describing a STEP-conformant B-spline curve. If the input conic is closed, the closed-flag of the B-spline curve will be set. The format of such a B-spline curve is also based on the Surface Application Protocol. Differences between the AP-format and the format used in this routine are pointed out in the technical description of this tool (chapter 3.3.6). Only one instance can be converted at a time. The input instance will not be modified.

CCIPAC

Category: basic; geometry conversion; C

Converts a point on a circular curve from parametric to cartesian. Input to this routine are parameters describing the circle (position and radius) and the location of the point in the parameter space of the circle. Output are the coordinates of the calculated cartesian point. Only one instance can be converted at a time. The input instance will not be modified.

Corresponding routines (clipac(), celpac()) have been developed for converting points on lines and ellipses.

Restrictions: This tool does not distinguish trimmed and untrimmed circles.

CELPAC

Category: basic; geometry conversion; C

Converts a point on a elliptic curve from parametric to cartesian. Input to this routine are parameters describing the ellipse (position, major and minor radius) and the location of the point in the parameter space of the ellipse. Output are the coordinates of the calculated cartesian point. Only one instance can be converted at a time. The input instance will not be modified.

Corresponding routines (clipac(), ccipac()) have been developed for converting points on lines and circles.

Restrictions: This tool does not distinguish trimmed and untrimmed ellipses.

CCICOC

Category: identifier; geometry conversion; C

Converts a conic curve of type circle or ellipse to a B-spline curve. Input to this routine is an IDS-identifier to an instance of type conic (circle or ellipse); output will be an IDS-identifier to an instance of type B-spline curve. An absolute geometry resolution value (recommended: 0.1) is required input. A local, relative value might be calculated that overrules the given value. Only one instance can be converted at a time. The input instance will not be modified or deleted. Input and output instances will be totally independent from each other. Thus, if the input instance is deleted, the output instance will not be touched. There is a parameter version of this tool called cccobc().

CCCOBC

Category: parameter; geometry conversion; C

Converts a conic curve to a B-spline curve. This routine is the corresponding parameter version to the identifier tool ccicoc(). Input to this routine is a set of parameters describing a STEP-conformant conic instance, i.e. a circle or an ellipse. The type of the conic must be given as well as its coordinate system. The coordinate system is described by three cartesian points, its origo, a point on the x-axis and a point on the z-axis. Each conic is defined by one or two parameters; these are

- for a circle   : radius

- for an ellipse : semi-axis 1 and 2.

For more explanations consult ISO 10303-205 version 1. May 1992, the underlying specification for this tool, the Surface Application Protocol. An absolute geometry resolution value (recommended: 0.1) is required input. A local, relative value might be calculated that overrules the given value. Output from this routine is a set of parameters describing a STEP-conformant B-spline curve. The format of such a B-spline curve is also based on the Surface Application Protocol. Differences between the AP-format and the format used in this routine are pointed out in the technical description of this tool (chapter 3.3.6). Only one instance can be converted at a time. The input instance will not be modified.

3.4.6.6 File CSIFUN.C - handling b-spline parameters

This file contains routines especially designed for handling the parameters of B-spline curves and surfaces. Among these routines the following ones are considered to be independent conversion tools.

GET_UP_IND_KNOTS

Category: basic; geometry conversion; C

Removes duplicated elements from a B-spline knot vector and returns the upper index of the resulting vector. The upper index is the number of elements in the vector; it is 1 both when the given vector is empty and when it contains one element only. Input to this routine are a knot vector, which might contain duplicated elements, and the size of this input array. The function returns as its value the size of the knot vector with counting duplicated elements only once. This representation is conformant to the definition of the STEP B-spline entities.

Restrictions: This tool is not callable from FORTRAN.

FILL_KNOTS_MULTS

Category: basic; geometry conversion; C

Converts a knot vector containing duplicated elements to two arrays: one containing not duplicated knots and the other multiplicities belonging to these knots. Input to this routine are a knot vector, which might contain duplicated elements, the size of this input array, and the upper index of the resulting arrays. The latter input value may be obtained by a call to get_up_ind_knots(). Output are one array holding the unique knot values and one array with the corresponding multiplicities. This representation is conformant to the definition of the STEP B-spline entities.

Restrictions: This tool is not callable from FORTRAN.

COMPOSE_MULT_KNOTS

Category: basic; geometry conversion; C

Composes a knot vector with duplicated elements from a unique knot vector and a corresponding multiplicity array. This tool is complementary to fill_knots_mults(). Input to this routine are one array holding unique knot values and one array with the corresponding multiplicities and the size of both these arrays (same size!). This representation is conformant to the definition of the STEP B-spline entities. Output is the exploded knot vector, which might contain duplicated elements.

DEFAULT_MULT_KNOTS

Category: basic; geometry conversion; C

Derives a default knot vector with duplicated elements from other information of the B-spline entity (uniform, degree, number of control points). Input to this routine are an indicator on the type of knot set (non-uniform, uniform, quasi uniform, piecewise Bezier) and the degree of the B-spline entity and the number of its control points (= upper index of control points + 1). These parameters are conformant to the definition of the STEP B-spline entities. Output is a knot vector, which might contain duplicated elements.

RAT_DESCR

Category: basic; geometry conversion; C

Decides whether a B-spline entity definition is rational or non-rational. Input to this routine are the weights-array of a B-spline entity and the length of this array. The function returns value 1 if the given description belongs to a rational B-spline entity, else 0.

Restrictions: This tool is not callable from FORTRAN.

COMPOSE_RAT_DESCR

Category: basic; geometry conversion; C

Derives an array of homogeneous control points from other information of a B-spline entity (arrays of control points and weights, dimension). Input to this routine are an array with IDS-identifiers to instances of type cartesian point that are the control points of a B-spline entity; an array with weights for these control points; the number of control points respectively weights (identical in size); the dimensionality of the cartesian points (2- or 3-dimensional). Output is an array with IDS-identifiers to instances of type cartesian point that is homogeneous according to the given rational description. The routine does not access the IDS.

Restrictions: This tool is not callable from FORTRAN.

3.4.6.7 File CSIIS2.C - parameter conversion from IDS to SISL

This file contains routines especially designed for the conversion of B-spline curves and surfaces from the IDS- to the SISL-format utilizing new and STEP-conformant versions of SISL curve and surface constructors (newCurve2 and newSurface2). Among these routines the following ones are considered to be independent conversion tools.

CCISBC2

Category: parameter; geometry conversion; C

Converts a B-spline curve from IDS- to SISL-format. This routine has no corresponding identifier tool. Input to this routine is a set of parameters describing a STEP-conformant B-spline curve. The format of such a B-spline curve is based on ISO 10303-205 version 1. May 1992, the Surface Application Protocol. Differences between the AP-format and the format used in this routine are pointed out in the technical description of this tool (chapter 3.3.6). Output is a pointer to a SISL curve-structure. A similar tool (ccisbc()) exists that utilizes the old SISL curve-constructor newCurve. SISL is the SI Spline Library. As many of the Conversion Tools are based on the SISL format, this tool is of general interest. Only one instance can be converted at a time. The input instance will not be modified.

CCISBS2

Category: parameter; geometry conversion; C

Converts a B-spline surface from IDS- to SISL-format. This routine has no corresponding identifier tool. Input to this routine is a set of parameters describing a STEP-conformant B-spline surface. The format of such a B-spline surface is based on ISO 10303-205 version 1. May 1992, the Surface Application Protocol. Differences between the AP-format and the format used in this routine are pointed out in the technical description of this tool (chapter 3.3.6). Output is a pointer to a SISL surface-structure. A similar tool (ccisbs()) exists that utilizes the old SISL surface-constructor newSurface. SISL is the SI Spline Library. As many of the Conversion Tools are based on the SISL format, this tool is of general interest. Only one instance can be converted at a time. The input instance will not be modified.

3.4.6.8 File CSIISC.C - b-spline curves from IDS to SISL

This file contains routines especially designed for the conversion of B-spline curves from the IDS- to the SISL-format. Among these routines the following ones are considered to be independent conversion tools.

CCIISC

Category: identifier; geometry conversion; C

Converts a B-spline curve from IDS- to SISL-format. Input to this routine is an IDS-identifier to an instance of type B-spline curve. Output is a pointer to a SISL curve-structure. SISL is the SI Spline Library. As many of the Conversion Tools are based on the SISL format, this tool is of general interest. The routine additionally requires a flag indicating whether a rational curve shall be converted to a non-rational one or not. The dimensionality of the input curve must also be given to indicate whether the curve is 2- or 3-dimensional. Only one instance can be converted at a time. The input instance will not be modified or deleted. There is a parameter version of this tool called ccisbc().

Restrictions: The conversion from rational to non-rational is not part of this Library.

CCISBC

Category: parameter; geometry conversion; C

Converts a B-spline curve from IDS- to SISL-format. This routine is the corresponding parameter version to the identifier tool cciisc(). Input to this routine is a set of parameters describing a STEP-conformant B-spline curve. The format of such a B-spline curve is based on ISO 10303-205 version 1. May 1992, the Surface Application Protocol. Differences between the AP-format and the format used in this routine are pointed out in the technical description of this tool (chapter 3.3.6). Output is a pointer to a SISL curve-structure. A similar tool (ccisbc2()) exists that utilizes the new SISL curve-constructor newCurve2 (STEP-conformant). SISL is the SI Spline Library. As many of the Conversion Tools are based on the SISL format, this tool is of general interest. Only one instance can be converted at a time. The input instance will not be modified.

3.4.6.9 File CSIISS.C - b-spline surfaces from IDS to SISL

This file contains routines especially designed for the conversion of B-spline surfaces from the IDS- to the SISL-format. Among these routines the following ones are considered to be independent conversion tools.

CCIISS

Category: identifier; geometry conversion; C

Converts a B-spline surface from IDS- to SISL-format. Input to this routine is an IDS-identifier to an instance of type B-spline surface. Output is a pointer to a SISL surface-structure. SISL is the SI Spline Library. As many of the Conversion Tools are based on the SISL format, this tool is of general interest. The routine additionally requires a flag indicating whether a rational curve shall be converted to a non-rational one or not. Only one instance can be converted at a time. The input instance will not be modified or deleted. There is a parameter version of this tool called ccisbs().

Restrictions: The conversion from rational to non-rational is not part of this Library.

CCISBS

Category: parameter; geometry conversion; C

Converts a B-spline surface from IDS- to SISL-format. This routine is the corresponding parameter version to the identifier tool cciiss(). Input to this routine is a set of parameters describing a STEP-conformant B-spline surface. The format of such a B-spline surface is based on ISO 10303-205 version 1. May 1992, the Surface Application Protocol. Differences between the AP-format and the format used in this routine are pointed out in the technical description of this tool (chapter 3.3.6). Output is a pointer to a SISL surface-structure. A similar tool (ccisbs2()) exists that utilizes the new SISL surface-constructor newSurface2 (STEP-conformant). SISL is the SI Spline Library. As many of the Conversion Tools are based on the SISL format, this tool is of general interest. Only one instance can be converted at a time. The input instance will not be modified.

3.4.6.10 File CSILIN.C - trimmed line to b-spline curve

This file contains routines especially designed for the conversion of trimmed lines to B-spline curves. Among these routines the following ones are considered to be independent conversion tools.

CCILBC

Category: identifier; geometry conversion; C

Converts a trimmed line to a non-rational B-spline curve. Input to this routine is an IDS-identifier to an instance of type line. Output is an IDS-identifier to an instance of type B-spline curve. The routine additionally requires start- and end-point of the line. These might be given either as IDS-identifiers to instances of type cartesian point or as parameter values of the line. A flag shall indicate whether trimmed curve and line have the same sense. An absolute geometry resolution value (recommended: 0.1) is required input. A local, relative value might be calculated that overrules the given value. Only one instance can be converted at a time. The input instance will not be modified or deleted. Input and output instances will be totally independent from each other. There is a parameter version of this tool called cctlic().

CCTLIC

Category: parameter; geometry conversion; C

Converts a trimmed line to a non-rational B-spline curve. This routine is the corresponding parameter version to the identifier tool ccilbc(). Input to this routine is a set of parameters describing a STEP-conformant line. The format of such a line is based on ISO 10303-205 version 1. May 1992, the Surface Application Protocol. An additional sense-flag must be provided. An absolute geometry resolution value (recommended: 0.1) is required input. A local, relative value might be calculated that overrules the given value. Ouptut from this routine is a set of parameters describing a STEP-conformant B-spline curve. Also here the Surface Application Protocol, Part 205, and chapter 3.3.6 of this document may be consulted for clarifications. Only one instance can be converted at a time. The input instance will not be modified.

CLIPAC

Category: basic; geometry conversion; C

Converts a point on a line from parametric to cartesian. Input to this routine are parameters describing the line (position and direction) and the location of the point in the parameter space of the line. Output are the coordinates of the calculated cartesian point. Only one instance can be converted at a time. The input instance will not be modified.

Corresponding routines (ccipac(), celpac()) have been developed for converting points on circles and ellipses.

Restrictions: This tool does not distinguish trimmed and untrimmed lines.

3.4.6.11 File CSIPAR.C - b-spline curve to pcurve

This file contains routines especially designed for the conversion of 3-dimensional B-spline curves to pcurves, i.e. 2-dimensional B-spline curves in the parameter plane of a surface. Among these routines the following ones are considered to be independent conversion tools.

CCBBPC

Category: basic; geometry conversion; C

Converts a 3-dimensional B-spline curve to a 2-dimensional B-spline curve that is defined within the parameter space of a surface. Input to this routine is a pointer to a B-spline curve and a pointer to a B-spline surface that the resulting 2D-curve shall be defined in. Output is a pointer to this 2D-curve. A value for the maximum deviation allowed between true and approximated curve must be provided. Curves and surface are defined by SISL structures. SISL is the SI Spline Library. As many of the Conversion Tools are based on the SISL format, this tool is of general interest. Only one instance can be converted at a time. The input instance will not be modified.

A corresponding routine (ccbpbc()) as been developed for converting the other way: from a 2D-curve in the parameter plane of a B-spline surface to a 3D B-spline curve. This tool is not part of this library.

3.4.6.12 File CSIPNT.C - point conversions

This file contains routines especially designed for the conversion of point representations. Among these routines the following ones are considered to be independent conversion tools.

CCIPOC

Category: identifier; geometry conversion; C

Converts a point on curve to a cartesian point. Input to this routine is an IDS-identifier to an instance of type point on curve. Output is an IDS-identifier to an instance of type cartesian point. Only one instance can be converted at a time. The input instance will not be modified or deleted. Input and output instances will be totally independent from each other. There is a parameter version of this tool called ccpobc().

Restrictions: The routine can only handle points that are defined in the parameter space of B-spline curves; analytic curves are not supported.

CCPOBC

Category: parameter; geometry conversion; C

Converts a point on curve to a cartesian point. This routine is the corresponding parameter version to the identifier tool ccipoc(). Input to this routine is a parameter describing the location of the point on a B-spline curve. Else the STEP-conformant parameter set for the B-spline curve that is the definition space of the point is provided. The format of the B-spline curve is based on ISO 10303-205 version 1. May 1992, the Surface Application Protocol. Differences between the AP-format and the format used in this routine are pointed out in the technical description of this tool (chapter 3.3.6). Output are the three coordinates x, y, z of the point. Only one instance can be converted at a time. The input instance will not be modified.

Restrictions: The routine can only handle points that are defined in the parameter space of B-spline curves; analytic curves are not supported.

CCIPOS

Category: identifier; geometry conversion; C

Converts a point on surface to a cartesian point. Input to this routine is an IDS-identifier to an instance of type point on surface. Output is an IDS-identifier to an instance of type cartesian point. Only one instance can be converted at a time. The input instance will not be modified or deleted. Input and output instances will be totally independent from each other. There is a parameter version of this tool called ccpobs().

Restrictions: The routine can only handle points that are defined in the parameter space of B-spline surfaces; analytic surfaces are not supported.

CCPOBS

Category: parameter; geometry conversion; C

Converts a point on surface to a cartesian point. This routine is the corresponding parameter version to the identifier tool ccipos(). Input to this routine is a parameter describing the location of the point on a B-spline surface. Else the STEP-conformant parameter set for the B-spline surface that is the definition space of the point is provided. The format of the B-spline surface is based on ISO 10303-205 version 1. May 1992, the Surface Application Protocol. Differences between the AP-format and the format used in this routine are pointed out in the technical description of this tool (chapter 3.3.6). Output are the three coordinates x, y, z of the point. Only one instance can be converted at a time. The input instance will not be modified.

Restrictions: The routine can only handle points that are defined in the parameter space of B-spline surfaces; analytic surfaces are not supported.

3.4.6.13 File CSISIC.C - b-spline curves from SISL to IDS

This file contains routines especially designed for the conversion of B-spline curves from the SISL- to the IDS-format. Among these routines the following ones are considered to be independent conversion tools.

CCISIC

Category: identifier; geometry conversion; C

Converts a B-spline curve from SISL- to IDS-format. Input to this routine is a pointer to a SISL curve-structure. Output is an IDS-identifier to an instance of type B-spline curve. SISL is the SI Spline Library. As many of the Conversion Tools are based on the SISL format, this tool is of general interest. Only one instance can be converted at a time. The input instance will not be modified or deleted. There is a parameter version of this tool called ccsibc().

CCSIBC

Category: parameter; geometry conversion; C

Converts a B-spline curve from SISL- to IDS-format. This routine is the corresponding parameter version to the identifier tool ccisic(). Input to this routine is a pointer to a SISL curve-structure. SISL is the SI Spline Library. As many of the Conversion Tools are based on the SISL format, this tool is of general interest. Output is a set of parameters describing a STEP-conformant B-spline curve. The format of such a B-spline curve is based on ISO 10303-205 version 1. May 1992, the Surface Application Protocol. Differences between the AP-format and the format used in this routine are pointed out in the technical description of this tool (chapter 3.3.6). Only one instance can be converted at a time. The input instance will not be modified.

3.4.6.14 File CSISIS.C - b-spline surfaces from SISL to IDS

This file contains routines especially designed for the conversion of B-spline surfaces from the SISL- to the IDS-format. Among these routines the following ones are considered to be independent conversion tools.

CCISIS

Category: identifier; geometry conversion; C

Converts a B-spline surface from SISL- to IDS-format. Input to this routine is a pointer to a SISL surface-structure. Output is an IDS-identifier to an instance of type B-spline surface. SISL is the SI Spline Library. As many of the Conversion Tools are based on the SISL format, this tool is of general interest. Only one instance can be converted at a time. The input instance will not be modified or deleted. There is a parameter version of this tool called ccsibs().

CCSIBS

Category: parameter; geometry conversion; C

Converts a B-spline surface from SISL- to IDS-format. This routine is the corresponding parameter version to the identifier tool ccisis(). Input to this routine is a pointer to a SISL surface-structure. SISL is the SI Spline Library. As many of the Conversion Tools are based on the SISL format, this tool is of general interest. Output is a set of parameters describing a STEP-conformant B-spline surface. The format of such a B-spline surface is based on ISO 10303-205 version 1. May 1992, the Surface Application Protocol. Differences between the AP-format and the format used in this routine are pointed out in the technical description of this tool (chapter 3.3.6). Only one instance per time. Input instance will not be modified.

3.4.7 User's guide of GfS tools

General description. The Common Tools provided by GfS are to be used as building blocks for STEP data manipulation processors. Therefore their description belongs to the chapter "Conversion Tools". The GfS tools can be grouped as follows:

- B-spline conversions

- Utilities including model conversions

- Fortran IDS.

In each of the three groups several files are provided and each file contains one or more modules (routines). The distribution of modules to the files was done in a way that allows the user to include almost arbitrary subsets of this software in his processor. This building block approach was considered to be more efficient than "one routine per file".

The arguments have only two different types:

1. integer*4 / long
2. real*8 / double

All arguments are passed by address!

Most of the modules are written in Fortran, a few in C. No problems occurred so far from mixing binary code from both sources. No special argument or compiler flag is necessary for compiling either of them. For service purposes a debug flag may be set.

In the following subchapters only those modules are included that are called from outside.

3.4.7.1 B-spline conversions

Module bsmacu. This package generates b-spline curves surrounding a given b-spline surface. A built-in dialog lists all occurring b-spline surfaces and asks which should be processed. The user specifies them one by one or all of them. Then the calculation is performed and the generated b-spline curves are listed. Those are collected together with the surfaces in a geometric_3d_surface_set.

Module bspocu. This conversion creates polylines from b-spline curves. An interactive control is provided to let the user decide on which curves should be converted. Moreover the user can select the method:

- derive number of line segments from curvature (prescribed tolerance)
- prescribe number of line segments directly
- derive number of line segments from number of knots (a relation must be specified)

Such a decision can be taken for all those conversions or can be queried case by case.

The first alternative uses modules cobcpo and p19503 (by SI) for conversion, the others use intbs which in turn branches to pobscu (polynomial case) or ponurb (rational case). The latter decision is again up to the user. After converting a curve the edges or geometric sets using it are updated.

Some arrays should be reserved before calling bspocu, they should be sufficiently large to process one curve at a time.

Modules checkn, conins, knoins. When evaluating a b-spline curve or surface at a specific parameter (pair), routines are called requesting b-spline parameters in a particularly prepared way. These three routines deal with

- parameter values (ensure they lie in the right interval and don't coincide with a double knot)
- control points (add their coordinates to an array of triples)
- knots (expand knot list according to the given multiplicities)

These routines are normally called from within the conversion processors. Their parameters comply with the lists and arrays available there.

Module mebssu. With this package a b-spline surface can be broken into smaller faces. Each of them inherits the b-spline geometry, but with sufficiently small patches this becomes less important. The procedure in this module can be used to generate a finite element mesh or a display grid or just a finer subdivision. Topology is always arranged properly: Each b-spline surface results in a face_based_surface_model.

An interactive control is provided asking for the subdividing method:

- prescribing number of patches in either direction

- deriving number of patches from number of knots

The decision can be taken for all surfaces or for a single surface.

Each patch can be bounded either by 4 straight lines (only two points) or by 4 polylines with three points each (to be understood as representing a quadratic spline function) or by 4 polylines with four points each (to be understood as representing a cubic spline function). Also this control can be specified for one or for all surfaces.

This routine calls bsflae which in turn branches either to pobssu (polynomial case) or prbssu (rational case). On request this branching can be controlled by the user. Then an appropriate topology is generated by modules fltopo and/or femwri.

For this the user can select between

- storing a face_based_surface_model in the IDS

- writing FEM data to an ascii file (using a very simple format)

- both

The decision can be taken for the present surface or for all coming surfaces.

Modules ponurb, prbssu. The evaluation of rational b-spline curves or surfaces at specific parameter values is the task of these two modules. All b-spline parameters must be input: control points (as coordinates, three per point), knots (expanded), weights (optional). If no weights are supplied, a non-rational (polynomial) b-spline is assumed. Some scratch area must be provided for the development of the basic functions.

The result are the global coordinates of a point on the curve/surface at a given parameter (pair).

Modules dinurb, dinurs. To draw a b-spline curve normally a limited number of line segments is drawn. To draw a b-spline surface normally a grid consisting of a limited number of parameter lines consisting of a limited number of line segments is drawn. These two routines subdivide a b-spline curve into those line segments respectively generate parameter lines of a b-spline surface and subdivide them accordingly. The resulting line segments are passed to a function "gralin" that receives the three coordinates of the beginning point and the three coordinates of the ending point of the line segment. Gralin should be user-supplied and can do with the coordinates whatever might be useful, e.g. draw with them a line on the screen.

Modules pobscu, pobssu. The evaluation of non-rational (polynomial) b-spline curves or surfaces at specific parameter values is the task of these two modules. All b-spline parameters must be input: control points (as coordinates, three per point) and knots (expanded). Some scratch area must be provided for the development of the basic functions.

The result are the global coordinates of a point on the curve/surface at a given parameter (pair).

3.4.7.2 Utilities and model conversions

Module stypes. For all Fortran programmers there is a limited access to the entity type variables. In C programs the definitions written by yacc to the file y_tab.h are used. The module stypes reads this file and puts the integer numbers into a set of entity type variables. These integer variables are all in the common block /TYPES/ which is contained in the include file "type.inc". All program units needing access to the entity types should include this file. All type variables begin with the characters "it" followed by 2 to 4 characters defined by the entity identification in the put and get routines. For example, a b-spline_curve has the get routine cgbscu, the put routine cpbscu and the entity type itbscu.

stypes should be called once at the beginning of the program, then the entity types are available throughout the running time. The tree name (and its length) has to be supplied by the calling program. After having read the file, stypes calls tytest to count how many entity types have been preset. This is regarded an indicator for successful operation.

Modules enumev, enumco. For all Fortran programmers there is a limited access to the enumeration type variables. In C programs the definitions written by yacc to the enum.h are used. The modules enumev and enumco perform the transformation between enumeration strings and enumeration integers independently from enum.h. enumev is fed with an enumeration class and a string and returns the integer number associated with the respective enumeration type. enumco is fed with an enumeration class and an integer number and returns the enumeration type as string.

The following list of enumeration classes and types is implemented:

```
1) logical
     1 = .T.
     2 = .F.
     3 = .U.
2) ap_enum_class
     1 = .BREP.
     2 = .SF.
     3 = .WF.
     4 = .CSG.
```

```
   5 = .CBR.
   6 = .UNKNOWN_AP.
3) enumeration_curve1_pcurves1
   1 = .CURVE_1.
   2 = .PCURVE_S1
4) bspline_curve_form
   1 = LINE_SEGMENT.
   2 = .CIRCULAR_ARC.
   3 = .ELLIPTIC_ARC.
   3 = .PARABOLIC_ARC.
   4 = .HYPERBOLIC_ARC.
5) bspline_surface_form
   1 = .PLANE_SURF.
   2 = .CYLINDRICAL_SURF.
   3 = .CONICAL_SURF.
   4 = .SPHERICAL_SURF.
   5 = .TOROIDAL_SURF.
   6 = .SURF_OF_REVOLUTION.
   7 = .RULED_SURF.
   8 = .QUADRATIC_SURF.
   9 = .SURFACE_OF_LINEAR_EXTRUSION.
6) intersection_enumeration
   1 = .BASIS_CURVE.
   2 = .PCURVE_S1.
   3 = .PCURVE_S2.
7) uniform_type
   1 = .NON_UNIFORM_KNOTS.
   2 = .UNIFORM_KNOTS.
   3 = .QUASI_UNIFORM_KNOTS.
   4 = .PIECEWISE_BEZIER_KNOTS.
8) trimmed_curve_enumeration
   1 = .CARTESIAN.
   2 = .PARAMETER.
9) curve_transition_code
   1 = .DISCONTINUOUS.
   2 = .CONTINUOUS.
   3 = .CONT_SAME_GRADIENT.
   4 = .CONT_SAME_GRADIENT_SAME_CURVATURE.
10) make_or_buy
   1 = .MAKE.
   2 = .BUY.
   3 = .NOT_KNOWN.
```

11) part_status
   1 = .COMMITTEE_DRAFT.
   2 = .DRAFT_INTERNATIONAL_STANDARD.
   3 = .INTERNATIONAL_STANDARD.

Module checker. This module was provided to ease a call to the checker and to have interactive control over what to check and where to write the results. The user can select between a single entity (name must be specified) or all entities. The result can be written to standard output or to a file (name must be specified).

Module creawf. This routine collects all entities with a dimensionality of up to 1 into a connected_edge_set or into a geometric_3d_curve_set (depending on availability of topology in original model) and creates a wireframe model. The user can decide, wether he wants to delete all wireframe entities.

The AP flags are set properly and representation entities are generated accordingly.

Module crefac. This routine creates a face_based_surface_model from a shell_based_wireframe_model by generating faces on top of all edge_loops. The interactive control first informs about the number of edge loops; then the user can select:

- only add a face to each edge_loop
- add the faces, collect them in a connected_face_set and create a face_based_surface_model
- add faces, produce connected_edge_set and face_based_surface_model, additionally delete the wire_shell previously holding the edge_loops.

The AP flags are set properly and representation entities are generated accordingly.

Module crtopo. This routine creates topology on top of geometric sets. A parameter controls the mode:

- process only curves (and check for common vertices)
- process only surfaces (and compute bounding edges)
- process curves and surfaces (and check bounding edges)

Prior to calling this routine, the IDS should contain a geometric_3d_curve_set or a geometric_3d_surface_set. The models generated here have non-manifold topology associated with them. Depending on the selected mode a face_based_surface_model or a edge_based_wireframe_model is created.

The AP flags are set properly and representation entities are generated accordingly.

Modules delent, clean. These modules provide an interactive control to delete and undelete single entities or a whole tree of entities.

A call to clean removes all entities that are not referenced. By first deleting a top entity (leaving their child entities without reference) the user can delete whole models. This method saves entities that are used in other models.

User's control is done by integer numbers:

- positive numbers ask for deletion of the entity having as name this number
- negative numbers ask for undeletion of the entity having as name this number
- -1 asks for deletion of all unreferenced entities
- 0 is given to leave this module

Module headit. Here all header entities can be viewed and changed. The user can select

- to view the contents of all header entities
- to edit the contents of header entities
- to call a list of top entities (to see which models exist)

In edit mode each header entity is presented and can be accepted by not giving it a new content. It can be changed by entering a new content. Enclosing single quotes may or may not be typed; composite texts like "author" or "organisation" can be subdivided by two consecutive single quotes.

Module showld. After a call to cgwrld the calling program has a list of top entities itpo(1:ntop). This list is given to showld to display it with the entity types spelled out. The following 10 top models are recognised:

```
MANIFOLD_SOLID_BREP              SHELL_BASED_SURFACE_MODEL
FACE_BASED_SURFACE_MODEL         GEOMETRIC_3D_SURFACE_SET
SHELL_BASED_WIREFRAME_MODEL      EDGE_BASED_WIREFRAME_MODEL
GEOMETRIC_3D_CURVE_SET           CSG_SOLID
TOPOLOGY_BASED_COMPOUND_MODEL    BODY_BASED_SOLID_MODEL
```

Module interr. For service purposes it may be interesting, which entity has which parameters (has which child entities). Knowing the name of an entity of interest (e.g. a top entitiy), the user can ask for its parameters and subsequently for the entities referenced there. Additional information is provided with each listed entity: entity type, scoping, AP flag.

This module can be used for navigating through the IDS and for debugging models (e.g. if errors in the model prevent the formatter to write a correct file.

Module parsers. This module avoids a lot of overhead in the program unit that wants to call the parser. All actions for initialising, presetting, selecting the right parser routine are done here. Normally the debug flag should never be set, the write_log flag can be set to generate a log file where eventual errors in the STEP file are marked.

The appropriate parsers are automatically selected. The selection is controlled by the use_ap flag in the parse_results structure:

- use_ap = 10 -> initial value to call the header parser (reading data until a new AP flag occurs)
- use_ap = 1   -> the brep_parser is called (reading data until a new AP flag occurs)
- use_ap = 2   -> the surface_parser is called (reading data until a new AP flag occurs)
- use_ap = 3   -> the wireframe_parser is called (reading data until a new AP flag occurs)
- use_ap = 4   -> the csg_parser is called (reading data until a new AP flag occurs)
- use_ap = 5   -> the compound_brep_parser is called (reading data until a new AP flag occurs)

When end-of-file is reached and no new AP flag was set, this module finishes.

Module sortop. The so-called top entities of the CADEX project are in fact only shape models. With version 6.0 of the Common Tool Kit, representation entities (described in STEP Part 41) are required. They must contain at least one senseful information: The Application Protocol. As this information controls the parser on reading the file, it must not be omitted.

When creating STEP data from a native model or when converting one model into another, the IDS contains a (shape) top entity not being referenced from a block of representation entities. sortop recognises these "orphans" and generates all necessary representation entities with an appropriate AP information. When writing out these models, the formatter can produce a correct file.

No user action is necessary to ensure a correct function of this module, all information is taken from the IDS.

Module statis. This module only provides a comfortable call to the statistics viewer and avoids passing of C file pointers over Fortran statements.

Modules heakfk, eval, putdat. To link any parser to the IDS, an interface was provided that is fed by a stream of tokens (in this case provided by calls to KfK parser routines), sorts the parameters needed for a complete entitiy, and then calls an appropriate put routine of the IDS.

The routine heakfk processes the header section, the routine eval processes the data section. All data belonging to an entity are collected and passed to putdat. Dependent on the entity type, putdat branches to the right IDS put routine.

This interface works with every stream of tokens being in the correct order and containing no errors. Therefore just scanning (i.e. dividing into tokens) the STEP file is not enough. A parser is necessary to ensure the correctness of entities.

This module was developed when the CADEX parser was not yet ready, mainly because of insufficient coverage of Application Protocols. With version 5.3 of the Common Tool Kit this problem was abolished and the KfK parser was not used anymore. (Having exactly the same functionality, the CADEX parser is 5 times faster !) Therefore the changes to version 6.0 were not done in this interface. It can be updated, if necessary.

3.4.7.3 Fortran IDS

This package was developed when the CADEX IDS had only an insufficient coverage of Application Protocols. With version 5.3 of the Common Tool Kit this problem did no longer exist. Nevertheless this package was updated continuously (present version 6.1) and is fully compatible with all other tools and with the Application Protocols. Being written in Fortran it allows for a quick dump of the whole data base on a file and - at a later time - a quick recall of the whole database from the file. This feature can be used for interruption of long interactive work, for saving of intermediate stages, for saving of models that preliminarily don't comply with STEP and therefore cannot be written by the formatter nor read by the parser.

The Fortran IDS has almost the same functionality (and approximately the same performance) as the CADEX IDS written in C. The complete set of get and put routines was provided, but no modify routines. Basic list functions are included, while the list management functions are excluded.

In order to enable the use of both data bases at the same time, all routine names of the Fortran IDS beginn with the letter "f" instead of "c".

The tool idscfo was provided to swap all data from one data base to the other (and back again).

3.4.8 User's guide of FEGS tools

General remark. Neutral data standard go a long way to enabling data transfer between modelling systems. However, fundamental differences in the type of modelling system cannot be resolved by data standards alone. As long as systems are not completely conformant to all classes of data model there will be a need for conversion tools.

True to the spirit of CADEX the Conversion Tools Group set about finding ways of minimising and centralising this challenge in data transfer. A plan was set (conversion tools document CADEX90) to utilise the intermediate data structure (IDS) as the place to operate common conversion tools. This allowed conversion between alternative standard representations enabling different kind of systems to communicate between them.

Consequently the conversion tools group went on to specifying access routines needed to enable conversion tools to operate on the IDS. FEGS were given the task of adding all the modify routines for the wireframe, surface and solid model entities of the IDS. Since then it has been policy within CADEX that all entities should have a put routine, a get routine, and a modify routine. On top of this layer a set of high level access routines were added. These could perform operations such as search of the tree structure for parents and children of a particular type.

These tools form the basis on which the integrated conversion tools from FEGS Ltd can operate.

Based on this design, FEGS Ltd has concentrated on developing particular generic integrated conversion tools which operate directly on the IDS. Whilst the work done to date is only the beginning, the concept has now been proved and several working integrated conversion tools are available. All these routines are self-contained and follow a generic form which makes it easy for any pre- or post-processor developer to link in.

3.4.8.1 File cvbspl.c - Module for b-spline to polyline conversions

The file contains a number of functions which together will convert a b-spline curve to a polyline within the formats defined in ISO 10303-42.

B-splines are essentially bounded curves and may exist on their own or be bounded by a trimmed curve in geometric 3d curve or surface sets. In topology bound models all curves must be referenced by edges. Because the use of trimmed curves as well could lead to ambiguous information, trimmed curves are assumed not to exist in topology bound models.

Functions included are:

void cvbspl (top,criterion,parameter,gtol,istat)

  Main call routine for this conversion tool. It divides each b-spline into a polyline according to set criterion.

void cvedbs (criterion,parameter,gtol,ed_name,num_par,istat)

  Converts curves of type b-spline into curves of type polyline.

void cvtrbs (criterion,parameter,gtol,trcu_name,num_par,istat)

  Converts a trimmed b-spline curve into a trimmed polyline with n cartesian points.

void cvbspo (criterion,parameter,gtol,bscu_name,istat)

  Converts an un-trimmed b-spline curve into a un-trimmed polyline with n points.

3.4.8.2 File cvcici.c - Module for circle to circular arc conversions

The file contains a number of functions which together will convert a circle to a number of circular arcs all smaller than a predefined tolerance. Both input and output conform to the formats defined in ISO 10303-42.

Circles are essentially unbounded curves and must therefore be bounded by either a trimmed curve or an edge. Trimmed curves are used to trim unbounded curves in geometrical 3d curve and surface sets, while edges are used to trim unbounded curves in topology bound models.

Functions included are:

void cvcici (top,maxang,istat)

  Converts all circular arcs of angle greater than maxang into arcs smaller than maxang and corrects all referencing entities.

void cvedci (maxang,ed_name,istat)

  Converts edges of type circle which span more than maxang into smaller ones
  and adjusts the edge_loops or connected edge set that reference them.

void cvtrci (maxang,trcu_name,istat)

  Converts trimmed curves of type circle which span more than maxang into
  smaller ones and adjusts the geometric 3d curve set that references them.

void cvcirc (maxang, cu, cusense, start_pnt, end_pnt, ndiv, angle, centre, vector1,
binormal, istat)

  Calculates the number of divisions and angle required to split a STEP circular
  arc into circular arcs <= maxang .

void cipaca (coor, cu, param, pnt, istat)

  Converts a point on a circular curve from parametric to cartesian.

3.4.8.3 File cvelel.c - Module for ellipse to elliptic arc conversions

The file contains the functions necessary to convert an ellipse to a number of
elliptical arcs based on a tolerance criteria.

Ellipses are essentially unbounded curves and must therefore be bounded by
either a trimmed curve or an edge. Trimmed curves are used to trim unbounded
curves in geometrical 3d curve and surface sets, while edges are used to trim
unbounded curves in topology bound models

  Functions included are:

void cvelel (top,maxang,istat)

  Converts all elliptic arcs of angle greater than maxang into arcs smaller than
  maxang and corrects all referencing entities.

void cvedel (maxang,ed_name,istat)

  Splits edges referencing curves of type ellipse which span more than maxang and
  adjusts the edge loops or connected edge sets that reference them.

void cvtrel (maxang,trcu_name,istat)

  Converts trimmed curves of type ellipse which span more than maxang into
  smaller ones and adjusts the geometric 3d curve sets that references them.

void cvelip (maxang, cu, cusense, start_pnt, end_pnt, ndiv, starang, angle, centre,
major, minor, istat)

  Calculates the number of divisions and the angle required to split a STEP elliptic
  arc into elliptic arcs <= maxang .

void elpaca (coor, cu, param, pnt, istat)

  Converts a point on an elliptic curve from parametric to cartesian point.

3.4.8.4 File cvorax.c - Module for axis2 placements correction

The function corrects an ill defined axis2 placement definition to avoid numerical
problems when using it.

Despite specification in ISO 0303-42, STEP, it is still probable that axis sets will be ill defined, i.e. parallel axis. This causes great problems in coding as several trigonometric functions are partial to bomb out when using the ill defined results from this routine. Consequently I have set a precedence whereby if the axis are incorrect the z-axis is taken as the master axis and the x-axis is redefined to be legitimate. This follows the philosophy used in Part 42 function Build Axis.

Function included is:

void cvorax (ax2, centre, xaxis, yaxis, zaxis, istat)

Get the normalised axis directions for an axis2_placement. The axis2 placement may be an ill defined set numerically, despite of the definition in ISO 10303-42, STEP.

3.4.8.5 File cvplpl.c - Module for polyline to polyline conversion

The file contains functions to convert a polyline to another set of points used to define a polyline.

Polylines are essentially bounded curves and may exist on their own or be bounded by a trimmed curve in geometric 3d curve or surface sets. In topologically based models all curves must be referenced by edges. Because the use of trimmed curves as well could lead to ambiguous information, trimmed curves are assumed not to exist in topologically based models.

Functions included are:

void cvplpl (top,criterion,parameter,gtol,istat)

Divides each polyline into n new polylines according to set criterion.

void cvedpl (criterion,parameter,gtol,ed_name,istat)

Converts edges of type polyline into several edges with polylines in and adjusts the edge loops or connected edge sets that reference them.

void cvplcu (criterion,parameter,gtol,plcu_name,istat)

Splits a polyline into n trimmed polylines and adjusts the geometric 3d curve set that references it.

void cvtrpl (criterion,parameter,gtol,trcu_name,plcu_name,istat)

Converts trimmed curves of type polyline into several trimmed_curves of type polylines and adjusts the geometric 3d curve sets that reference them.

3.4.8.6 File cvspbs.c - Module for sphere to b-spline surface conversions

The file contains the functions necessary to convert an elementary surface of type sphere to a b-spline surface.

The generated b-spline sphere is a full rational b-spline closed in u and v. Consequently faces on the original sphere may end up lying across the seam of the generated b-spline. I hope this does not cause to much of a problem.

Functions included are:

void cvspbs (top,istat)

Converts all spheres below the top entity in the IDS to b-spline surfaces.

void cvsphere (sphere_name,istat)

Converts a sphere to a b-spline surface.

3.4.8.7 File cvsurf.c - Module for b-spline surface to point set conversions

This file contains c conversion function to conver a b-spline surface to point set.

Functions included are:

void cvsurf (top,criterion,parameter1,parameter2,gtol,istat)

Divides b-spline surface into patches according to set criterion.

void cvbsps (criterion,parameter1,parameter2,gtol,bssu_name,istat)

Converts an untrimmed B spline surface into a grid of n points, which can be dumped to a file for reading in a native form.

3.4.8.8 File cvtobs.c - Module for torus to b-spline surface conversions

The file contains the functions to convert an elementary surface of type torus into a b-spline surface.

The choosen b-spline torus is a full rational b-spline closed in u and v. Consequently faces on the original torus may end up lying across the seam of the generated b-spline. I hope this does not cause to much of a problem.

Functions included are:

void cvtobs (top,istat)

Converts all torii below the top entity in the IDS to b-spline surfaces

void cvtorus (torus_name,istat)

Converts a torus to a b-spline surface.

3.4.8.9 File cvtopo.c - Module for topological operators

This file contains topological operators for the IDS.

Functions included are:

void adedlo (elo_name, seg_name, nedges, edges, istat)

Replaces an edge with several edges in a loop in the correct order.

void adedces (ceds_name, seg_name, nedges, edges, istat)

Replaces an edge with several edges in a connected_edge_set.

void adcug3cs (g3cs_name, trcu_name, ncurves, trcu, istat)

Replaces a trimmed curve with several trimmed curves in a geometric 3d curve_set.

void adcug3ss (g3ss_name, trcu_name, ncurves, trcu, istat)

Replaces a trimmed curve with several trimmed curves in a geometric 3d surface_set.

3.4.8.10 File cvvect.h - Module for vector operators

This file contains a set of vector operators.

Functions included are:

void crossv (v1,v2,v3)

Calculates the cross product of two vectors of dimension 3.

void diffv (v1,v2,idim,v3)

Calculates the difference of two vectors.

void sumv (v1,v2,idim,v3)

Calculates the sum of two vectors.

double vdist (v1,v2,idim)

Calculates the distance between two vectors.

double dotv (v1,v2,idim)

Calculates the dot product of two vectors.

void scalev (s1,v1,idim,v2)

Scales a vector.

double magv (v1,idim)

Finds the magnitude of a vector.

void vnorm (v1,idim,v2)

Normalises a vector.

4. Development of Processors

Organisation of processor descriptions. This part of the report deals with the task "Development of STEP Pre- and Postprocessors" which is the biggest workpackage in the project. The CADEX partners represent ten different CAD sytems (including modelers of analysis systems) and have provided them with bidirectional STEP interfaces. The processors developed for this are described here. As the systems of the CADEX partners work with different kinds of data thus representing different applications, the idea of Application Protocols (basically subsets of the STEP resource model) was promoted by CADEX. The processors are always designed for one or several Application Protocols (APs). The partners' contributions in the following subchapters will state explicitly which APs are covered by their processors.

4.1 Det Norske Veritas Research

4.1.1 Introduction

DNV has developed processors for the finite element modeller PREFEM. PREFEM is a part of the analysis package SESAM.

PREFEM uses the geometry model for the creation of a finite element model. It is necessary to have geometry corresponding to different types of analysis. In STEP terms the necessary geometry corresponds roughly to solid models, surface models, wire frame models and combinations of these. The primary use of data exchange involving finite element modellers is the importing of models from CAD systems. This enables re-use of the already existing models. Often the exchange may be in order to use the powerful modelling capabilities of CAD systems for the modelling, and the specialised finite element modelling capabilities for that part.

The processors have been developed for a beta version where the SI APS package is used for sculptured surface representation. For this reason the surface AP has been in the focus of the processor development. The compound b-rep application protocol is the CADEX answer to the need for models which are needed in the finite element modellers. This application protocol has also been in focus in the processor development for DNV.

4.1.2 Processor architecture

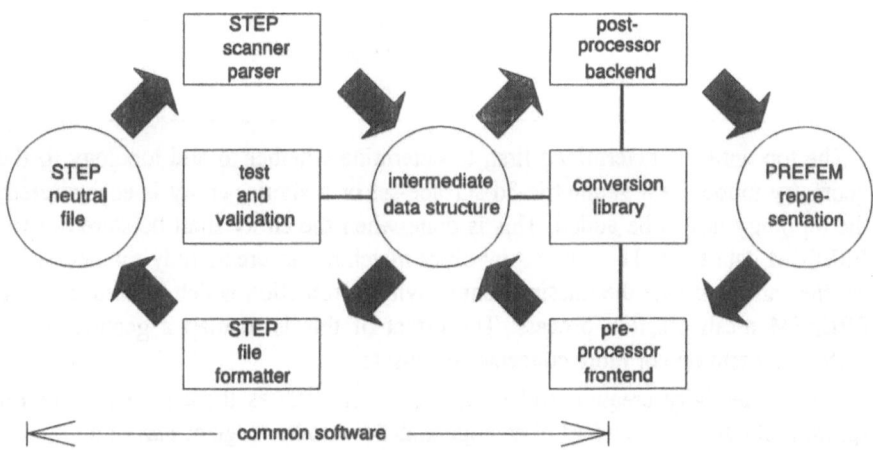

Fig. 4.1.1: Architecture of DnV processors

Figure 4.1.1 shows the overall architecture of the processors for PREFEM. The common tools takes care of the interface between the intermediate data structure and the STEP file. The pre-processor front-end translates from PREFEM data structure to IDS and the post-processor back-end translates from the IDS to PREFEM data structures.

4.1.3 Post-processor backend

The post-processor back-end identifies the top entities, traverses the IDS, and translates all information to PREFEM information. The mapping of STEP entities into PREFEM entities depend on whether topology is present or not.

```
----------------------------------------------------------------
STEP entity        PREFEM entity        PREFEM entity/entities
                   topology present     no topology present
----------------------------------------------------------------
Point              -                    -
                                        Vertex (future extension)
```

```
------------------------------------------------------------------
B-spline curve    B-spline curve    B-spline curve
                                    Vertex (at each end)
                                    Edge
------------------------------------------------------------------
B-spline surface  B-spline surface  B-spline surface
                                    Vertex (at each corner)
                                    B-spline curve
                                      (from the surface boundary)
                                    Edge (for each curve)
                                    Edge loop
                                    Face
------------------------------------------------------------------
```

The top entity is determined first, to determine whether to add topology to the geometry model. If a geometric-3d-surface-set or a similar entity is encountered, the topology has to be added. This is done when the entity shall be stored in the PREFEM data base. The process involves matching, to create only the necessary entities, and to create the missing connectivity information which is needed in the PREFEM mesh creation process. The effect of this is to map a geometric-3d-surface-set into one or more connected-face-sets.

The connectivity creation still needs improvements, as there is at present no splitting of edges when a part of an edge matches another edge or part of it.

4.1.4 Post-processor conversion

The geometries encountered can not be used directly in PREFEM. PREFEM is used for the creation of finite element models an approximative method. Approximations of curves and surfaces are thus sufficient for internal use in PREFEM. The only mathematical representation in PREFEM is polynomial b-spline curves. All other curves, including non-uniform rational b-spline curves (NURBs), have to be converted into PREFEM b-spline curves. This conversion is not available in the common tools of CADEX.

Analytical curves are also approximated as b-spline curves. In this case the conversion routines are the CADEX common conversion tools.

4.1.5 Pre-processor frontend

The pre-processors all creates connected sets. There are two calculations needed.

1. Calculation of connected set.

2. Calculation of correct logical structures.

The PREFEM model is used to create a set, from which the connectivity is checked. All entities that are connected are put into one set which is then processed by the PREFEM to IDS translator, and made as a connected set of some kind. The logical structure calculation is at present done for the edges, in which case it is essential also for PREFEM to have the information in the post-processing process.

There are topology in PREFEM which can not be mapped into even the compound b-rep application protocol. An example is a part of a cylinder. In PREFEM this is one surface and one loop. The loop have edges for the ends of the cylinder, but also one or more edges connecting the cylinder ends. The connecting edges are included more than once in the loop. These type of topologies can not currently present be translated to STEP.

4.1.6 Available processors.

Application protocol	Model Type	Pre Processor	Post Processor
Surface	Geometric 3d surface set	Yes	Yes
	Face based surface model	Yes	Yes
	Shell based surface model	No	No
Compound b-rep	Edge based wire frame model	Yes	Yes
	Face based surface model	Yes	Yes
	Body based solid model	Yes	Yes
	Topology based compound model	Yes	Yes
Wire-frame	Edge based wire frame model	Yes	Yes
	Shell based wire frame model	No	No
B-rep	Manifold solid brep	Yes	Yes

Note that the geometry coverage is not complete.

4.1.7 Future

The processors developed in the CADEX project are prototypes. This is necessarily so as STEP itself is not an official ISO standard at the end of the project. Commercial processors have to be developed when that is the case.

There will be other application protocols in addition the those developed in the CADEX project that will be important for DNV. These include finite element related application protocols, and ship and offshore application protocols.

4.1.8 Examples

The figures 4.1.2 and 4.1.3 show two models received from Siemens Nixdorf as surface level 1 models. The bottle in figure 4.1.2 has had topology added, and finite element mesh created. The finite element mesh is shown. The other model is shown with constant parameter curves of the b-spline surfaces.

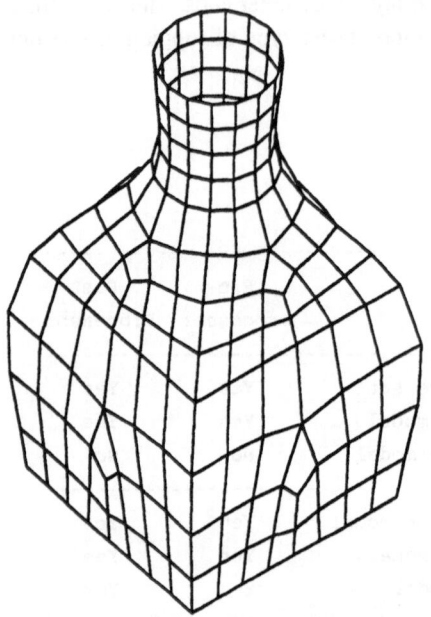 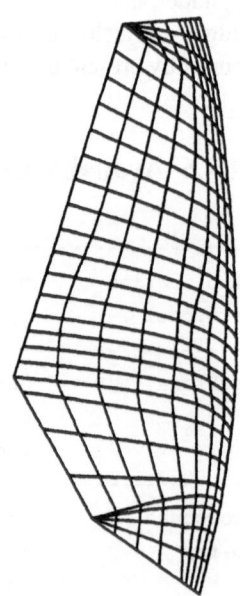

Fig. 4.1.2: Bottle, with topology added and finite element mesh generated

Fig. 4.1.3: Surf sail with b-spline parameter lines shown

4.2 DISEL

4.2.1 Introduction

DISEL has participated in the CADEX project generating processors for the CAD system CATIA. CATIA has been developed by DASSAULT SYSTEM and is distributed by IBM.

Since 1991 IBM markets a new workstation with RISC architecture and the new CATIA version for AIX operating system. Our initially foreseen development environment was changed from a 6150 workstation to the new one with AIX system. The new equipment allowed us to take advantage in a better way of the

CADEX common tools, and the use of C language, but we encountered significant problems in the new CATIA version.

This report is a description of the CATIA processors developed during this project.

4.2.2 Development environment

4.2.2.1 System description

DISEL's processors have been implemented in an IBM RISC System/6000 POWERstation 530 with AIX operating system.

We have used CAD system CATIA version 3 for AIX and we have worked with the following two layers:
1. Geometry and calculation comprising drafting, 3D design, advanced surfaces and solid modelling capabilities.
2. Interfaces with the CATIA base module.

Some specific development environment characteristic are required:
- AIX version 3.1.5 or later is essential to run the application; the processors could not run in a previous version.
- CATIA Version 3.2 or later.

4.2.2.2 CATIA geometry description

The geometric elements in CATIA can be defined in two ways:
1. an exact definition of elements which respect to certain number of constraints
2. an approximate definition using facets (solids and surfaces).

Exact definition. Mathematically, elements with an exact definition have a canonical or polynomial form:
- Canonical form: The elements are defined by their geometric characteristics; these elements are curves and simple surfaces such as circles, ellipses, parabolas, spheres and cylinders
- Polynomial form: The elements are defined in a precise and unique manner (one parameter for curves, two parameters for surfaces, and so forth); an exact volume can be defined using a topological restriction of surfaces.

Approximate definition. Elements with an approximate definition have a poly-hedral form. A polyhedron is an assembly of planar polygonal facets that are joined and form a surface which is not self-intersecting. This surface may be closed (polyhedral solid) or open (polyhedral surfaces).

These elements are the base to build up the geometric object in CATIA.

The CATIA geometry can be split in three levels:

1. Wireframe models: Wireframe design using points, lines and curves; two types of curves are considered:
   * analitycal curves
   * b-spline curves
2. Surface models: Surfaces are extensions of the wireframe geometry in surfaces, and faces are supported by the wireframe object; faces are topological restrictions of a surface limited by curves lying on the surfaces and forming one or more closed contours; the types of surfaces used are analytical surfaces, interpolated surfaces, evolute surfaces and net.
3. Solid models: Solids can be defined by using an approach based on facetted volumes or can be directly defined and then combined by means of boolean operations. Volumes constitute an intermediate step between surfaces and solids; they can be modeled in two ways: exact volume with a precise geometrical definition and facetted volume in which the geometric definition is aproximate.

4.2.2.3 CATIA interface description

CATIA incorporates an internal interface that allows the interrogation of the CATIA database for differents applications. Two of this modules have been used for our processors:
- CATGEO routines
- CATMSP routines

The CATGEO routines - CATIA Geometry Interface - are the interface for programming application used to read, write, delete and modify the CATIA data base. They are a set of routines written in FORTRAN but they can be linked into C programms. CATMSP (CATIA Mathematical Subroutine Package) provides the capability to perform operations on geometric elements (intersection, transformation, boolean operations...), analyses and basic mathematical operations.

4.2.3 Processors description

4.2.3.1 The processors that have been developed

The processors to be developed were stablised according to two characteristics of the CATIA system: the CATIA topology and the CATIA solid representation.

The CATIA topology. In CATIA, the topology is basically used to have an exact representation of a defined domain in two or three dimensions. Therefore, it is based on three single entities that, in fact, only have a geometric description. These entities are: edge, face and volume.

EDGE
An edge in CATIA is a curve on a surface. When a curve is projected onto a permanent surface, the program creates two elements: an isolate curve and an edge (polynomial definition linked both surface $(n,v) = f(w)$, where u and v are

the values of the surface parameters. The edge is however associated to the curve in the model.

FACE

A face is a topological restriction of a surface (or plane) limited by edges forming a closed contour.

VOLUME

A volume is a topological restriction of space, limited by adjacent faces. It offers the possibility of having holes in its domain.

CATIA solid representation. The solid representation used by CATIA is based on the method CSG. The solids normally obtained have two representation: an exact one through its CSG definition kept in a historical file and an approximate one that consists on a facetted description (type brep-facetted) of the limits of the solid, obtained through a polyhedric approach to the exact limits of the solid (its tolerance is controled by the user).

According to these characteristic, the processors that we have developed are:

Preprocessor:	CSG models	
	Brep models level 1.:	facetted brep
	Advanced surfaces level 2:	geometric set
		face based models
Postprocessor:	CSG models	
	Brep models level 1:	facetted brep
	Advanced surfaces level 2:	geometric set
		face based models
	Wireframe models level 1.	

4.2.3.2 Processors architecture

The architecture of the processors has been designed integrating the CADEX common tools developed in the project: IDS, scanner/parser, formatter and IDS data checker (figure 4.2.1).

Preprocessor and postprocessor for STEP data transfer are considered two separate programs.

The operations during the preprocessing are:

- identify the models to be sent and open the corresponding working files

- analysis of entities; test the conformance of the entity to a particular STEP entity

- conversion of entities (if it is required) using CATMSP routines or our local conversion library routines,
  and once the entities are STEP conform

- write the entities into the IDS and execute the formatter to produce the STEP file

- close all files.

The operations during the postprocessing are:

- identify the STEP file to be sent and open the corresponding working files

- the scanner/parser is executed and the IDS is created
- use the corresponding IDS routines to read STEP entities
- analysis of entities, test the conformance of the STEP entity to a particular native-entity
- conversion of entities using CATMSP routines or the local conversion library routines
- write the entities in native format into the native database by means of CATGEO routines, and
- close all files.

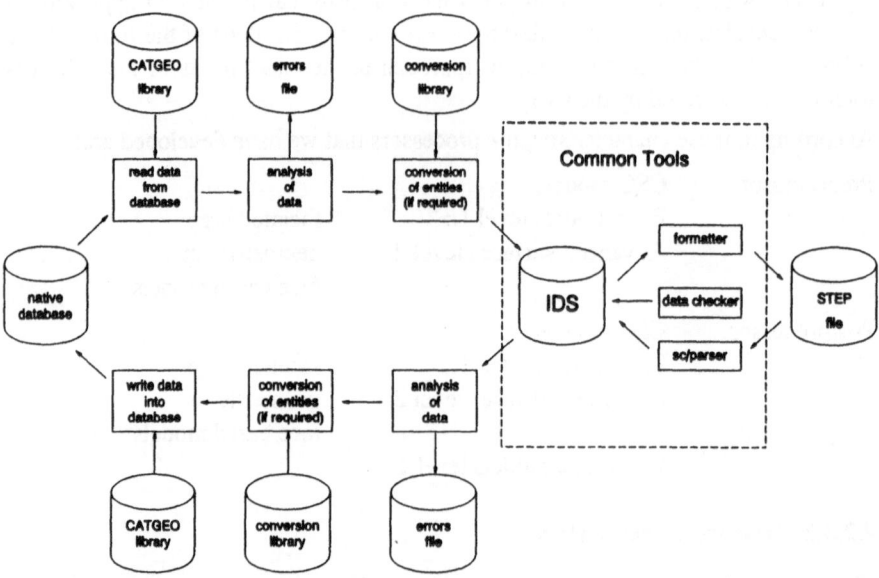

Fig. 4.2.1: DISEL processor architecture

4.2.3.3 Functionality of the processors

The Application Protocol defined in the project provided us the entities to be covered in each geometric models.

The documents ISO 10303 Part 204, 205 and 206 and the CSG Applications Protocol developed by CADEX Project have been used as base to establish the funtionality of the processors.

4.2.3.3.1 CSG processors

The functionality of the CSG processor is described in the following list:

ENTITY	PREPRO	POSTPRO
CSG_SOLID	X	X
UNION	X	X
INTERSECTION	X	X
DIFFERENCE	X	X
SPHERE	X	X
RIGHT_CIRCULAR_CYLINDER	X	X
RIGHT_CIRCULAR_CONE	X	X
TORUS	X	X
RIGHT_ANGULAR_WEDGE		X
BLOCK	X	X
SOLID_OF_REVOLUTION	X	X
SOLID_OF_LINGAR_EXTRUSION	X	X
HALF_SPACE		X
BOX_DOMAIN		X
SOLID_REPLICA	X	X

The entities RIGHT_ANGULAR_WEDGE, HALF_SPACE and BOX_DOMAIN do not exist in CATIA. The entity SOLID_REPLICA (not provided in processor planning) was included in the final version.

4.2.3.3.2 Wireframe processor

For wireframe models only a postprocessor was developed.

The entities covered are:

    GEOMETRIC_3D_CURVE_SET
    CARTESIAN_POINT
    POINT_ON_CURVE
    DIRECTION
    AXIS2_PLACEMENT
    CARTESIAN_TRANSFORMATION
    LINE
    CIRCLE
    ELLIPSE
    HYPERBOLA
    PARABOLA
    POLYLINE
    BSPLINE_CURVE
    TRIMMED_CURVE
    COMPOSIVE_CURVE
    D3_OFFSET_CURVE
    CURVE_REPLICA

4.2.3.3.3 BREP processors

The functionality of the brep processor is described in the following list:

ENTITY	PREPRO	POSTPRO
FACETTED_BREP	X	X
SHELL	X	X
FACE	X	X
POLY_LOOP	X	X
CARTESIAN_POINT	X	X

4.2.3.3.4 Advanced surfaces

The functionality of the surfaces processors is describe in the following list:

ENTITY	PREPRO	POSTPRO
GEOMETRIC_3D_SURFACE_SET	X	X
B_SPLINE_CURVE	X	X
B_SPLINE_SURFACE	X	X
CARTESIAN_POINT	X	X
DIRECTION	X	X
FACE_BASED_SURFACE_MODEL	X	X
POLYLINE	X	
CIRCLE	X	
ELLIPSE	X	
PARABOLA	X	
HYPERBOLA	X	
PLANE	X	
CYLINDRICAL_SURFACE	X	
CONICAL_SURFACE	X	
SPHERICAL_SURFACE	X	
TOROIDAL_SURFACE	X	
CONNECTED_FACE_SET	X	X
FACE	X	X
LOOP_LOGICAL_STRUCT	X	X
VERTEX_LOOP	X	X
VERTEX	X	X
EDGE_LOOP	X	X
EDGE_LOGICAL_STRUCT	X	X
EDGE	X	X
SURFACE_LOGICAL_STRUCT	X	X

4.2.3.4 Error processing

The error that could occur during the execution of CATIA processors can have four different sources:

1) Errors caused in the routines of IDS or FORMATTER. When an error occurs the operating system takes the control and a short message is displayed with the characteristics of the error.

2) Errors caused in the SCANNER/PARSER software when it is testing (syntactically and semantically) the physical file. An error occurs when the physical file is incorrect (it is not in accordance with the AP specifications); that returns the control to the operating system and provides an error file whose name is made from the physical file name plus an extension ".l".

3) Errors caused in the routines specific to the CAD system (CATGEO) that return the control to the operative system and show a display with the CATIA code and description of the error.

4) Errors caused in the specific processor software, that returns the control to the operating system, show a display with the description of the error occurred and provide a file whose name is made from the physical file name plus an extension ".p", and that collects all this type of errors occured during the processing.

4.2.3.5 Testing Results

The processors have been tested with all models availables in the library provided by FIAT. Two significant problems have been found:

- Different accuracy in two differents systems could cause problems with tolerances.
- The system could have big problems generating the IDS if the models are composed by a great number of entities (more than 50000 entites make problems).

4.3 FEGS Ltd

4.3.1 Introduction

The development of processors is based on the use of an extensive collection of basic software modules, the common tools software. This software is developed around a system architecture with a central data repository, the intermediate data structure, IDS, an in core C data structure which can hold all the entities defined in the integrated resources in STEP, parts in the 40 series and 100 series. The common tools software includes the scanner/parser, the formatter, and the IDS itself. Access routines to modify the content on the IDS have been developed, together with conversion tools, verification tools, statistical tools, etc.

The application protocols defined in the project represent a grouping mechanism for entities to be used in the transfer of particular model classes between dissimilar

CAD and FEM systems. The application protocols, APs, are supported by the scanner/parser, formatter and IDS on entity level, ie each entity included in the different APs must be specifically implemented in the system.

This common tool kit is used in the development of the processors in FEGS Ltd.

4.3.2 The status of FAM

FEGS Ltd market an analysis modeller for field analysis called FAM. This is a traditional mesh generator with a restricted topological coverage, but with the extended functionality to hold non-manifold topology models. FEGS Ltd has seen the increase in user awareness of the benefits of geometry modelling in a CAD environment, and reacted to this by embarking on a major program for geometry interfacing using STEP for data exchange.

The target is to receive, manipulate and hold the models engineers make in their CAD systems, maintaining the mathematical representation of the model as well as the topology. This can only be achieved by extending the geometric and topological coverage of the FAM analysis modeller in parallel with the development of STEP data exchange technology.

The functionality of FAM at the outset of the project included the ability to hold one or more wire frame, surface and solid parts as individual models or combined in the same model. A special case is a collection of boundary representation solid models which share a common face, edge or vertex.

The models that can be transferred between FAM and other systems are limited in some respects, but comprehensive in others. The functionality added to FAM in parallel with the development in the CADEX project includes the introduction of NURBS curves and surfaces and experimental extensions to the topology entities towards general definitions as found in STEP Part 42. The functionality of the prototype set of STEP processors will reflect this fact:

- The manifold solid b-rep models originating in FAM will be rather simple, but complete. The manifold solid b-rep models FAM can receive and hold are presently limited to the domain of meshable topology. The manifold solid b-rep models originating from other systems are generally more complex than this. It is therefore necessary in some cases to reduce manifold solid b-rep models into lower topology levels e.g. a face based surface model. As a result of this reduction these models will not pass a loop test.

In order to compensate for this data degeneration, it is possible to utilise special grouping mechanisms in FAM to hold the non-meshable topology. This option has been used in the first implementations of solid model transfer.

- FAM is able to hold compound boundary representation models. This however is not a feature of two-manifold modelling systems. Therefore collections of meshable topologies may be sent between systems for analysis modelling, but not be exchanged with traditional solid modellers. It is hoped that this restriction will be lifted when analysis modellers and CAD solid modellers are based on the same technology.

- circle tests can only be performed on models covered by the topological domain in FAM

A similar situation exists with geometry, as the coverage there is limited to wireframe geometry and elementary surfaces, planes, cylinders, cones, and spheres. However, the entities describing free form curves and surfaces have been introduced into FAM. This development has enabled transfer of sculptured surface models according to part 205 and Part 204 level 3.

4.3.3 The processors that have been developed

FEGS Ltd has developed processors for the transfer of CAD models between STEP external files and the FAM data base. The processors were developed in two stages. This is indicated in the table below. The following application protocols are implemented:

processor type:	pre	post
the boundary representation application protocol, level 2	x	x
the boundary representation application protocol, level 3	x	x
the surface model application protocol, functional level 1	x	x
the surface model application protocol, functional level 2	x	x
the surface model application protocol, functional level 3	x	x
the wire frame application protocol, functional level 1	x	x
the wire frame application protocol, functional level 2	x	x
the wire frame application protocol, functional level 3	x	x
the compound boundary representation AP, functional level 1	x	x
the compound boundary representation AP, functional level 2	x	x
the compound boundary representation AP, functional level 3	x	x
the compound boundary representation AP, functional level 4	x	x

The entire set of processors has been updated following the completion of upgrading the common tools software to support the latest versions of resource models and the current versions of the ISO APs.

The processor implementation in a many-to-many implementation, allowing conversion between the different functional levels in the different APs in both the pre-processor and the post-processor. The conversions are concentrated in the transfer between the IDS and FAM data base, and the transfer between the FAM data base and the IDS. However, some of the conversions are made directly on the IDS to enable alternative model representations to be sent to and from different systems conforming to different APs.

4.3.4 Requirements to the processors

The functional requirements to the processors are:
- the transfer between STEP and FAM should be symmetrical, ie the model content should be the same before and after the transfer into FAM or the STEP physical file
- the entity coverage should be 100% conformance
- if model information is degenerated, the resulting model should conform to another AP
- the accuracy of the model should be maintained or improved during the transfer

These requirements could not all be met up front, but were met towards the end of the project.

4.3.5 System architecture

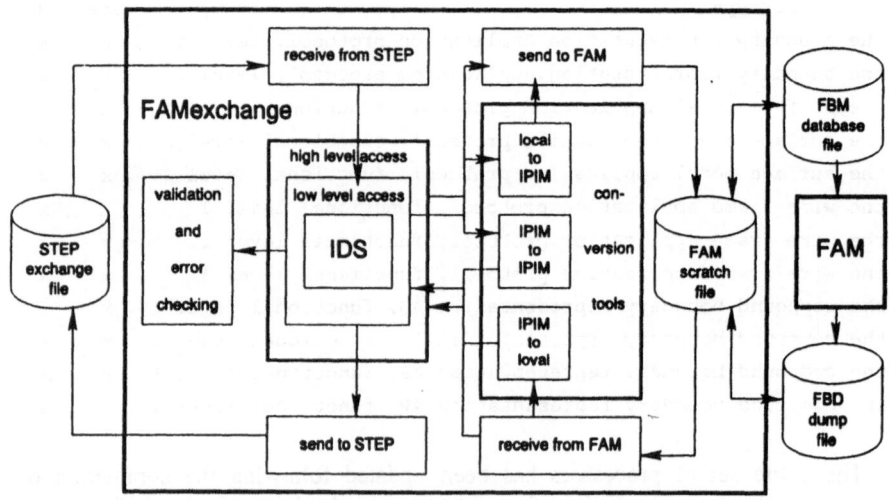

Fig. 4.3.1: The architecture of FAMexchange

The pre- and post-processors for STEP data transfer will form a separate program from the main product FAM, and are incorporated into a single program called FAMexchange in the prototyping phase. All the pre- and post-processors will be operated from within the same program, as all will work on the same data structures and be operated in a similar way. The system architecture is given in figure 4.3.1.

The processing functions developed in the project have been integrated within an infrastructure called the FAM shell. This is the command processor, the graphics, the in core data structures and the data base used in FAM. The system is based on the concept developed in the project, having a data structure between the physical file and the native database for storing the model during transfer both to and from the physical file. This data structure will co-exist with the internal data structure in FAM during the processing session. The chosen architecture preserves the functionality of FAM and includes the prototype in a family of programs with the same look and feel. The software will be as portable as any of the other modules in FAM and can be used on any computer where FAM is used today, including the VAX range, UNIX machines, etc, utilising the X-windows graphics implemented in FAM. The only restriction imposed at the moment is the need for a C compiler during porting, bearing in mind that the communication between FORTRAN and C may be different on different machines.

4.3.6 User interface and software operation

The FAMexchange program is an interactive, graphical program. The user can run it as any other modules in the FAM suite, maintaining the same flexibility as with any of the other programs in FAM.

The user communicates with the system through the standard FAM user interface. A number of the traditional FAM commands are available in FAMexchange, enabling a user to manipulate and view the model as part of the transfer to and from STEP.

The user will be able to run a complete transfer session from within the FAMexchange program. The operations during pre-processing include:
- to identify the model or parts of a model that should be sent
- to extract those entities from the native data structure, convert the information to STEP entities and store them in IDS
- to convert the entities within the IDS to suite a particular application protocol or receiving system
- to test for conformance to a particular AP, run statistics and whatever
- to write the model into the physical file

The operations during the post-processing are:
- to read the physical file
- to test for conformance to a particular AP, run statistics or whatever
- to convert entities to other entities from the Integrated Resources to suit FAM as the receiving system
- to extract the entities from the IDS, convert the information to local entities and store them in the native data base.

The basic processing commands are shown in figure 4.3.2.

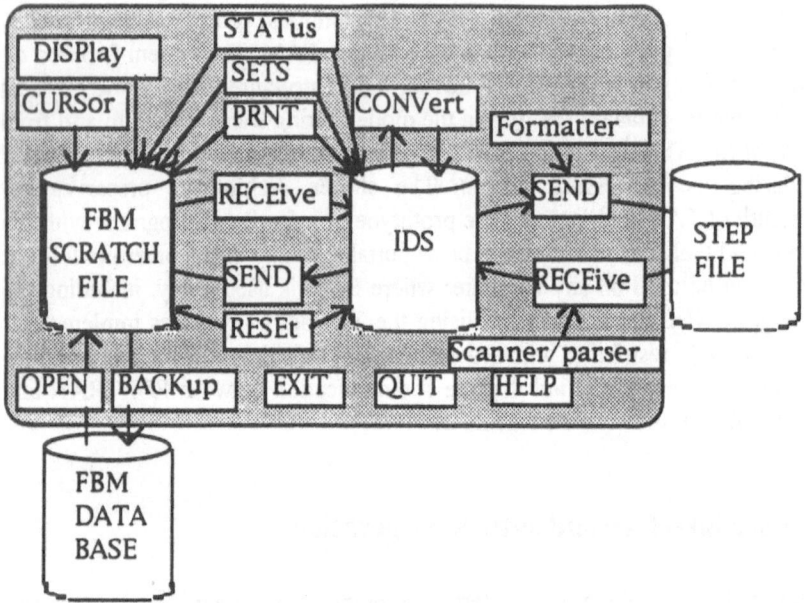

Fig. 4.3.2: The commands of FAMexchange

4.3.7 Transfer between IDS and FAM

The transfer between the IDS and the FAM data base will be done during the interactive session. The transfer is a read and a write operation between two different in-core data structures, one in FORTRAN and one in C. As part of the transfer, the conversion between the native representations and the STEP definitions will take place. The transfer to the native data base will be controlled by the Receive command, while the transfer to the IDS will be controlled by the Send command.

These operations will be the same for all model types, independent of the AP.

4.3.8 Conversions between IDS and FAM

The transfer from the FAM native data base to IDS includes a conversion process as well. The user specifies what AP and level the model should conform to as part of the transfer from FAM to IDS. In the process, the entities are converted according to the AP rules and restrictions that exist for the specified AP.

The conversion included in the transfer between the IDS and FAM includes the mapping from FAM entity format to STEP and vice versa. The mapping of FAM geometry onto STEP entities will be as follows:

- the straight line will be an unbounded curve with associated topology
- the arc will be an unbounded curve with associated topology
- the intersection curve will be sent to IDS as a polyline
- the conic arcs do not exist in FAM explicitly, but will be held as a special case of intersection curves, and they will be sent to the IDS as conics
- the elementary surfaces will be elementary surfaces
- the NURBS curve will be a NURBS curve
- the NURBS surface will be a NURBS surface

The mapping from STEP to FAM entities will be as follows:

- the line will be a bounded line, ie topology will be used to bound the entity
- the circular arc will be one or more circular arcs
- the intersection curve will be an intersection curve
- the polyline will be a intersection curve
- the conic arc will be a special case intersection curve
- the elementary surfaces will be elementary surfaces
- the NURBS curve will be a NURBS curve
- the NURBS surface will be a NURBS surface

It is worth noting that FAM cannot hold conic arcs explicitly, but will be able to hold them as intersection curves. The post-processing should therefore include a conversion of them into intersection curves, where the elementary surfaces are defined to be a plane and a cone. This will work well during post-processing, but there will be an overlap between genuine intersection curves between a plane and a cone during pre-processing. This will violate the requirement for symmetrical conversions, but is an acceptable solution, as long as the model improves:

- the intersection between a plane and a cone is a genuine conic arc, but may be represented as a polyline (not a intersection curve in the b-rep AP) in the model
- the post-processing will create an intersection curve in FAM
- the pre-processing will create a conic arc in the model stored in IDS
- the model will now have a conic where there was a polyline in the past

This will work well, as the accuracy is maintained by the intersecting elementary surfaces and not by the curve representations themselves.

Topology will be used in the definition of the bounded FAM entities and included as implicit information in the FAM model. When the model is converted from FAM to IDS, the information will be separated into the different STEP entities.

4.3.9 Conversions between STEP entities on IDS

The implementations are based on the fact that a traditional FEM analysis modeller like FAM does not have the functionality to receive and hold a general boundary representation solid model or a surface model.

The models received into the IDS will conform to a particular AP, but still be incompatible with the data formats used in the receiving system. A conversion between alternative representations within the STEP entities is possible directly on IDS. This will allow model conversions between APs and enable transfer between systems of different kinds.

The conversions on IDS will typically be mapping between alternative geometry representations from within STEP. The mapping is mostly between alternative approximations used in different contexts.

Examples of conversions on the IDS:

- the breaking up of circular arcs into arc segments smaller than 180 degrees
- the breaking up of ellipses into ellipse segments
- the breaking up of a polyline into two or more polylines
- the conversion of conics to intersection curves
- the conversion of NURBS curves to polylines
- the conversion of elementary surfaces to NURBS surfaces

The transformation into alternative representations and data degeneration is caused by incompatibilities between the model representations in the CAD systems in question. This is a prevailing problem in the transfer between CAD and FEM systems in particular. The model degeneration that takes place during transfer from a CAD solid modeller to an FEM analysis modeller, reduces the content of the model to such an extent that the transfer back is made impossible. For FAM, the models conforming to the b-rep AP must be represented as a surface or a wire frame model. For surface models, elementary surfaces and NURBS surfaces can be represented exactly.

There is also a possibility to reverse the geometry conversions, by identifying what approximations together form a complete entity. With the development of these conversion tools, the transfer of models into a wire frame representation and back will be possible. However, the critical conversion is on topology, as there is no general method for automatic re-introduction of face and shell topology in a wire frame model. Certain special cases can be handled with some degree of success, but user interaction is necessary to re-introduce topology into a wire frame which initially was a solid model. The development of conversion tools for two way transfer will therefore not lead to a complete solution.

4.3.10 The support of the boundary-representation AP, ISO 10303-204

The AP for the data transfer of STEP boundary representation solid models via the physical file is based on unbounded geometry and a separate and complete topology for solid models. There is no provision for alternative representations.

The geometry is limited to elementary surfaces and the curves that will be intersections between these surfaces. The intersection curve is represented as an explicit curve in 3D space and the intersection curve entity in Part 42 is not used. If a system requires the analytical form of the intersection curve, it may be calculated by finding the underlying surfaces through the topology relationships. This implicit representation of the intersection curve reduces the space requirement in the transfer and ensures that the geometric accuracy is carried by the surfaces.

FAM has an intersection curve similar to the definition in Part 42, and a conversion will be necessary to support the boundary representation AP. This conversion will be made on the IDS.

The geometric entities in Part 204 functional level 2, analytical b-rep, and level 3, advanced b-rep, are supported in full, ie 100% conformance, with the exception of the torus. The torus has been implemented as a NURBS surface in FAM while an implementation of a surface of rotation is planned.

The topological entities in the boundary representation AP are supported in full, with a restriction to the complexity of the face, the closed shell and the manifold solid b-rep entities. The current version of FAM restricts the face, the closed shell and manifold solid b-rep entities to be the entities used for meshable topology which is restricted to three or four sided faces and five or six faced manifold solid b-rep. As long as the boundary representation models satisfy those criteria, a complete transfer will be possible.

Most b-rep models will be more complicated, however. If the model is more complicated in face topology than can be held in FAM, the face, the closed shell and manifold solid b-rep entities will be lost, storing the relationships in a connected edge set and a topological entity similar to the geometric 3D curve set and the geometric 3D surface set respectively. If the model is more complicated at the solid level but not at the face level than can be held in FAM, the closed shell entities will be lost, storing the relationships in a connected face set. Any attempt to capture partly the topology may be useful, but will not represent a significant improvement, as the received topology must be replaced by meshable topology anyway.

The sending of a solid model using the boundary representation AP has to obey the same rules. If the solid model in FAM is a closed shell, there will be no problems. If the model has lost its topology, it cannot be transferred back, as the boundary representation AP does not provide for the use of alternative representations, as do both the surface AP and the wire frame AP. An analysis modeller cannot take part in a closed loop test because the topology of the model will disappear.

4.3.11 The support of the surface AP, ISO 10303-205

The surface AP has three functional levels where the difference is purely topological. The first level is without topology, while level 2 accepts non-manifold topology and level 3 is restricted to manifold topology.

The implementation in FAM supports the geometry entities in all levels, while the topology is supported within the limitations found in FAM, as described for the b-rep AP support.

When receiving surface models, the information is mapped onto the NURBS entities in FAM. For geometry bound surface models this is sufficient for complete model transfer. For topology bound surface models and shell based surface models the topology must also be transferred. If the topology is within the range FAM can handle, the topology is mapped directly to FAM faces. If the topology is too complex, the information is mapped to a set mechanism, allowing the relationship to exist, but not as a face.

When a surface model is sent out of FAM to a STEP file, the geometry is mapped directly to the respective geometry entities. No attempt is made to recreate topology which is lost in the receiving process at this stage. All models are therefore degenerated to geometry bounded surface models.

4.3.12 The support of the wire frame AP, ISO 10303-206

The wire frame AP has three functional levels where the difference is purely topological. The first level is without topology, while level 2 accepts non-manifold topology and level 3 is restricted to manifold topology.

The implementation in FAM supports the geometry entities in all levels, while the topology is supported within the limitations found in FAM, as described for the b-rep AP support.

When receiving models from the STEP file into IDS, the geometric information is mapped directly to the equivalent FAM entities. FAM maintains topology on wire frame level and the geometry bounded wire frame model will be automatically upgraded to a connected edge set. The edge based wire frame model will map directly to the representation in FAM, while the shell based wire frame model will create face definitions for all edge loops in the model that are compatible with a FAM face.

When sending models to the STEP file from FAM, a model is converted to the format which the users selects or the system finds most appropriate for the model at hand. The wire frame model in FAM has topology by default, and a geometry bounded wire frame model will loose information which is already there. A topology bound wire frame model will go directly across to the STEP file, while a shell based wire frame model can only be created if it is a manifold model with faces defined. The edge loops in the shell definitions represent faces without underlying surfaces and are therefore a simplified representation which lends itself to interpolating surface definitions like Coon's blend.

4.3.13 The support of the compound b-rep AP

The compound b-rep AP has four functional levels where the difference is purely topological. The levels are wire frame models, surface models, solid models and topologically mixed models.

The implementation in FAM supports the geometry entities in all levels, while the topology is supported within the limitations found in FAM, as described for the b-rep AP support.

The reading and writing of STEP files follow the same patterns as described for the other three APs.

If the solid model is a collection of solid models connected together (a cellular model), the individual cells may be re-defined as a separate b-rep and the model sent as a sequence of independent two manifold solid b-rep models. The receiving system must be able to put them together again to make this kind of transfer worthwhile. The boundary representation AP should not be used to send cellular models. The compound boundary representation AP has been developed for this purpose.

This situation is not specific to STEP, but the status for transfer of solid models into an analysis modeller using any file format. The problem is the incompatibility between the two technologies solid modelling and analysis modelling and can only be properly solved when the systems can communicate the same kind of information.

Successful transfer can be demonstrated within the common domain the systems can cover, e.g. the union of their functionality.

4.3.14 Current status

The development of FAMexchange has progressed to the point where a FAM model can be viewed in the graphical environment. The scanner/parser, formatter and IDS are all operational, and an integration with the defined transfers and conversions is possible for wireframe models, surface models, boundary representation models and compound boundary representation models. The existing software, FAMexchange, is a prototype from which a product can be developed.

4.3.15 The experience gained in the project

The AP for the data transfer of STEP wire frame models via the physical file is specified to satisfy the following criteria:

- to cover the same entities as existing file formats
- to cover the entities used in the APs defined in CADEX in a complementary way
- to cover the entities used in CAD and FEM systems interested in wire frame transfer
- to make use of the improved functionality that STEP can offer

The pre- and post-processors for FAM demonstrate that STEP can be used to achieve a higher level of model integrity than existing wire frame transfers based on IGES and VDA/FS. So far, the circle tests on FAM that have been performed show that the model content can be maintained 100% during the transfer.

The transfer of surface models is successful only when the implementation of b-spline curves and surfaces is completed. So far, the transfer of surface models to FAM will result in a collection of points, as all the control points for the curves and surfaces are translated.

An experimental mapping from the b-spline curve to a parametric spline in FAM has shown that it is possible to get an approximation across, but also that substantial conversion tools are required to have a complete mapping.

The problems of data exchange between dissimilar systems ought to be solved in a different way, e.g. by making the systems compatible. This is now going on in FEGS Ltd. The knowledge of the model representations used in STEP has enabled FEGS to specify entity definitions that are compatible with traditional CAD systems, reducing the need for complex conversion tools.

4.3.16 Future plans, FAM and FAMexchange

The ability to receive and send models based on STEP is determined by the functionality of the FEM analysis modeller. The development of FAM as a product will result in a number of changes, both in topology and geometry coverage.

It has been realised that STEP represents the accumulated knowledge of how different entities used to describe geometry and topology ought to be. The entity coverage is complete with respect to what industrial applications use, and well structured. The definitions are context free, ie they do not reflect the functionality of the systems where they are implemented. Therefore, it makes sense to look at the Integrated Resources in Part 42 when extensions to the existing functionality of a CAD system are planned.

The new entities introduced to FAM include the definitions of geometric and topological entities from STEP Part 42. The definitions must be augmented with application specific attributes that will allow the new entity definitions to carry the additional information needed for the application.

The topological coverage will be extended to the topology used in solid modelling, allowing the reception of complete boundary representation models. This will be done in two stages:

- the meshable topology will be extended to cover a larger range of regular topologies
- the structure of the topology entities in STEP Part 42 will be implemented

This will be done in the commercialisation phase after the project.

The geometric coverage is currently being extended to cover the NURBS curve and surface as defined in STEP allowing the reception of sculptured surface models, both the geometry and the topology, if any.

The reception of models from boundary representation modellers and sculptured surface modellers will therefore be staged, based on the functionality of FAM at anyone time. Initially the reception will be slightly better than for IGES, ie some of the topology will be lost and the geometry the system can receive will be converted, the rest will be lost. The development of FAM will increase the number of entities transferred and eventually enable complete transfer. The projects to achieve this are scheduled.

These changes to FAM will improve the reception of CAD models conforming to the given APs. The use of STEP entities will reduce the need for conversions and approximations, increase the chances for successful transfer and be a step towards the goal of 100% conformance to the various APs.

The goal is to bring to the market a product which is STEP compatible, e.g. able to receive, hold and send any model conforming to the application protocols developed in CADEX, in particular those which are progressed inside the ISO/STEP community: ISO 10303 Part 204, Part 205 and Part 206. The processor development within the CADEX project is a good start for achieving this high level of functionality in product model data transfer.

4.4 Gesellschaft für Strukturanalyse

4.4.1 Introduction

GfS is selling a FEM system with extensive pre- and post-processors. Essential for the quality of systems is the completeness in functionality, beginning with the input of the describing geometrical data and ending with the evaluation of results. GfS has entered the CADEX project to enhance its FEM pre-processor PROLOG. For this a general interface for geometrical data described in an international standard is needed. Thus GfS provided PROLOG with a STEP interface for the exchange of CAD geometrical data. Having brought to the market the new FEM modeler DIAMOS, GfS uses the STEP processor with almost no change for this successor product DIAMOS.

4.4.2 Models to be covered

The GfS STEP interface covers all model types relevant for CAD to FEM data transfer. The entities to be covered come from four different Application Protocols (APs) including Boundary-Representation (B-rep), Surface and Wireframe models, which are described in the STEP Parts 204/205/206. In addition to these ISO APs there was a need to handle models with meshable geometry (e.g. for geometry exchange between different FEM systems) and models with special geometry occuring in structural analysis. Therefore the Compound B-rep AP was designed to cover all models containing internal boundaries and/or parts of different dimensionality (e.g. a solid with a flat part attached to it). The CSG is explicitly excluded as defined in the technical annex, because these models are incompatible with geometry representation in FEM modelers.

As models in the analysis domain do not happen to be limited to only one AP, the GfS processor must also be able to deal with a mixture, i.e. to combine entities from different APs in one model. The Compound_Brep AP can overcome this difficulty, but compound models will be written only by other analysis partners, not by normal CAD users. Thus the GfS STEP processor is held open to include a (future) Mechanical Design AP that covers entities from several presently existing APs.

At the present GfS has a STEP processor that can read models of the four APs listed above. All entities being used there are included. All changes in the STEP standard until the beginning of 1992 have been payed regard to. The concept is flexible enough to deal with further changes.

Entities that are not usual in FEM modelers must undergo conversions; such functions are built in and are partly operated interactively, e.g. meshing of b_spline surfaces. As the native representation in DIAMOS is similar to a compound representation, also topological conversions are necessary to use models coming from other CAD systems. As far as conversions belong to the Common Tools, these processor parts were based on those. Conversions that are specific for the particular system were developed separately.

As output from the GfS modeler the workplan requires only Compound Brep models. This CADEX AP comprises entities from B-rep, Surface and Wireframe models plus a few entities necessary to describe compoundness. The STEP pre-processor of GfS supports all these entities and is able to write files according to the Compound B-rep AP. Additionally Wireframe models can be written. The processor is designed to be switched to a comprehensive Mechanical Design AP (replacing all others) when it becomes available.

4.4.3 Functionality

As stated in previous reports, the architecture of GfS's STEP processor is almost the same as for all CADEX members. The processor is designed as stand-alone program, i.e. on post-processing it writes a native file for the GfS FEM modeler and on pre-processing it reads that file, which reflects the data structure of the modeler. The following table shows the main menue of the STEP processor.

```
1 = initialise IDS
2 = read new STEP file              scanner/parser
3 = write STEP file                 formatter

4 = write data to native file       \ native
5 = read data from native file      / interface

6 = display structure               on screen
7 = call checker                    test tools
8 = report statistics
9 = list top entities
a = interrogate data base           to trace contents

b = entity conversions              e.g. b_spline
c = model conversions               e.g. WF
d = store IDS in data base file     dump
e = read IDS from data base file    for resume
f = delete entities

g = other and service functions
0 = end of program
```

Less important branches shown in the previous reports have been removed or organised in a different way. As shown in figure 4.4.1 the main functions are:

- reading and writing STEP files
- reading and writing native files
- reading and writing internal files (for save/recall)
- displaying and checking
- converting

All functions are linked to the IDS which forms the central data base. In almost all cases input and output of these functions are IDS-conform. Only a few conversions must be performed in the native interface. These cover entities which are local to the native system and not available in the IDS.

The program is used interactively to keep conversions under user's control and to show the model to the user. An enhanced user interface can be provided later without affecting the main body of the program. As sophisticated models need a lot of conversions, a single session can last long. That is why the save/recall tool was kept to store the present contents of the data base (IDS) on a (binary) file. This also allows to interrupt processing and to backup any intermediate step.

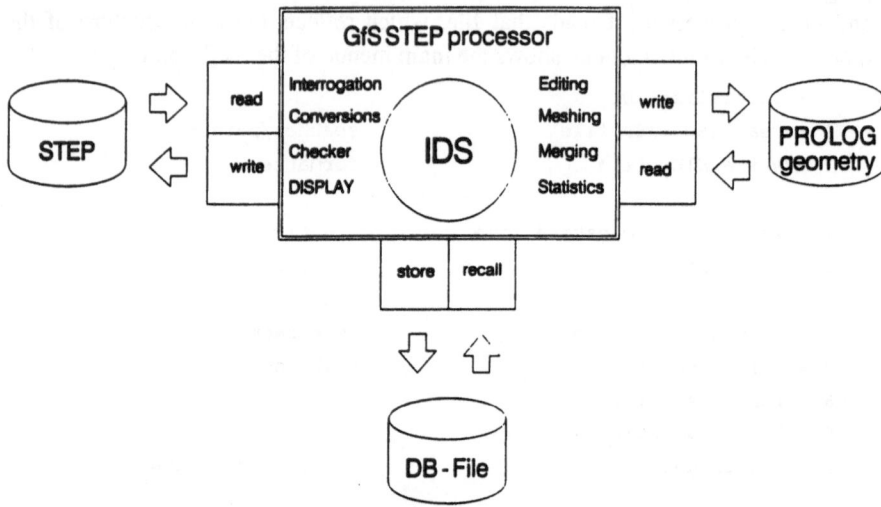

Fig. 4.4.1: Architecture of GfS STEP Processor

4.4.4 Implementation and Test

Implementation was done on three hardware platforms:
- Apollo workstation DN3000 with operating system AEGIS 10.3 (for the parser generation software a Unix subsystem was invoked)
- MIPS C2035 with UNIX (system 5.3)
- PC 386/486 with MSDOS using Extender (SALFORD DBOS)

To test the functionality of the processor, the test file library (distributed by FIAT) was used. All models could be read. Most of them could be transferred successfully. Problems occur when models unsuitable for FEM systems are encountered. In such cases only part of the information can be used. As FEM modelers internally use Compound B-rep models, this kind of file guarantees a maximum exploitation of information in both directions.

Experience with all test models showed that the most important advantage (compared with previous "neutral" solutions) is the completeness and standardisation of given information.

Figures 4.4.2 and 4.4.3 show some examples taken from the CADEX test file library in their state on input (before any conversion) and after processing.

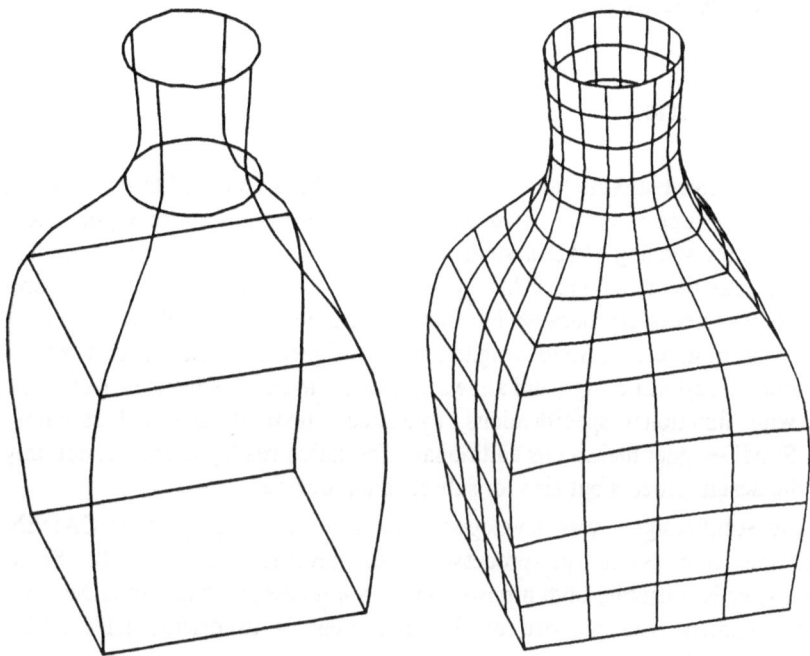

Fig. 4.4.2: CAD model "bottle" before and after conversion

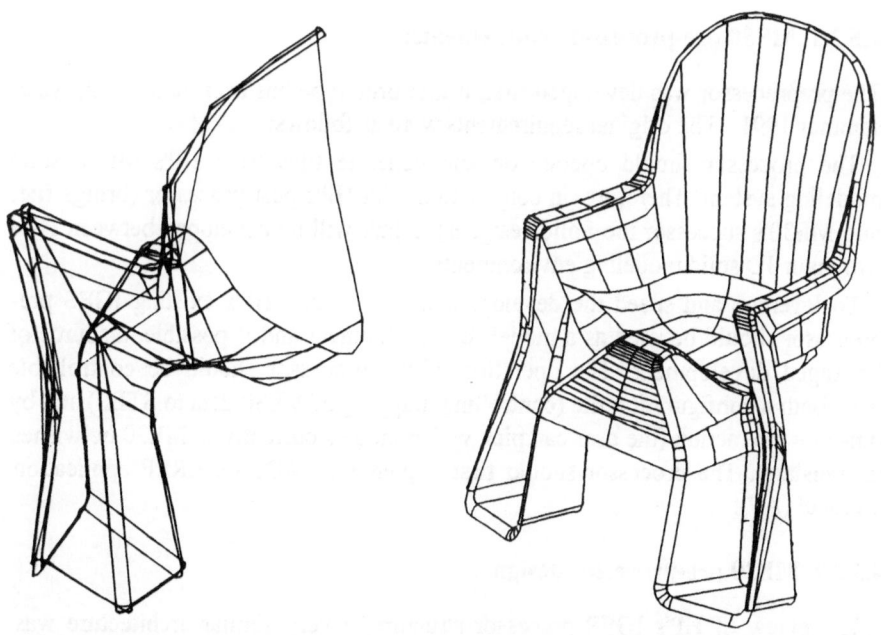

Fig. 4.4.3: CAD model "bicycle chair for children" before and after conversion

4.5 Hewlett-Packard

4.5.1 Introduction

As part of its contribution to the CADEX project, HP has produced three processor prototypes that support STEP. Two of them are pre-processors (they generate a STEP file from sending CAD system) and one is a post-processor that reads a STEP file into the receiving CAD system. The decision to develop two pre-processors was necessary because HP introduced a new 3D modeling system the HP Precision Engineering SolidDesigner. SolidDesigner is the successor of ME30 and supports free-form design and allows a designer to easily enhance a geometric model with functional specifications, tolerances, textural data and assembly details. SolidDesigner makes this additional information readily available, not only during the design process but also for subsequent processes.

Because SolidDesigner was developed in the same time frame as the CADEX project, the first STEP pre-processor was developed for ME30. Since SolidDesigner is operable, both a post- and a pre-processor for the SolidDesigner was implemented. The purpose of this document is to discuss HP's STEP processors, their design, implementation, and comments regarding the work.

4.5.2 HP's ME30 STEP pre-processor

4.5.2.1 ME30 pre-processor requirements:

The preprocessor was developed first; a first prototype has been functioning since summer 1991. The original requirements were as follows:

The processor should operate on and generate files from HP's ME30 solid modeling system. Therefore, in conjunction with HP's post-processor (brings files into ME30's successor the SolidDesigner), a link will be developed between HP's two main 3D solid modeling environments.

To leverage and speed the development, the existing HP's existing IGES pre-processor should be used as a model for development and if possible, a source of leveraged development. The operation of the processor should be controllable from both a configuration file (controlling mapping of ME30 data to STEP) and by run-time interaction (the user can pick which models currently in ME30 he wishes to translate). The processor should first implement CADEX's BREP application protocol (AP).

4.5.2.2 ME30 pre-processor design

After review of HP's IGES processor structure, a very similar architecture was chosen. Please refer to the drawing "Reference Model for a STEP Pre/Post

Processor" (figure 3.5 in chapter 3.3.2.1) for a block diagram. The Preprocessor in this diagram shows the Intermediate Data Structure (IDS). This IDS is not required for HP's ME30 pre-processor since the model data can be extracted directly from the native model in memory.

Once the processor is called, its configuration file is read to set variables to determine what "style" of BREP AP information should appear in the eventual file. The user then specifies what bodies (3D solids) he wishes to translate. This list is then scanned, with the list of shells of the first body being fetched, followed by the list of faces associated with the first shell, etc. until vertices and their corresponding points are returned from ME30.

Then the translator is invoked. As there is a clear mapping between ME30 and STEP entities, the translator takes ME30 information and converts it into a form that is identical to what is in STEP.

Once the translator is finished converting a cartesian point (or vertex, curve, or any other ME30 entity) into its STEP equivalent, the HP pre-processor formatter is called. The formatter gives the new STEP entity a name, and writes this name with the entity keyword and attributes to the STEP file. It then returns the chosen entity name back to the translator, which saves this entity name by attaching it as an attribute to the ME30 entity itself. This is done for several reasons; to remember if an entity has already been visited, and to convert memory pointers in ME30 to entity references in STEP. Once the STEP name has been stored, the translator returns control to the pre-processor scanner.

The scanner then continues by seeking out the next entity in ME30's BREP tree, calling the translator with it, the translator translates it, and the formatter formats its, etc. In this way the entire ME30 BREP data structure is traversed and represented in the STEP physical file. At the very end the software writes a log file showing which ME30 entities were translated, and which STEP entities were written.

4.5.2.3 ME30 pre-processor implementation

The implementation of the software is entirely in the programming language "C". No CADEX tools were used in the implementation (the formatter was not ready). It's hard to see other than the formatter what tools might be useful.

Geometry entity mapping

point, ME_TYGEPT
A ME30 points is directly mapped to the STEP CARTESIAN_POINT entity (3 coordinates).

Curve Types:

line, ME_TYCUST
A straight line in ME30 is represented by a vector used as a point on the line and a vector describing the direction of the line. In the pre-processor the STEP LINE

entity is directly derived from the geometry of the vertices building an ME30 edge. The start point of the ME30 edge will be the location placement. The vector difference between end and start point is used for a STEP DIRECTION describing the direction of the line.

circle, ME_TYCUCI

In ME30 the placement of the circle in 3D space and the axis direction to describe the orientation and positive direction of the circular curve can directly accessed with the curve entity. For STEP an AXIS2_PLACEMENT entity is used to describe location and orientation. The STEP CIRCLE entity uses this placement and the circle radius. The x-axis direction of the AXIS2_ PLACEMENT is arbitrary (but of course perpendicular to the z-axis).

ellipse, ME_TYCUEL

In ME30 the placement of the ellipse in 3D space and two vectors to describe the direction of the minor axis and major axis can accessed with the ellipse entity. For both axis the corresponding radii are given (majorradius and minorradius). To create a STEP ELLIPSE entity the outer product of the ME30 axes is necessary to describe the orientation and positive curve direction of the ellipse. The axis for the majorradius can be used directly.

intersection curve, ME_TYCUIN

The ME30 intersection curve is represented by an STEP POLY_LINE. To compute the points necessary to describe the poly line an internal ME30 function is called to march along the intersection. If the returned points are connected with straight lines ME30 guarantees that the chordal error of this polygon with respect to the true curve varies with respect to the curvature of the curve according to r *0.002 where r is the local radius of the curvature of the curve.

Surface Types:

planar surface, ME_TYSUPL

In ME30 a planar surface is defined by two vectors. One builds the location, the other the orientation of the surface normal. This vectors are used to build a STEP AXIS2_PLACEMENT, which is referenced by a STEP PLANE entity.

cylindrical surface, ME_TYSUCY

In addition to the AXIS2_PLACEMENT (see planar surface), which represents the inner axis of the cylinder the radius of the generating circle is inserted to build the STEP CYLINDRICAL_SURFACE entity. The direction of the x-axis is arbitrary.

spherical surface, ME_TYSUSP

In addition to the AXIS2_PLACEMENT (see planar surface), which represents the location of the center of the sphere the radius of the sphere is inserted to build the STEP SPHERICAL_SURFACE entity. The direction of the z- and x-axis is arbitrary.

conical surface, ME_TYSUCO

In ME30 a conical surface is defined by the semi-convergence angle, one vector for the placement of the center of the central swept generating circle in 3D space with its radius at the position and a second vector describing the sweep direction. The semi-convergence angle can be negative if the cone gets wider when looking along the axis direction.   To build a STEP CONICAL_SURFACE a translation has to be done because in STEP first the semi_angle should always be positive and second looking along the axis direction should show the cone getting wider.

toroidal surface, ME_TYTOSU

ME30 describes a toroidal surface by sweeping a circle of minorradius about an axis through an center placement and an axis direction. The swept circle traces a circular path of radius majorradius. As all surface types the STEP TOROIDAL_SURFACE uses an AXIS2_PLACEMENT to describe the sweeping axis and both radii. In spite of the current version of Part42 [2] a majorradius smaller than minorradius will be written to the STEP file if this constellation is found in the ME30 data structure. Considering the balloting process on Part42 [3] it is likely that this will be allowed in the future.

Topology entities mapping

vertex, ME_TYTOVX

A ME30 vertex is mapped directly to the STEP VERTEX

edge, ME_TYTOED

A ME30 edge is mapped directly to the STEP EDGE

loop, ME_TYTOLO

To convert a ME30 loop it is to be distinguished whether the loop consists of a list of edges (STEP EDGE_LOOP) or a single vertex (STEP VERTEX_LOOP).

face, ME_TYTOFA

A ME30 face is mapped directly to the STEP FACE

shell, ME_TYTOSH

A ME30 shell is mapped directly to the STEP CLOSED_SHELL

body, ME_TYTOBY

A ME30 body is mapped directly to the MANIFOLD_SOLID_BREP

assembly, ME_TYTOAS

A ME30 assembly is not processed because at the moment the BREP AP does not support assembly structures

STEP entities derived implicitly from ME30 entities

AXIS2_PLACEMENT

STEP and ME30 places Curve and surface entities in 3D space (location and two vectors representing the z- and x-axis direction). Whereas ME30 connects this information direct to the entity, STEP uses this intermediate entity, having a more common understanding on placements in space. For each entity to be placed in space, the two DIRECTION and one CARTESIAN_POINT entity are created. This three entities are referenced by the AXIS2_PLACEMENT.

DIRECTION

In addition to the use of DIRECTION in the environment of an AXIS2_PLACEMENT the STEP LINE entity is defined by a point and a direction entity. In this case the direction is created by building the vector connecting both points of the corresponding edge.

EDGE_LOGICAL_STRUCTURE

The direction of the edge respecting to the curve direction is accessed form the kernel of ME30   .T. means that the topological direction of the EDGE in a given loop fits with the geometric representation of the curve linked to the edge and the direction given by start and end vertex. If start and end vertices are identical only the geometric curve direction is used as reference.

CURVE_LOGICAL_STRUCTURE

Conforming to the BREP AP the logical value of this entity is in any case .T. (for true).   By setting a flag in the configuration file the writing of this redundant information can be skipped.

FACE_LOGICAL_STRUCTURE

Conforming to the BREP AP the logical value of this Entity is in any case .T. By setting a flag in the configuration file the writing of this redundant information can be skipped.

SURFACE_LOGICAL_STRUCTURE

The direction of the surface respecting to the face direction is accessed form the kernel of ME30

LOOP_LOGICAL_STRUCTURE

Conforming to the BREP AP the logical value of this Entity is in any case .T. By setting a flag in the configuration file the writing of this redundant information can be skipped.

SHELL_LOGICAL_STRUCTURE

Conforming to the BREP AP the logical value of this Entity is in any case .T. By setting a flag in the configuration file the writing of this redundant information can be skipped.

An example log file

To supply the user with a short description of the translation process a log file is created which describes the configuration of the pre-processor and a table of the entities written. The following example shows the output after the translation of a simple cylinder.

```
STEP translation log file ( 04 Dec 1991  12:42:36 )

STEP 3D Translator  Rev. preprocessor_version: 2.0, 01-Oct-1991

Writing file 'cylinder.step'

Configuration file
-------------------
```

$$ HP Precision Engineering Systems STEP 3D Translator Rev. 0.1

```
$$ STEP configuration file STEPo.con
$$
$$ Each line has the format 'Value range  ** comment : 'value'
$$ You may change comments and values, but the ':' may
$$ not be removed.
$$
$$ Output coordinate system  G-Global W-Workplane
$$ Use LOGICAL structures or not : T - true,  F - false
$$
$$ File output format    STEP-No line delimiter HP-UX-<LF>
$$ File output format    DOS-<CR><LF> PAWS<CR>
$$
$$ For a detailed explanation of each parameter refer to the
$$ Interfacing manual.
$$
   String    ** Destination path              : .
   G,W       ** Output coordinate system       : G
   T,F       ** use LOGICALs                   : T
   String    ** Product ID (originating system) : ME30 4.10/t3 25-Jan-91
   String    ** Author          : Hermann Ruess, Peter Schild
   String    ** Organization    : Hewlett-Packard GmbH, MDD, R&D,
                                   Herrenberger Str. 130,
                                   D7030 Boeblingen,
                                   Phone: 49-7031-14-2329,
                                   Fax: 49-7031-14-3930
   String    ** File output format             : HP-UX
STEP entities written:

Entity Type                   Count
------------------            -----
Man. Solid Breps ...................   1
Shells ............................    1
Faces .............................    3
Surf. Log. Structs ................    3
Edge Loops ........................    4
Edge Log. Structs .................    4
Edges .............................    2
Vertices ..........................    2
Planes ............................    2
Cylindrical Surfs .................    1
Circles ...........................    2
Cartesian Points ..................    7
Directions ........................   10
Axis2 Placements ..................    5
```

File organization

```
FILE              FUNCTION
----              ---------
step.c            Control program.
step_conf.c       Routines that read the config file and write the
                  log file.
step_scan.c       Routines that scan the ME30 data structure.
step_trans.c      Routines that translate the ME30 entities into
                  STEP entities
trans_sup.c       Translation utility routines (from IGES).
trans_form.c      Formatter routines.
```

4.5.2.4 Comments on ME30 pre-processor

The ME30 pre-processor was not very complex because of the similarities between ME30's data structure and STEP's definition of analytic BREP entities. Note that it is not planned to incorporate CADEX's IDS and formatter into the ME30 to STEP pre-processor software. It was viewed as inappropriate because of HP's ability to write the STEP file 'on the fly'. This ability has been admittedly easier by allowing only backward referencing in the file. As you will find in the section about the SolidDesigner STEP pre-processor the CADEX common toolkit was used for that implementation because it is a more flexible approach which simplifies the adoption after changes in the syntax of STEP files and allows long term supportability.

The ME30 STEP pre-processor was not updated to the latest versions of the CADEX AP's. The actual version is very stable and the produced STEP files had a very good quality. So HP was able to supply other CADEX partners with valid BREP files in a very early stage of the CADEX project, which allowed testing their post-processors. Since the SolidDesigner STEP pre-processor is operational, and because it is easy to create complex analytic and also free-form surfaces with SolidDesigner the ME30 pre-processor can be replaced completely.

4.5.3 HP's SolidDesigner STEP post-processor

4.5.3.1 SolidDesigner post-processor requirements

The post-processor was developed with the following requirements in mind:

The post-processor should be integrated in HP's SolidDesigner to read in STEP BREP files. In conjunction with HP's ME30 pre-processor a link was developed between HP's two main 3D modeling environments.

The development of the SolidDesigner STEP post-processor was done in parallel with the developing SolidDisigner itself. HP is convinced that there is a need from the HP customers for a STEP interface very soon after ISO STEP is stable. So in a short time frame after the introduction of SolidDesigner HP will be able to supply the customers with a complete STEP post- and pre-processor interface.

During the development of a totally new product like SolidDesigner it is useful to have a neutral file output for the CAD models created by the system. The idea to use STEP as neutral file in the development phase can only carried out if there exits a SolidDesigner STEP pre-processor, too. Because both processors are available SolidDesigner contain a real bidirectional link to the STEP neutral file.

To develop the post-processor in a very short time frame all available tools to support the development process should be used. Therefore unlike in the case of the ME30 pre-processor, the Scanner/Parser and IDS Part of the CADEX Common Toolkit (CTK) are used for the scanning of the physical STEP file. In addition to the advantage in the development time the post-processor back end will not be influenced by possible changes in the STEP definition of the physical file. For more information about the Scanner/Parser and the IDS see [1] and the related chapters in this report.

Because of the close integration of the post-processor in SolidDesigner the processor is written in C++ to conform to HP's modeling system environment. Because the CADEX Common Toolkit is written in C a special C++/C interface was necessary to combine both modules.

4.5.3.2 SolidDesigner post-processor design

The main aspect of the design of the post-processor was to create a supportable architecture which is open for future improvements in efficiency (speed and memory) and to allow the easy addition of new STEP entities. Aspects of code sharing with other HP links products and the use of functions of the CADEX conversion library should be considered.

To fulfill this design goals and the general post-processor requirements stated above a modular and layered architecture was developed. Because of the use of the Scanner/Parser and of the IDS from the CTK the way to process a STEP file is split into two sequential steps. After the name of the STEP file which is desired to read in is supplied from the SolidDesigner user interface the scanning and parsing of the physical file is triggered. If the STEP file is syntactically correct this step results in an IDS data structure filled with the model(s) read in. In a second step the individual STEP entities which build the whole model(s) can be accessed directly via the reference identifier. This identifier is the same integer number which is used in the STEP file to identify a single entity.

Conversion of STEP entities. The global concept to build a SolidDesigner model from the STEP model is a top down rebuild of the STEP model installed in the IDS. This means that first as precondition the top level entities of the STEP model must be accessed. In the case of a BREP model the only possible top entity is the MANIFOLD_SOLID_BREP entity. Nor does the top down parsing of the STEP model mean that for each STEP entity must exist a corresponding SolidDesigner entity, neither that if there exist such a SolidDesigner entity that this entity must be created within SolidDesigner before the creation of other lower level SolidDesigner entities. This counts only for entities which are leafs of the STEP entity tree. For example a STEP CARTESIAN_POINT entity will cause a immediate creation of a the SolidDesigner point entity. On the other hand, for instance before a SolidDesigner face is created not only the STEP FACE but also the referenced surface geometry must be accessed. So if there is information about the structure of entities on a lower reference level necessary to build a corresponding SolidDesigner entity, this entities are scanned and translated before a build of the actual entity itself.

The object oriented concept is used to allow an easy addition of new STEP entities and to allow the maintenance of small portions of the existing software without affecting most of the other parts. For each STEP entity there exists a corresponding object class containing the internal variables necessary to represent the STEP entity and a conversion function which describes the process necessary to translate this STEP object to a SolidDesigner object. All defined STEP objects are derived form a general STEP_ENTITY class. To simplify the build of STEP classes with a common structure of their data some entities are not directly derived from STEP_ENTITY but from a common entity which itself is derived from STEP_ENTITY. For example all entities which need a list of references inherit this data structure form the STEP_TOPOLOGY entity.

This concept simplifies the addition of a new entity. The only thing which must be done is the addition of a new object class derived form STEP_ENTITY and to define the conversion function. Adding improvements this way the Conversion routines of other object classes can not be influenced.

As mentioned before the only possible top level entity in an BREP environment is the MANIFOLD_SOLID_BREP STEP entity. By the end of the CADEX project it will be possible to read in entities conforming to the CADEX CSG AP too. Also Surface AP entities will be readable by future versions of the post-processor. These improvements are easy to implement because the general design is open for all possible STEP entities. If thea additional C++ class as container for dependent attributes and the conversion routine is added the new functionality is operational.

The following description of a typical translation is based on the CADEX BREP AP. As stated above the described conversion process should only be seen as a general example.

A STEP MANIFOLD_SOLID_BREP entity consists of one or more STEP CLOSED_SHELL entities. This translation knowledge for the entity is contained in the MANIFOLD_SOLID_BREP conversion function. In this case the conversion task is simple: To build a complete SolidDesigner body there is at least one complete SolidDesigner shell entity necessary (the outer shell). So the only thing to be done is to build a SolidDesigner body, trigger the complete conversion of the shell entities and at least to connect the computed SolidDesigner shell with the created SolidDesigner body. So the return of a conversion of a STEP MANIFOLD_SOLID_BREP is a complete SolidDesigner body.

This conversion example shows the general structure of all conversions. The only STEP entities of interest beside the entity which is to be converted are the entities directly referenced by the actual entity. In this way STEP FACES, LOOPs, EDGES, and VERTICEs, are converted into HP's data structure. Eventually the bottom level of the scan (usually a CARTESIAN_POINT) is reached, a SolidDesigner point is created and returned to the next higher entity level conversion function.

The example of recursive conversions of all STEP entities shows in addition that in the design no distinction between topological, geometric or shape STEP entities is done. Each entity knows its own conversion and triggers the conversion of all directly needed sub entities. For instance the SolidDesigner face needs the topological bounding loop entities and the geometric surface reference to be complete. It depends on the internal decision of the conversion function whether the conversion of the geometric or the topological sub entities are done first. In the view of a higher level entity both topology and geometry are build in parallel.

Interface to the CTK. To hide the functions of the CADEX Common Toolkit and to create an interface between the "C" functional oriented structure of the CTK and the object oriented "C++" world a modul STEP_ENTITY_MANAGER is necessary. The function of this modul is to deliver the corresponding STEP entity object to a given identifier.

Another type of access to the CTK is necessary to get the precondition for the translation in form of a list of all top entities found in the STEP file. This can be done for instance by a direct call to the IDS or within the concept of the STEP_ENTITY_MANAGER by defining a dummy STEP world entity, which consists of a list of references to all found top entities. Because both concepts only need simple implementation differences this decision is not fixed in this design analysis. The call via the STEP_ENTITY_MANAGER is much more structured and fits better in the general layered concept and supports a more common process flow. On the other hand it should be considered that the introduction of dummy data structures, which cannot be found in real STEP files, is not only a simple addition to the existing data structures (to have a consistent structure one of the consequences of such a step can be the need to add a conversion function...).

Some STEP entities are referenced twice or more times in one STEP file. For instance in the BREP environment each EDGE is a part of two different LOOPs or a surface description can be used by more than one STEP FACEs. Because of this fact it is not meaningful to process such an entity more than once. Mention that this is not only an aspect of speed and memory efficiency, because in the case of an SolidDesigner body each individual EDGE entity should only be represented once as a SolidDesigner edge entity. Therefore for the post-processor it is necessary to know whether a STEP entity was processed before the actual call. In this case the SolidDesigner entity from the first pass should be returned to a upper layer entity conversion function. To allow this SolidDesigner ENTITY_MANAGER module is added above the STEP_ENTITY_MANAGER layer. Each call to the SolidDesigner ENTITY_MANAGER should return a valid

SolidDesigner object. At the first call of a unique entity the SolidDesigner ENTITY_MANAGER should ask the STEP_ENTITY_MANAGER for a first creation of the STEP entity and trigger the conversion to an SolidDesigner entity. A repeated call with the same reference should be answered by the same SolidDesigner entity.

4.5.3.3 Post-processor implementation

The modular structure shown in the previous chapter was directly converted to the object structure and as well the file structure of the post-processor system. HP's short term goal concerning the post- processor was to test the concepts of the generated CADEX application protocols and to detect missing functionality, ambiguous definitions and general problems of BREP input which will still be present in spite of the use of STEP as neutral file for CAD model exchange.

Therefore only small work was done on aspects like tuning of efficiency, error handling and configuration control. In a long term planning, but not before STEP is stable, a SolidDesigner customer will be able to handle bidirectional STEP data exchange. At the time when the STEP processors will be official part of HP's SolidDesigner product the aspects above will become important.

Because of the strong similarities between STEP BREP and HP's modeler data structure conversions for topology or analytical geometry are minimal and therefore, neither special tools or complicated conversion routines were neither written nor used up to now. The work while running the post-processor is mostly to copy elements between the data structures. An important exception to this rule is the processing of intersection curves. In an CAD system intersection curves represent non analytic curve geometry which results from a lot of surface types sharing a common topological edge (e.g. the two circular arcs which are the result of boring a hole into a cylinder). In the BREP-AP intersection curves are represented by the POLY_LINE entity. This entity references a list of CARTESIAN_POINTS which lie on the curve. The resolution and amount of the points depend on the accuracy of the sending system. In the SolidDesigner Postprocessor it was decided to recalculate this intersection. Because this recalculation needs the surfaces which participate in the intersection the information contained in the POLY_LINE entity is not sufficient. Therefore the surface information must be derived from the entities accepted before. Because a general intersection of surfaces results in more than one curve the points given in the POLY_LINE are used to select the corresponding curve.

In the case of the CSG input the conversion process is not so simple as described above for BREP entities. Because SolidDesigner is no CSG modeler, it is necessary to translate the CSG description of the STEP file into the more efficient BREP structure. Therefore the conversion routine of a CSG primitive has to create a "BREP-primitive" entity which is parametrized by the parameters given in the CSG primitive. Because CSG boolean like operations can be handled also in SolidDesigner the generated "BREP-primitives" can be united, intersected or subtracted as normal BREP models. This functionality is done by the conversion routines of the CSG_UNITE, CSG_INTERSECT and CSG_SUBTRACT C++ classes derived from STEP_ENTITY. CSG entities representing partial infinite segments of 3D object space (e.g. HALF_SPACE) need special conversion functions to build "pseudo-infinite" objects large enough to allow the same operations like on the CSG infinite objects.

4.5.3.4 Comments on post-processor

The S/P and IDS software was used mainly to allow rapid prototyping of the post-processor and to be independent of changes in the syntax of the physical file, which is regular using an emerging standard. The architecture to distribute the conversion knowledge within the STEP_ENTITY objects and the concept of the SolidDesigner_ENTITY_MANAGER will allow a direct call of constructors and conversions of the entities as well. Such a change of the control flow of the conversion will be easy as long as entities only reference other entities read in before. This can not be guaranteed in the future because for instance it can occur, that e.g. display or presentation attributes can exist as independent entities which reference later to come entities. To handle such entities, modules for a partial build up of some kind of "IDS" are necessary to allow a delayed conversion. For the moment it is not planned to change the concept of using the CADEX toolkit, so the IDS will be a component of next versions of the post-processor, however, it should be noted that the IDS consumes a lot of system memory and the implemented memory management (allocation of many small pieces of memory) could cause trouble with very large files.

Because the BREP AP was one of the first AP's defined by CADEX, and the implementation of the toolkit was based on this AP, an update of the toolkit was carried out in April 1992 to reflected the large amount of changes necessary to allow a common structure of all CADEX APs. Also all new entities for advanced and elementary BREP files conforming to the current BREP AP were added. Because of the architecture shown above HP was not forced to introduce large changes on the post-processor necessary to cope with the additional entities of the new AP. Indeed the new updated AP's include the mandatory product structure information of part 41 and 43. These entities above the level of topology and geometry need more work on the mapping to the entities of the SolidDesigner CAD system. Syntactical changes in the STEP file only affected the Scanner/Parser and the changes on the post-processor are only required to add the STEP_ENTITY classes and their conversions for the new entities.

Current work is involved in the area of adding log files and improve error handling and the integration into the SolidDesigner user interface. All geometric and topological types of the BREP AP level 3 are supported (this is dependent on SolidDesigners own development). The extension to interpret a STEP file conforming to the actual CSG AP is in progress.

An issue to cope with is the experience that the number of IDS routines will grow drastically when more and more STEP entities need to be supported (for each STEP entity at least a create -, a retrieve -, a modify function is required). This growth in number of functions expands the amount of software support effort.

4.5.4 HP's SolidDesigner STEP pre-processor

4.5.4.1 SolidDesigner pre-processor requirements

The SolidDesigner STEP Preprocessor completely replaces the ME30 STEP pre-processor. The schedule for HP was to complete the SolidDesigner STEP Preprocessor as the last of the three processor prototypes contributed to the CADEX project. So the experience with the other pre-processor could be used and the "old" STEP files generated with the ME30 pre-processor existed as reference for testing the SolidDesigner Preprocessor. The pre-processor is based on the latest stage of the CADEX BREP AP and its first release completely supports the actual definition of BREP level 3.

The SolidDesigner pre-processor should support both the BREP AP in all three functional levels as well as the Surface AP level 1. Whereas the users should is not be forced to select a different pre-processor level for writing models containing analytic or free-form geometry (level 2 and 3) there is a user intervention necessary if the generated STEP file should contain only entities conforming to the Surface AP or the BREP AP level 1 (facetted Breps).

As the ME30 pre-processor the operation of the processor should be controllable from both a configuration file (controlling mapping of ME30 data to STEP) and by run-time interaction (the user can pick which models currently in the SolidDesigner she wishes to translate).

4.5.4.2 SolidDesigner pre-processor design

In contrast to the architecture chosen for the ME30 pre-processor, the SolidDesigner Preprocessor completely takes the advantages of using the CADEX common toolkit.

Because at HP there exists a generalized and configurable scanner to traverse the SolidDesigners data structure it was only necessary to select the SolidDesigner entities which should be scanned and to implement a set of translation routines which convert the SolidDesigner data into a representation conform to STEP. Afterwards a call to one of the "put" routines of the IDS is necessary to store the generated STEP entities.

After the complete data structure which describes the SolidDesigner model is scanned and translated the control of the pre-processor is transfered to the formatter of the CADEX toolkit to do the actual writing of the physical file.

One additional task of the translator module is to generate unique identifiers for the generated STEP entities and control the translation process that the same

SolidDisigner entities are neither translated nor scanned twice. If a SolidDesigner entity is used more than once the identifier generated during the first translation should be used as reference in the STEP structure.

This general architecture is also valid for the translation conforming to the Surface AP and BREP AP level one. The only difference is that for this functionality of the pre-processor a different scanner (in most cases only a different scanner configuration) is set up, which may call different translator functions as in the standard BREP case. For facetted models a scanner having access to the (facetted) graphical representation of a model in SolidDesigner will be used.

4.5.4.3 SolidDesigner pre-processor implementation

The implementation of the processor is strongly related to the design described above. The main modules are scanner, translator and the toolkit modules for the IDS and the formatter.

As the SolidDesigner post-processor, the pre-processor modules are written in "C++". The interface between the "C" functions of the IDS and the pre-processor modules is implemented in the translator module by a direct call to the IDS "put" functions.

To demonstrate the functionality of the scanner one rule of the scanner configuration file is discussed:

```
// Rule for FACE
config.set_config (FACE::type, WS_TRANS::test_if_already_known,
                   surface::type,  NULL,  LOOP::type,
                   WS_TRANS::face, config.END);
```

The rule describes that if the type of a scanned SolidDesigner entity is a FACE it should be first checked whether the actual FACE has been already translated in the past. The result of that test may be that the actual scanning in suspended because it does not make any sense to search for entities which are connected to an already known one. In such a case also the connected entities have been scanned (and translated) before. If the SolidDesigner FACE was already know the following part of the rule describes that first a connected surface and afterwards all LOOPs connected to the actual FACE should be scanned. Because the parameter after the surface type is NULL, no special action should be executed after the surface is scanned and translated. The parameter after the loop type is not NULL but references a function which should be called after all LOOPs (and the surface scanned before) are evaluated. The function referenced in this rule is exactly the translator function for the SolidDesigner FACE itself.

The translator function has access to the results (the assigned STEP entity identifiers) of the scanning and translation of the connected entities. It can use these results to call the IDS "put" routine to insert a STEP face into the STEP data structure. The described scanning process results in a top down scanning but the translation order is that terminal entities are translated before top level entities. So entities which are referenced from an other entity are always translated first. This creates a backward referencing STEP structure.

4.5.4.4 Comments on the pre-processor

As with the ME30 pre-processor, because of the similarities between the SolidDesigner's data structure and STEP's definition of BREP entities the translation task for the SolidDesigner pre-processor is not very complex. Because of the use of the generalized scanner for SolidDesigner and the use of the CADEX common toolkit a fully working BREP level 3 pre-processor was implemented without large effort. As result of the working processor a library of complex models containing free form geometry can be build up with the easy to handle user interface of SolidDesigner.

4.5.5 Examples of model exchange

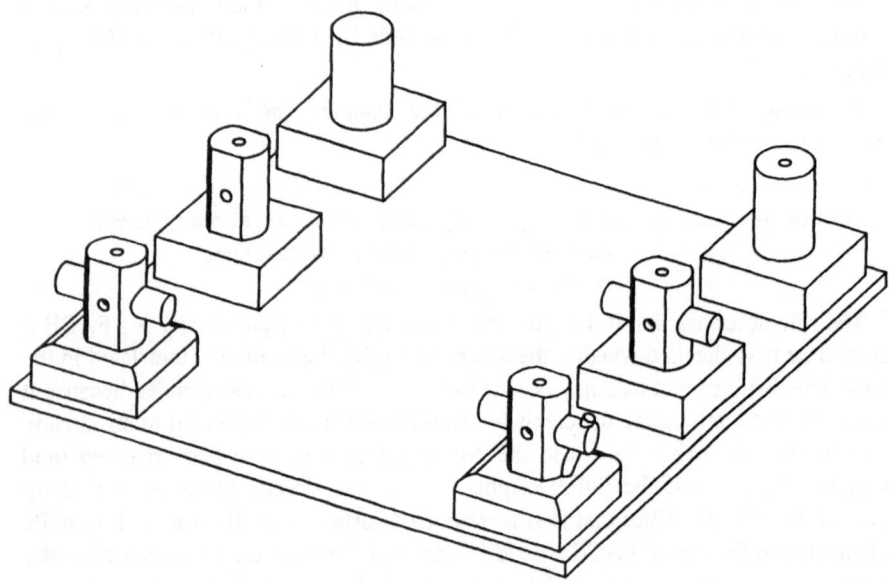

Fig. 4.5.1: Test model

Figures 4.5.1 and 4.5.2 show the contents of two example test files written by the ME30 pre-processor and which were exchanged with other partners. Figure 4.5.1 shows bodies which combine cylindrical and planar surfaces with increasing complexity. Especially the intersection of the cylindrical surfaces needs elliptical as well as non analytic intersection curves. Test STEP files containing these bodies were offered at the ISO Houston meeting to the ISO community. The files are on the 'road' now around the world, distributable through the Implementation

Working group (WG7). HP hopes to get feedback from different system vendors and implementors on those files.

Figure 4.5.2 shows an exhaust pipe which was designed with ME30 from descriptions of model number 7 from the test and validation working group of the CADEX project [4]. The model was successfully exchanged with several CAD and Application Systems especially the CADEX partner systems SIGRAPH-CAD-3D (Siemens-Nixdorf), PROREN (PROCAD), HP's Precision Engineering SolidDesigner and FAM (FEGS). The exchange of this model was part of the demonstration of the CADEX project on the ESPRIT exhibition in Brussels at November 1991. There are other more complex models which have been successfully exchanged.

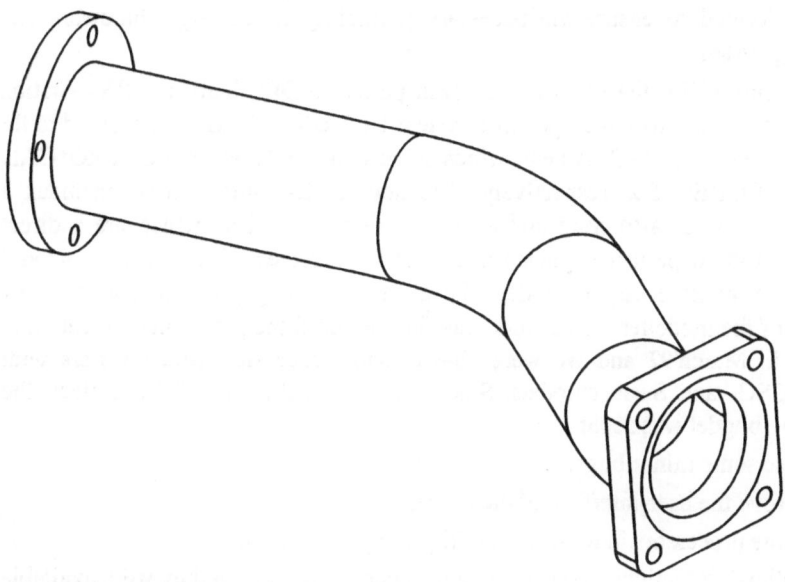

Fig. 4.5.2: Exhaust pipe

4.5.6 References

[1] Esprit Project 2195 CADEX, Deliverable 6 Report, pg. 14-94, November 1991
[2] Part 42 Geometric and Topological Representation, ISO TC184/SC4/WG3/ P1,N45, May 1991
[3] Balloting results in Committee Draft ISO CD 10303 - 42, October 1991
[4] Model Generation Description of "Industrial" Test Cases CADEX document no. WG3F105C, Model Number 7, June 1991
[5] WG3FI04C Analysis and Choice of a set of "Industrial" Test Cases
[6] WG3FI06C Model Generation Description of "Industrial" Test Cases

4.6 ISYKON and PROCAD

4.6.1 Introduction

The basis for the common processor development of ISYKON and PROCAD is the CAD system PROREN, which is a 3D system with BRep modeling technique and an integrated surface modeller dealing with b-spline mathematics.

PROREN works with different hardware platforms like DEC or IBM and supports different operating systems like UNIX or VMS. For the first test and implementation phase the UNIX version on CADMUS hardware available at PROCAD was chosen. Many tests, before and during the implementation have been performed to ensure the necessary portability to all target hardware and operating systems.

At the end of '91, the software has been ported to DEC hardware (Dec-Station 5000/xxx) running with the operating system ULTRIX and has been adapted to the new release of the CAD-System, which is now available as revision PRO*CAD 5.31 or PROREN 5.2, respectively. The new CAD-system release includes a revised internal datastructure and a revised kernel interface, which are a direct result from the experiences gained in this ESPRIT project. Consequently, several mandays were necessary to update the already working processors to the new features of the modeller. This work was finished until the presentation meeting at Munich in March'92 and we were able to show sucessfull data transfers with BRep-, CSG- and Surfacemodels. Since March and the end of the project, the remaining bugdet was spent

- to delete some minor bugs,
- to improve the user-interface of the processors,
- to test the processors in terms of stability and performance.

The following sections describe the developed processors as they were available at end of May 92.

4.6.2 The pre-processors

4.6.2.1 General remarks on pre- and post-processors

The pre- and postprocessors are designed as separate external operators which can be handled independent of the CAD-System. If necessary, they can be easily integrated into the CAD-system because of their modular structure and well defined interfaces. The benefits of having them independent are:

- smaller size of the CAD-System
- batch capability
- easier maintainance

The communication between the CAD-system and the processors is done by writing/reading of native datafiles. The programming languages used are C and Fortran. All processors work in case of errors, warnings or other messages to the user with texts retrieved from ascii-tables for an easier adaption to foreign languages.

4.6.2.2 Description of the pre-processors

The actual preprocessors within CADEX are able to handle the following CADEX AP's and related models:

- Brep-AP: (elementary) manifold solid BRep and facetted BRep
- Surface-AP: Geometric_3D_Surface_Set

Other models, like the advanced BRep or the topological oriented surface models will be possible with the next version of the CAD-system. The preprocessors for surfaces and breps are logically two distinct software moduls (at the frontend side) but use a lot routines commonly, so that it was feasible to combine them into only one operator and to select the needed functionality by a user given parameter.

Below is a listing of the entities which are currently supported by the preprocessors:

Group	B-rep-AP-Entity	Surface-AP-Entity
Shape	MANIFOLD_SOLID_BREP FACETTED_BREP	GEOMETRIC_3D_SURFACE_SET
Topology	SHELL_LOGICAL_STRUCTURE CLOSED_SHELL FACE_LOGICAL_STRUCTURE FACE LOOP_LOGICAL_STRUCTURE EDGE_LOOP POLY_LOOP VERTEX_LOOP EDGE_LOGICAL_STRUCTURE EDGE VERTEX	no topology available
Geometry	SURFACE_LOGICAL_STRUCTURE PLANE CYLINDRICAL_SURFACE CONICAL_SURFACE SPHERICAL_SURFACE CURVE_LOGICAL_STRUCTURE LINE	B_SPLINE_SURFACE B_SPLINE_CURVE CARTESIAN_POINT

(continued)

```
| POLY_LINE            |
| CIRCLE              |
| ELLIPSE             |
| HYPERBOLA           |
| PARABOLA            |
| AXIS2_PLACEMENT     |
| DIRECTION           |
| CARTESIAN_POINT     |
```

4.6.2.3 Functionality and environment

The function of the preprocessor is to read the PROREN native data from a given file and write the translated entities onto a STEP physical file.

Starting with a native database file of the CAD-system the preprocessor frontend selects the related data and stores it into an internal data structure PDS (PROREN Data Structure) which is equivalent to the CAD internal data structure (see figure 4.6.1). After this storage, the frontend validates, that the data read is the correct input for the logical preprocessor selected by the user.

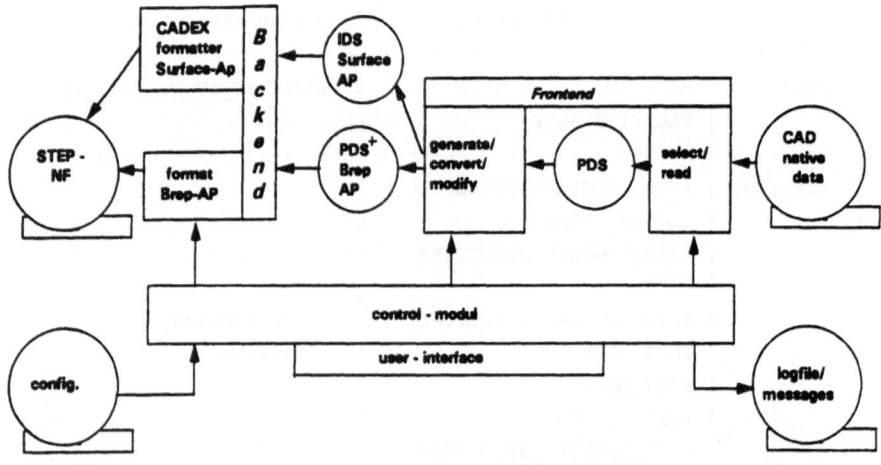

Fig. 4.6.1: The pre-processor

Assuming accordance, the next step is to convert this data into a datastructure called PDS+ which is a combination of the original PDS and a few new structure elements (e.g. elements to hold shell information).

The conversion process can be logically divided in three different functions:

a) modifying of existing data (e.g. splitting of edgecurves)

b) conversion of mathematical representations of datatypes (e.g. all conical curves)

c) generation of entities from implicit representations (e.g. shell elements).

This modified internal data structure (called PDS+ on the figure) can now be written to the STEP physical file format by the preprocessor back-end which is not identical with the CADEX formatter for effiency reasons.

In the area of the surface models, the preprocessor uses low level access routines to fill the IDS instead of producing an PDS+. Therefore, the related backend is identical with the CADEX formatter.

Another preprocessor output is a statistic file which can be used for a first test for completeness of processing, showing the original PROREN elements and the created STEP entities.

The three main blocks(selection of data, transfer from PDS to PDS+ and the format process) are controlled by one modul which is also responsible for the interface to the operating system (log-files, configuration files etc.) and the interaction with the user (e.g. Switching between BRep-AP and Surface-AP).

If the processor works in batch mode the necessary control information is stored in tables, which are read by the processor at starting time. In case of errors, the software distinct between warnings and severe errors and writes a message into the logfile. In case of severe errors the whole process is terminated. The messages itself are stored in an external file for easy changes and language independencies.

4.6.3 The post-processors

4.6.3.1 Functionality

The postprocessor is mainly written in C for an easy utilization of Scanners/Parsers and the IDS, except of the interface-routines to the CAD system which are written in FORTRAN. Any access to the native database is done by interface routines to the kernel (see figure 4.6.2).

The postprocessor reads the STEP file and transfers the data to a PROREN native file.

The main program calls the language-interpreter from which the user can control the way of processing. The interpreter consists of a LEX and YACC based software which reads commands from the standard input or from a script file (see section 3.4.2 for a description of the available commands and the language syntax).

By using the related commands the scanner/parser software from the common toolkit reads the NEUTRAL FILE and fills up the IDS. Based on this IDS, there are now a lot of possibilites for the user:

1) the back-end routines can produce the native data or

2) the test software can be used to examine the IDS

3) the user can get detailed informations about entities

4) the formatter can create a new STEP file (only available in the test environment)

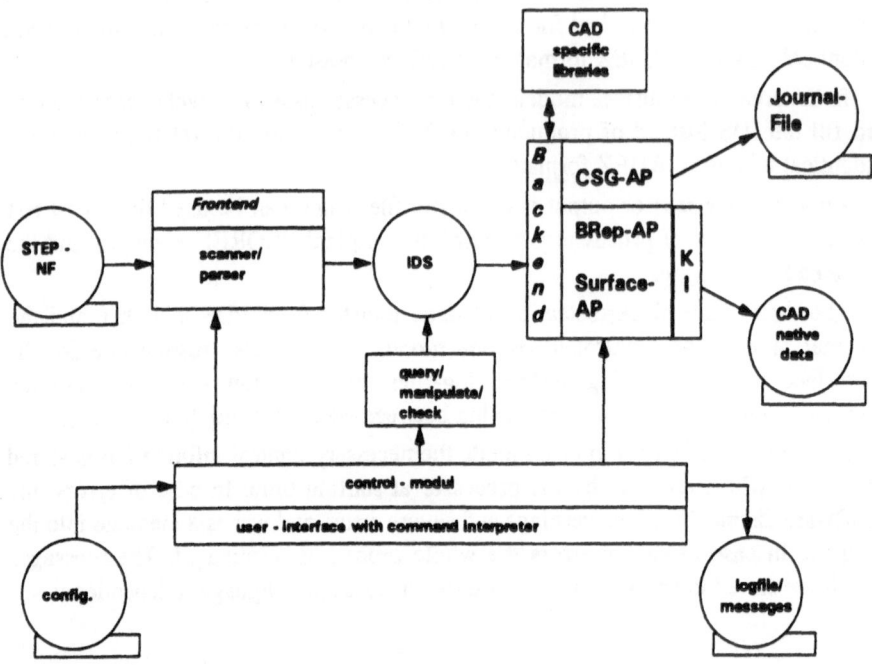

Fig. 4.6.2 Post-processor

All these activities are controlled by this language-interpreter called the dialogue module. It's an environment for postprocessing, testing, modifying and all the other activities which work with the data in different phases and in different matters. In a real product, however, the usage may be limited to the first two items on the list above.

As mentioned, the main task is to transfer the data from IDS to the native format of the CAD-system. This can be done in the following ways:

a. the user wants to process any data in the IDS.

In this case, the IDS is scanned for topentities (BRep's, CSG-Solids or single Curves and Surfaces from the 3D_GEOMETRIC_SET) and transfers them step by step to native data. The corresponding command is "PROCESS ALL ;" at the command level.

b. the user wants to process only selected entities.

In this case, the user has to select the corresponding entity type and its identifier (e.g. "PROCESS BREP 4711 ;").

The generated native data is stored in different types of datafiles, according to the different types of the topentities and the related AP's. The following sections show the different postprocessor backends in more detail.

4.6.3.2 B-rep post-processor

Currently all entities of the Brep-AP 1.2, except the TOROIDAL_SURFACE entity, are supported by the postprocessor. Equivalent to the preprocessor, the capability to handle Surface-models is reduced to GEOMETRIC_3D_SURFACE_ SET at this time. Facetted_Breps from the newer AP204 is supported, too. A complete list of supported entities can be found in the table above.

4.6.3.3 CSG post-processor

CSG-models can be handled in the postprocessor even PROREN bases on a true BRep-modeller. The reason is, that PROREN offers a mighty user interface, which is partwise very close to the definition of CSG-primitives and the boolean operators. All commands and actions of the user during a work session are stored in a so called journalfile, which is used to reproduce or to modify CAD-models. This facility is used by the CSG-postprocessor in the following way:

The frontend of the CSG-postprocessor reads in the STEP file via the CADEX tools into the IDS. The backend produces now a journal-file for the CAD system by transfering the boolean expressions from STEP into the corresponding boolean expressions according to the PROREN command language. This file can now be read by the command interpreter of PROREN and produces the respective CAD-Model in PROREN. Please note, that the resulting model will always be a BREP model, because the CSG-syntax is only available at the user interface, represented by the command language but inside the datastructure itself. Therefore, a CSG-preprocessor will not be possible with the current CAD-system philosophy.

Handling and userinterface of the CSG-postprocessor is identical to the BREP postprocessor.

The CSG-postprocessor supports the following entities:

```
BOOLEAN_EXPRESSION
    UNION
    INTERSECTION
    DIFFERENCE
CSG_PRIMITVE
    SPHERE
    RIGHT_CIRCULAR_CYLINDER
    RIGHT_CIRCULAR_CONE
    TORUS
    RIGHT_ANGULAR_WEDGE
    BLOCK
SWEPT_AREA_SOLID
    SOLID_OF_LINEAR_EXTRUSION
    SOLID_OF_REVOLUTION
GEOMETRY
    CARTESIAN_POINT
    DIRECTION
```

```
        AXIS1_PLACEMENT
        AXIS2_PLACEMENT
        LINE
        CIRCLE
        PLANE
  TOPOLOGY
        VERTEX
        EDGE
        EDGE_LOOP
        FACE
```

Please note, that the TORUS will be approximated by cylinder segments, because a true torus entity is not available in PROREN.

4.6.4 The dialogue language

The purpose of the dialogue software is to provide the processors with an powerfull interactive user interface. After starting the program, a prompt appears on the screen identifying the dialog mode and the user can now input the necessary commands. There is also a possibility to read all commands from a script file, so that a standardized processing or a batch behaviour can be provided.

The interpreter itself is writtem with lex and yacc and has its own grammar rules, which are oriented at the SQL-style. The commands are used to

- include testroutines
- control the logfiles
- start the parsersoftware
- start the formatter (only for test purposes)
- start the postprocessor backends
- set parameters
- show different aspects of the IDS and the current environment

The present implementation supports the following commands, shown in terms of grammar rules:

```
CHECK
  ( ALL              (* the ids_checker checks the whole IDS    *)
  | BREP name        (* the ids_checker checks the named brep   *)
  | G3DSS name       (* the ids_checker checks the named        *)
  | ENTITY  name     (* the ids_checker checks the named entity *)
  ) [ TO file ] ;


CLEAR
  ( ALL              (* deletes the IDS and removes the logfile *)
  | IDS              (* deletes the IDS *)
  | LOG              (* removes the logfile *)
  ) ;
```

```
CLOSE LOG ;              (* closes the logfile *)

EXIT ;                   (* ends the program  *)

FORMAT ;                 (* calls IDS formatter in testenvironment *)

HELP  [ command ] ;      (* Help information; general or for a  *)
                         (* given command                      *)

HOST  command ;          (* executes a system command *)

PARSE  file ;            (* reads a new stepfile and fills up the  *)
                         (* IDS with this data *)

PROCESS
  ( ALL                  (* calls back-end to generate the native  *)
                         (* data of the whole STEP file            *)
  | BREP  name           (*  of the named BRep                     *)
  | G3DSS name           (*  of the named geometric_3d_surface_set *)
  | SBSM  name           (*  of the named shell_based_surface_model *)
  | FBSM  name           (*  of the named face_based_surface_model *)
  | CSG   name           (*  of the named CSG model                *)
  ) [ TO file ] ;        (*  output to file                        *)

QUIT ;                   (* ends the program and lets you return to *)
                         (* your system *)

SET
  ( LOG [ file ]         (* activates a logfile   *)
  | OUTPUT file          (* sets a new output file name  *)
  | SIZE number          (* sets the size of the lookup_table *)
  | TORUS number         (* sets number of generic elements of *)
                         (* the csg torus *)
  | TOLERANCE number     (* sets computing tolerance  *)
  ) ;

SHOW
  ( ( ALL                (* calls every single show command      *)
    | ENTITIES name      (* shows the whole data of the IDS      *)
    | BREP name          (* shows the whole data of the named BRep *)
    | G3DSS name         (*  of the named geometric_3d_surface_set *)
    | SBSM name          (*  of the named shell_based_surface_model *)
    | FBSM name          (*  of the named face_based_surface_model *)
    | CSG name           (*  of the named CSG model       *)
```

```
| VIEWER              (* calls statistic and structure viewer    *)
| PARAMETER           (* shows the settings of the parameters    *)
| HEADER              (* shows the data of the header      *)
| RELEASE             (* shows the release of the postprocessor  *)
| TOEN                (* shows the top_entities       *)
| STAT                (* writes a statistic        *)
) [ TO file ]
| SYMTAB              (* writes the entities in order of their   *)
                      (* appearance *)
| SCAN                (* scans the ids       *)
) ;
```

As the grammar shows, a lot of the CADEX partners' tools were combined to build a interactive tool with a large scope of functionality. The experience with LEX and YACC showed, that it will be very easy to include further tools (e.g. modify routines) into this environment.

4.6.5 Integrated test tools

As mentioned, there are several tools for testing the STEP file or the IDS respectively used from the partners. Nevertheless, there are also a few test routines from PROCAD embedded, which were found necessary in the development process:

```
- show()              shows the whole IDS
- ccbrep(*name)       shows the whole data of the named BRep
- chg3ss(*name)       shows the whole data of the named
                      geometric_3d_surface_set
- chsbsm(*name)       shows the whole data of the named
                      shell_based_surface_model
- chfbsm(*name)       shows the whole data of the named
                      face_based_surface_model
- chcsg(*name)        shows the whole data of the named CSG model
- cdstoe()            shows all top_entities of the IDS
- cdshead()           shows the data of the header
- cstats()            output a table with a statistic of all the
                      the entities in the IDS
```


4.6.6 Summary

The development of the processors during the CADEX project was a very interesting task. It was influenced by
- an always moving target called STEP

- many different views on technical aspects caused by the amount of project partners
- new revisons and updates of the CAD-system
- the increasing KnowHow gained in all the specification and standardization efforts
- experiences made by testing and exchanging data

The processors itself seem to have with a good quality in terms of robustness, less errors and performance. With a few exceptions, all data, the CAD-system can handle, can be transferred via the neutral file. All models, which do not include a torus element, can be postprocessed into the CAD-system. Therefore, the result of the project can be summarized as follows:

1) The processors of the CADEX-project are still prototypes and can't be marketed because of a still non-available standard STEP and some lacks of the CAD-system.

2) The KnowHow gained in the project allows our companies to build real products based on the CADEX processors in a short time frame.

3) Results from CADEX will be used indirectly to improve the CAD-system itself.

Overall, we see a positive summary and hope to get STEP available soon and hopefully close to the specifications we used in CADEX.

4.7 ITALCAD

4.7.1 Introduction

ITALCAD's S7000 CAD system includes a sculptured surfaces modeler GSM (Generalized Surface Modeler) and a CSG solid modeler; ITALCAD's pre- and post-processors for STEP data transfer are designed for this CAD system.

At present ITALCAD supports a new solid modeler called 3D-PSM (Parametric Solid Modeling); however, the CSG and B-rep processors , developed by Italcad, will be completly compatible with 3D-Psm.

Any type of application based on S7000 can be developed using the Application Interface (AI), which is a Fortran 77-callable interface with S7000.

4.7.2 Processor architecture

Processors have been developed integrating the CADEX common tools, the S7000 Application Interface and the software developed to map information between different data structures. Figure 4.7.1 shows the components of this architecture.

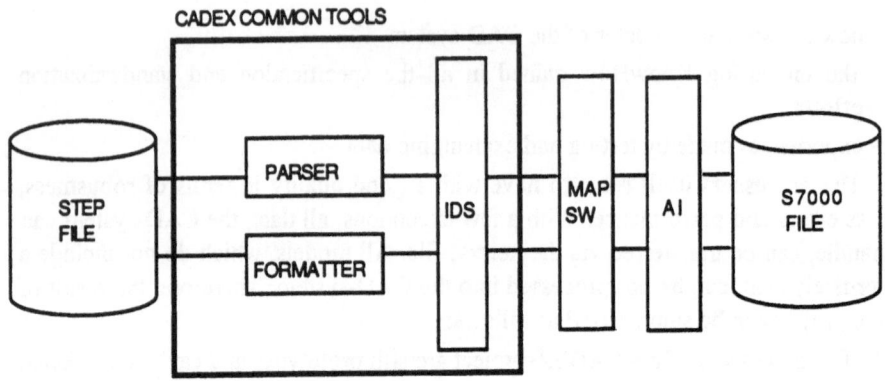

Fig. 4.7.1: ITALCAD's STEP processors for S7000

The major effort in the development of ITALCAD's processors was the software mapping between the Intermediate Data Structure (IDS) and the S7000 data structure, in particular the software converting the mathematical description used by S7000 surface modeler and the Nurbs entities (see CADEX Status Report 29.8.91).

At present the S7000-STEP interface can run directly from S7000 (in graphical session) or out of the system as external application.

The main steps in pre-processing are:

1-identification of S7000 model (name of external file or screen selection)

2-choice of STEP file name

3-identification of Application Protocol

4-reading entities information from S7000 data structure

5-convertion

6-writing converted data in IDS

7-writing STEP file by FORMATTER

The main steps in post-processing are:

1-identification of STEP file

2-choice of S7000 file name

3-identification of Application Protocol

4-reading entities information from STEP file by S/P

5-reading data from IDS

6-convertion

7-writing converted entities in S7000 data structure

4.7.3 Processors built

4.7.3.1 SS_AP processors

The processors for Sculptured Surfaces models are based on SS_AP implementation level 1. The following table shows the main supported entities.

STEP entities	Pre	Post
3D_GEOMETRIC _SURFACE_SET	Y	-
B_SPLINE_CURVE	Y	Y
B_SPLINE_SURFACE	Y	Y
3D_CARTESIAN_POINT	Y	Y

The S7000 Surface Modeler does not support topological information.

4.7.4.2 CSG_AP processors

The processors for CSG models are based on CSG_AP implementation level 1.
Both pre- and post-processing is covered. In figure 4.7.2 an example taken from
the FIAT test part library is shown: Model 1, a nozzle holder, was created in
S7000 and exported by the CSG pre-processor.

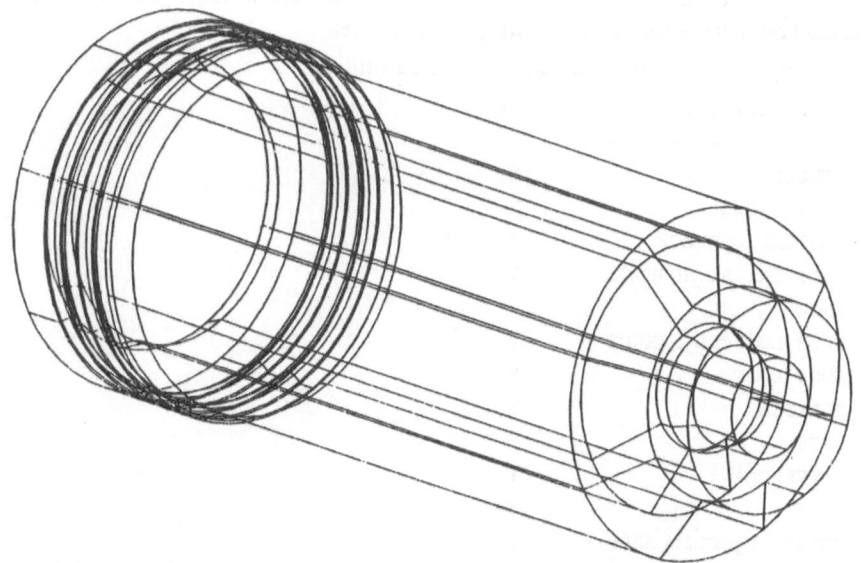

Fig. 4.7.2: Model 1 from FIAT, example for CSG pre-processing

The supported entities are listed in the following table. ITALCAD's solid
modeler uses HALF_SPACE and BOX_DOMAIN only for internal computation.

Solids constructed by REVOLUTION or EXTRUSION are described only by CSG trees.

STEP Entities	Pre	Post
CSG_SOLID	Y	Y
UNION	Y	Y
INTERSECTION	Y	Y
DIFFERENCE	Y	Y
SPHERE	Y	Y
BLOCK	Y	Y
RIGHT_CYRCULAR_CYLINDER	Y	Y
RIGHT_CYRCULAR_CONE	Y	Y
RIGHT_CYRCULAR_WEDGE	Y	Y
TORUS	Y	Y
SOLID_OF_REVOLUTION	-	Y
SOLID_OF_LINEAR_EXTRUSION	Y	Y
HALF_SPACE	-	-
BOX_DOMAIN	-	-

4.7.3.3 B-REP_AP processor

The pre-processor for B-rep models is based on BREP_AP implementation level 1. ITALCAD does not support the B-rep post-processor, because S7000 cannot process boundary information to build a solid model.

The following table shows the main supported entities.

STEP Entities	Pre
CIRCLE	Y
CLOSED_SHELL	Y
CONICAL_SURFACE	Y
CYLINDRICAL_SURFACE	Y
EDGE	Y
EDGE_LOGICAL_STRUCT	Y
EDGE_LOOP	Y
ELLIPSE	Y
FACE	Y
HYPERBOLA	Y
LINE	Y
MANIFOLD_SOLID_BREP	Y
PARABOLA	Y
PLANE	Y
SPHERICAL_SURFACE	Y
SURFACE_LOGICAL_STRUCT	Y
TOROIDAL_SURFACE	Y
VERTEX	Y

4.7.4 Implementation and test

The software was developed on HP 400, with operating system AEGIS 10.3. The test library distributed from Fiat was used to test the processors: all models could be read and trasferred. An example of an exported model is shown in figure 4.7.2.

4.8 Kongsberg 3D Partner A.S.

4.8.1 Introduction

Due to restructuring of Norsk Data (ND) with the effect that ND no longer will be active in the CAD/CAM business, Kongsberg 3D Partner (K3DP) has taken over ND's role in the CADEX project. K3DP was established february 1st. with the same staff previously working at the ND development department in Kongsberg, Norway. ND's engagement in the CADEX project was taken care of by the staff now employed by K3DP. This report must therefore be read and compared to what has previously been reported by ND.

Kongsberg 3D Partner/ND has within the CADEX project developed STEP processors for the prototype geometric modeller TECHPRO. The goal for the work within CADEX was to implement pre- and postprocessors for level 1 of the Sculptured Surface Application Protocol version2 (SS-AP level 1). These processors are finished. Version 2 handles b_spline geometry only, and without topology. SS-AP level 1 version 3 will contain more geometric elements than only b_splines. It will be necessary to develop the processors such that they conform to the latest and hopefully at last ISO approved version of the SS-AP, This will have to continue after CADEX has finished.

TECHPRO is the prototype of what was intended to be Norsk Data's next generation 3D geometric modelling system. K3DP will not continue the development of TECHPRO in order to provide it to CAD end-users. It is implemented in C++, the prototype is running on Silicon Graphics computers. The operating system environment is IRIX4.01, and version 2.0 of the C++ compiler is used.

The modeller is based on:

1) ACIS kernel.

  The core solid modeller within TECHPRO. This is a new generation 3D brep modeller delievered by Spatial Technology Inc., and developed by Three Space Ltd. in Cambridge. It is implemented by utilising object oriented techniques, using C++ as programming language. ACIS is now licensed by a lot of CAD vendors and users, who intend to use it as a kernel modeller in next generation CAD systems.

2) SISL (SI Spline Library).

A C function package containing functions for definitions of, and operations on spline geometry. SISL is from Center for Industrial Research (SI) in Oslo, and is extendinng the ACIS kernel and Techpro to also handle spline geometry.

Together, SISL and the ACIS kernel give potential for developing integrated modellers covering surface and wire modelling as well as solid modelling.

4.8.2 System architecture

The system architecture of Techpro with STEP processors is sketched in the figure 4.8.1.

Extensions which add functionality to the modeller are called husks or applications. The STEP processors are implemented as such husks. We also see from figure 4.8.1 that the CADEX Common Tools are extensively used. The Common Tools are described in chapter 3.

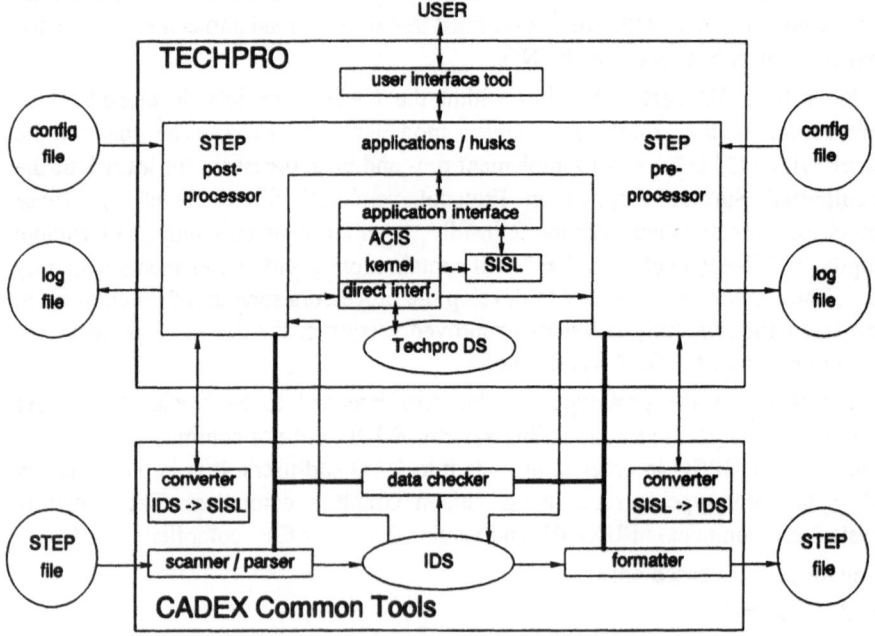

Fig. 4.8.1: K3DP system architecture, thick lines: control, thin lines: data

The STEP configuration file contains data for setting up the STEP processing sessions.The name of the logfile is defined here, and so is the different data fields going into the header section of the STEP file. Such as: Sending company, sending system, step version, author, etc.... (Only used by preprocessor). In addition values needed for initialization of IDS are defined here.

The configuration file gives the user easy access to modify parameters for his STEP processing sessions, without being bored with too many input requests during processing and without needing to change software. The configuration file also enables running the STEP processors as standalone programs in batch mode.

The STEP logfile will after a STEP processing session contain various inform-ation on success and failure for each of the elements processed. At the end of the file a log summary will be written. More details are given in the section 'Error and log handling'.

The implementations of the STEP processor husks are described in more detail in the next two sections.

4.8.3 Pre-processor implementation

This section describes the subroutine structure of the preprocessor, and gives a short verbal description of how the preprocessor works. The most important routines are sketched in the subroutine hierarchy below. Routines from the CADEX common tools are marked with grey boxes.

The preprocessor works as follows:

First the user must input the wanted geometry (BODY) to be processed, and then give the STEP file name. STEP logfile name may be given in by user or the default name from the configuration file may be used. Refering to the subroutine hierarchy in figure 4.8.2, the processing goes on with

* `step_logfile`
- A global struct 'step_conf_table' is initialised with values from the STEP configuration file.
- Initialise error handling and logging. (Using values from 'step_conf_table')
- Initialise IDS (cinids).

* `step_head`
- Write STEP header data into IDS, taking values from 'step_conf_table'.

* `step_out_body`
- Traverse the native datastructure given the BODY (top node in native datastructure) as input, find all instances of b_spline_curves and b_spline_surfaces. Convert the b-splines from SISL format to IDS format using Common Tools conversion routines (ccsibs, ccsibc) and store them in IDS (cpxxxx). Each BODY will become a geometric_3d_surface_set / half_space in STEP. Each b_spline curve and surface will be written to IDS each time they occure in the native datastructure, without notifying if ithey already are written to IDS.

Only b_spline geometry will be written to IDS, i.e. no conversions from analytical to b_spline is done in the preprocessor.

- Write contents of IDS to the STEP file by using the CADEX formatter (cfmt).

- Free memory used by IDS (cfrmem).

* `cnd_logsum_pre`

- Write a log-summary of the preprocessor session at the end of the STEP logfile. Refer to section 'Error and log handling'.

More detailed description of the preprocessor is found in the document: 'System Documentation ND Preprocessor'

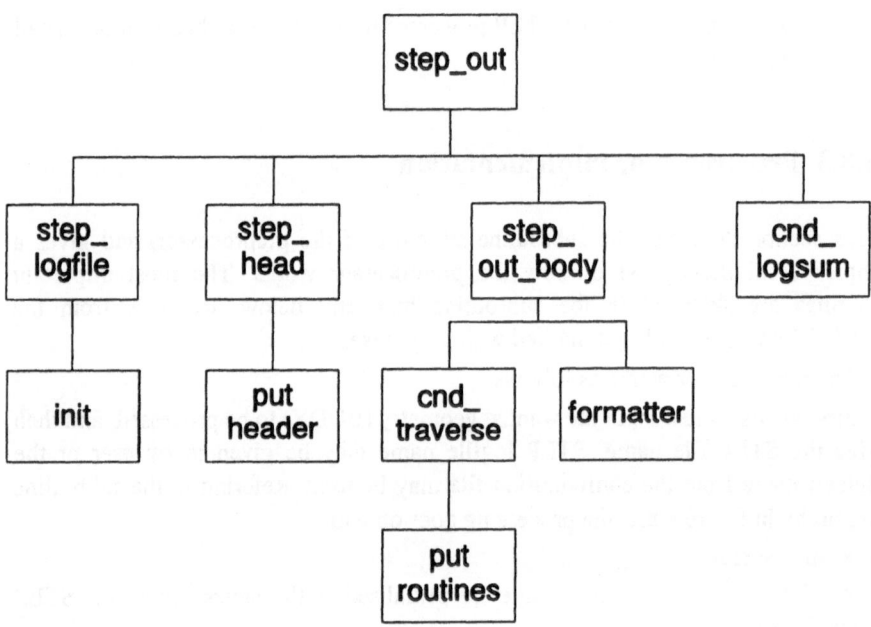

Fig. 4.8.2: Subroutine tree for pre-processor

4.8.4 Post-processor implementation

This section describes the subroutine structure of the postprocessor, and gives a short verbal description of how the postprocessor works. The most important routines are sketched in the subroutine hierarchy below. Routines from the CADEX common tools are marked with grey boxes.

The postprocessor works as follows: First the user must give as input the name of the STEP file he wants to be postprocessed. The logfile name may be given in, or the default name from the configuration file may be used.

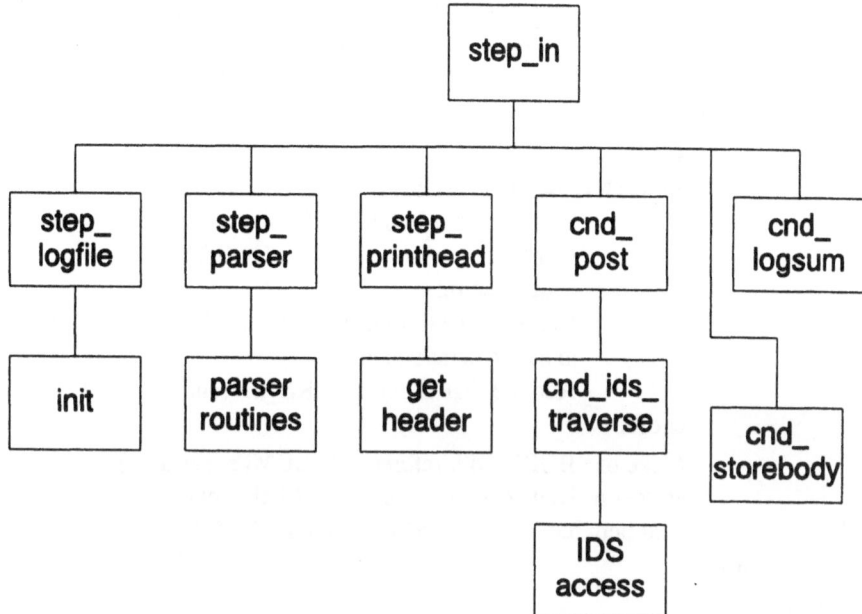

Fig. 4.8.3: Subroutine tree for post-processor

Refering to the subroutine hierarchy in figure 4.8.3, the processing goes on with:

* `step_logfile`
- Same as for pre-processor, but the header data in the configuration file is not used.

* `init_parser`
- Initialize the CADEX scanner/parser.

* `step_parser`
- Parse the STEP file and write into IDS using the CADEX scanner/parser.

* `step_print_head`
- Read STEP file header from IDS and write it to stdout for user information.

* `cnd_post`
- Traverse IDS and collect the data in this way:

> for each HALF_SPACE
>> for each B_SPLINE_SURFACE
>>> - Convert from IDS to SISL format. (csi_isbs)
>>> - Surfaces in TECHPRO need to have boundaries, so generate 4 b-spline curves along the min and max parameter values of the surface.
>>> - Connect the b-spline surface to a FACE element in

native datastructure, and the boundary curves to
LOOP, EDGE and COEDGE elements for this FACE.
- Connect FACE to SHELL in native datastructure.
- Connect SHELL to LUMP in native datastructure.
- Add the LUMP into a LUMP_list (cnd_lu_add).
endfor

for each B_SPLINE_CURVE
- Convert from IDS to SISL format. (csi_isbc)
- Connect b_spline_curve to EDGE in native
datastructure.
- Connect EDGE to COEDGE and COEDGE to WIRE
in native datastructure.
- Add the WIRE into a WIRE_list (cnd_wi_add).
endfor

Make one BODY with reference to LUMPs (surfaces).
Make one BODY with reference to WIREs (curves).
Add both bodies into BODY_list (cnd_bl_add).
endfor

* cnd_store_body
- Go through the BODY_list created in cnd_post, and store each of the bodies into the Techpro model.

* cnd_logsum_post
- Write a log-summary of the postprocessor session at the end of the STEP logfile.
Refer to section 'Error and log handling'.

* display the Techpro model generated from the STEP file

The fact that additional b-spline boundary curves are generated during postprocessing, will give the result that when the model is written to STEP file again, the new and the original STEP file will not be equal. There will be additional b-spline curves on the new STEP file. The model is not changed, but additional information is created. This special solution will not be necessary with level 2 of SS-AP, but is necessary now because TECHPRO needs topology and SS-AP level 1 does not support topology.

The current implementations of 'cnd_post' (postprocessor) and 'cnd_prtbody' (preprocessor) and the routines below them in the diagrams above, give the framework which easily enables extension of the processors to cover more STEP elements, and to support other Application Protocols. Hence, much of the basic work for implementing version 2 and 3 of the SS-AP, and the Brep-AP is done.

More detailed description of the postprocessor is found in the document: 'System Documentation ND Postprocessor.

4.8.5 Error and log handling

In the processors, six routines for error and log handling are used. All interaction with the STEP logfile goes through these routines. Both SI and K3DP are using the same concept for error and log handling.

A short description of the routines are given below:

* cnd_errinit

- Initialising error and log environment.

* cnd_error

- Analyse a reported error status and print an error/warning message onto the logfile.

* cnd_logent

- Print entity related logging information onto the logfile (entity name, entity type, action).

* cnd_logtext

- Print any logtext onto the logfile.

* cnd_logsum_pre

- Print summary of all accumulated logging info from preprocessing Techpro elements onto the logfile, and close the logfile.

* cnd_logsum_post

- Print summary of all accumulated logging info from postprocessing Techpro elements onto the logfile, and close the logfile.

Cnd_errinit is called at the beginning of the processing session, and cnd_logsum_pre/post are called at the end of the different sessions. The other three routines are used all places in the code where errormessages and logging information are senseful.

More details on the error and loghandling routines are given in the document: 'System Documentation ND Preprocessor'.

An example of a logfile created by the postprocessor would look like this:

```
Mon Jan 13 12:47:59 PST 1992
Version 0.1 of the ACIS SS-AP FL1 post-processor
LOG:   initialize_error_handling_-_OK
LOG:   Initialize_ids_lut_-_OK
1 (B_SPLINE_CURVE): Detected in IDS.
1 (B_SPLINE_CURVE): Conversion started (IDS -> ACIS).
1 (intcurve): Stored in ACIS.
6 (B_SPLINE_CURVE): Detected in IDS.
6 (B_SPLINE_CURVE): Conversion started (IDS -> ACIS).
6 (intcurve): Stored in ACIS.
11 (B_SPLINE_CURVE): Detected in IDS.
```

```
11 (B_SPLINE_CURVE): Conversion started (IDS -> ACIS).
11 (intcurve): Stored in ACIS.
16 (B_SPLINE_CURVE): Detected in IDS.
16 (B_SPLINE_CURVE): Conversion started (IDS -> ACIS).
16 (intcurve): Stored in ACIS.
21 (B_SPLINE_CURVE): Detected in IDS.
21 (B_SPLINE_CURVE): Conversion started (IDS -> ACIS).
21 (intcurve): Stored in ACIS.
26 (B_SPLINE_CURVE): Detected in IDS.
26 (B_SPLINE_CURVE): Conversion started (IDS -> ACIS).
26 (intcurve): Stored in ACIS.
31 (B_SPLINE_CURVE): Detected in IDS.
31 (B_SPLINE_CURVE): Conversion started (IDS -> ACIS).
31 (intcurve): Stored in ACIS.
36 (B_SPLINE_CURVE): Detected in IDS.
36 (B_SPLINE_CURVE): Conversion started (IDS -> ACIS).
36 (intcurve): Stored in ACIS.
41 (B_SPLINE_CURVE): Detected in IDS.
41 (B_SPLINE_CURVE): Conversion started (IDS -> ACIS).
41 (intcurve): Stored in ACIS.
46 (B_SPLINE_CURVE): Detected in IDS.
46 (B_SPLINE_CURVE): Conversion started (IDS -> ACIS).
46 (intcurve): Stored in ACIS.
51 (B_SPLINE_CURVE): Detected in IDS.
51 (B_SPLINE_CURVE): Conversion started (IDS -> ACIS).
51 (intcurve): Stored in ACIS.
56 (B_SPLINE_CURVE): Detected in IDS.
56 (B_SPLINE_CURVE): Conversion started (IDS -> ACIS).
56 (intcurve): Stored in ACIS.
61 (B_SPLINE_SURFACE): Detected in IDS.
61 (B_SPLINE_SURFACE): Conversion started (IDS -> ACIS).
61 (spline): Stored in ACIS.
78 (B_SPLINE_SURFACE): Detected in IDS.
78 (B_SPLINE_SURFACE): Conversion started (IDS -> ACIS).
78 (spline): Stored in ACIS.
95 (B_SPLINE_SURFACE): Detected in IDS.
95 (B_SPLINE_SURFACE): Conversion started (IDS -> ACIS).
95 (spline): Stored in ACIS.
112 (B_SPLINE_SURFACE): Detected in IDS.
112 (B_SPLINE_SURFACE): Conversion started (IDS -> ACIS).
112 (spline): Stored in ACIS.
129 (HALF_SPACE): Detected in IDS.
129 (body): Stored in ACIS.
LOG:  Transfer_SS_AP_data_from_IDS_to_ACIS_-_OK
LOG:  Free_memory_from_ids_-_OK
```

```
LOG:   Save_body_-_OK
LOG:   Save_body_-_OK


Log-summary post-processor
*************************
Number of entities identified in IDS  : 17
     12 B_SPLINE_CURVE
      4 B_SPLINE_SURFACE
      1 HALF_SPACE
Number of entities not post-processed : 0
Number of IDS-entities to be converted: 16
Number of IDS-entities converted      : 16
Number of entities stored in ACIS     : 17
      1 body
      4 spline
     12 intcurve
Of these post-processed with warnings : 0
Of these post-processed with errors   : 0
Number of unknown actions performed   : 0
```

The error and log handling environment described in this section will report all errors/warnings registered in the TECHPRO husks (preprocessor front end and postprocessor back end). Errors discovered in the formatter and scanner/parser will be reported by these modules, but the error handling described here will report that something has gone wrong in the formatter or scanner/parser. For the postprocessor, syntactical errors in the STEP file will be discovered by the scanner/parser, but semantical errors will be discoverd in the back end of the postprocessor and reported on the STEP logfile. The errormessage may not allways give the exact place to find the error, but rather give an indication on where the error is. The error and log handling has proven to be efficient especially for semantical errors, errors in data conversions and for finding errors in implementations of the processors.

4.8.6 Testing

ND/K3DP has exchanged a number of STEP files with SI, DnV SNI, and one with ITALCAD. No crucial problems are discovered during these exchanges. Some of the test objects from the test and validation group has been used. These models and the results of the processing is given with the report from the test group.

Since Techpro is a solid modeller with upto now few applications working on freestanding surfaces and curves, the only checking of imported geometry from STEP is done by visual inspection (display, rendering). From our experience with geometry imported from VDAFS we have learned that irregularities in the imported geometry are not discovered before heavy applications use the geometry (intersection algorithms, machining applications). Therefore, an extensive testing of the processors and the imported geometry is not possible with Techpro today.

Due to the fact that STEP turned out to be less stable than expected, the testing suffered from the excessive time that had to be spent in order to update code for latest versions of the AP's and Common Tools.

This section describes two models which we have received from SI and used our STEP postprocessor to read into the TECHPRO modeller. They are shown in figure 4.8.4. The STEP file "Half of a bicycle chair for children" contains 65 b-spline surfaces in 1 half_space. The STEP file "Half of a snow sledge for children" contains 36 b-spline surfaces in 1 half_space. No errors occurred in the post-processing.

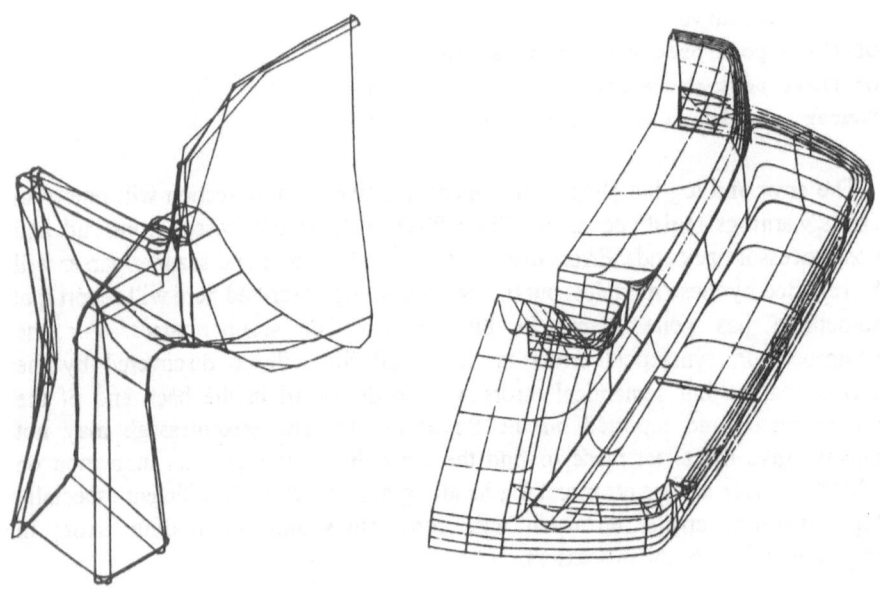

Fig. 4.8.4: Bicycle chair and snow sledge for children (transfer of one half each)

Two models which are written to STEP with the preprocessor are also shown. These models are generated in TECHSURF, where all geometry is converted to b-splines and sent via a dedicated link to TECHPRO. Version 2 of the Sculptured Surface Application Protocol level 1 is used in the processors.

The first one of these models is a graphite electrode used for making a drill. housing. Figure 4.8.5 shows the result in TECHPRO after writing the model to a STEP file, and reading it in again. The STEP file contains 132 b-spline curves and 33 b-spline surfaces in 1 half_space. No errors in the processing.

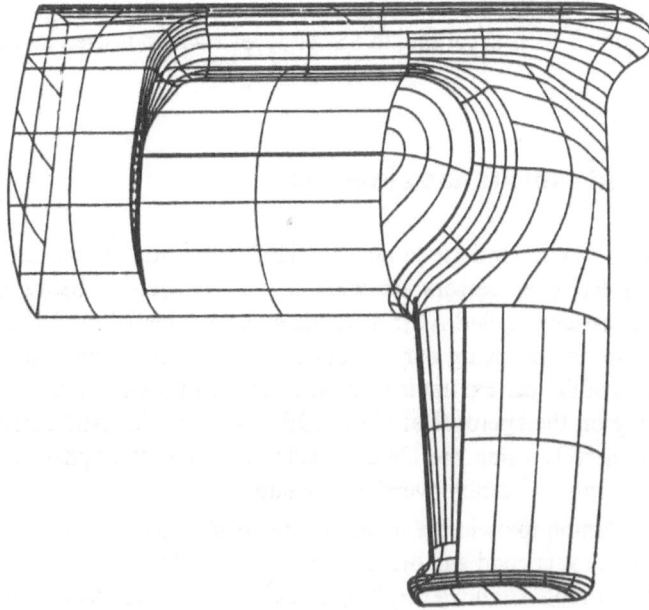

Fig. 4.8.5: Model of a graphite electrode used for making a drill housing.

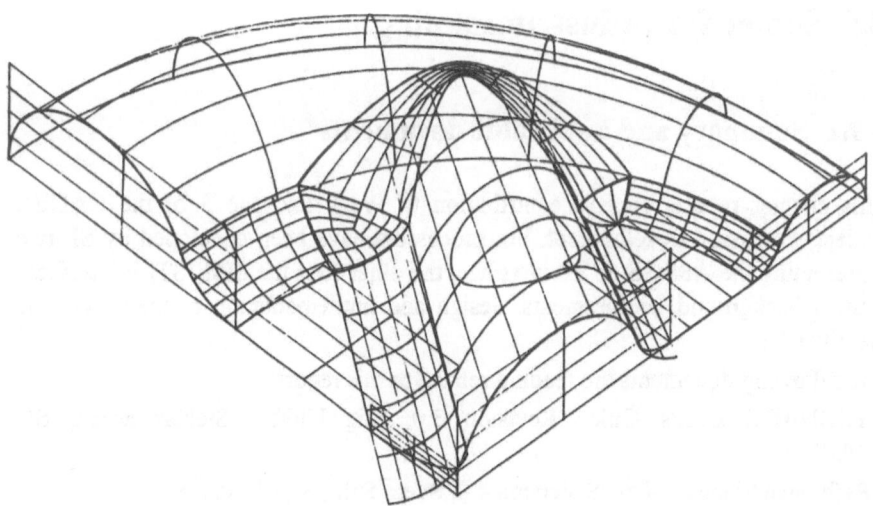

Fig. 4.8.6: Demo part after being written to STEP file and read in again.

The second model is a demo part consisting of 71 b-spline curves and 17 b-spline surfaces in 1 half_space. We notice that the surfaces are not trimmed against boundary curves anymore. This is according to SS-AP level 1 which is used in the processors.

4.8.8 Comments and exploitation of results

Kongsberg 3D Partner will not continue to develop TECHPRO into an end-user product. But since our STEP processors are working as a convertion between STEP and the ACIS datastructure, and ACIS is a widely spread kernel modeller, we see a potential market in enhancing our processors to cover all the brep- and SS-AP levels. The value of the current implementations as basis for enhancements are of course depending on the approval of the CADEX AP's in ISO. And since our processors are extensively using the CADEX common tools, the update of these tools to allways comply to latest AP versions are important.

Using the CADEX common tools in the processor implementations has been a nice experience. The concept is good and the implementations seems to be stable. The value of the common tools are however, as previously mentioned, depending on the updates and maintenance of these tools, and that they are accepted in the market as a good basis for building STEP processors.

4.9 Senter for Industriforskning

4.9.1 Summary and applicable documents

This chapter reports on SI's contribution to Work Package 5 of the CADEX project, Processor Development. Six processors have been developed by SI, two representing the APS/SS-STEP interface, the other four the SISL-STEP interface. System background, requirements, design and implementation of the processors are given below.

The following documents are reading related to this report:

1 TORNADO User's Guide Revision 3.0, Stig Ulfsby, Steinar Meen, SI, 10.05.1983

2 Reference Manual APS-SS Version 4.0, Svein Solli, SI, 01.10.1987

3 Application Protocol for the Data Transfer of STEP Sculptured Surface Models via STEP Physical File (version 1.1), Per Evensen/Jochen Haenisch, Oslo, 21.5.90

4 CADEX Status Report, 30.01.91

5 SISL Reference Manual Version 2.1, Arne Laksî, Mike Floater, SI, 05.04.1991

6 CADEX Status Report, 14.02.92

7 ISO TC184/SC4/WG3/P1 N132: 10303-205 Industrial Automation Systems - Product Data Representation and Exchange - Part 205: Application Protocol for Mechanical Design Using Surface Models, Per Evensen, Jochen Haenisch, Oslo, 01.05.1992

4.9.2 The SISL system

4.9.2.1 History

SISL, the SI Spline Library, is a subroutine library based on B-spline mathematics developed by SI. The library is a mathematical tool box for the modelling of, and operations on, curves and surfaces of high complexity.

The B-spline activities started at SI in 1976, and through 1978-1981 the development was enhanced within the inter-Nordic GPM project (Geometric Product Models). The concepts and results from these sculptured surface activities were offered to and accepted by the APS project. APS (Advanced Production System) was a joint German - Norwegian research and development project over the period 1981-1987. Within the APS project a FORTRAN package for defining B-spline curves and surfaces and for intersecting B-spline described geometries and B-spline represented geometries and analytical surfaces was developed. It was called the APS B-spline library, and is used in several European CAD/CAM and FEM systems.

Instead of upgrading the APS B-spline library, SI decided in 1988 to make a new product based on the latest technological results, experience and ideas from the use and development of the B-spline based software. This product is the SI Spline Library (SISL). SISL is written in C and made to satisfy the demands for functionality and accuracy expected from the market in the 1990s. The first version of SISL was based on non-rational B-splines. However, in 1990 NURBS (Non-Uniform Rational B-splines) based algorithms were introduced into SISL, e.g. NURBS intersection functionalities. Today almost all SISL functionality is applicable also to NURBS.

4.9.2.2 The functionality of SISL

The functionality of the spline library can be grouped as follows:
CONSTRUCTION
Curves and surfaces can be defined by interpolation or other approximation methods. Some keywords here are: Hermit interpolation, lofting, offset, fillets and blending.

MANIPULATION

Manipulations of many kinds are possible. Functions exist for splitting and linking of geometry. NURBS can be converted to a non-rational B-spline or Bezier representation.

DATA REDUCTION

Most interesting here is the data reduction functionality that reduce the amount of data, but keep the geometry within a guaranteed tolerance. In a similar way the polynomial degree of a B-spline represented geometry can be reduced, controlled by a user specified tolerance.

CALCULATION

Calculation of position, derivative up to wanted order, tangents and normal vectors, as well as calculation of area properties.

INTERSECTION

Finding intersection points and curves between two geometries is central to SISL. These functions are frequently used and have high precision. Examples of intersection functions available are curve/curve, curve/conic surface, curve/surface, surface/conic surface and surface/surface. Also calculations of extremals, and of "closest points" between geometries exist. The surface/surface intersection is divided into two parts, first the topology of the intersection is found, then the intersection curves are marched with a user specified tolerance.

ANALYSIS

Adjacency analysis, a method to establish adjacency relations among B-spline surfaces, is one type of analysis that is available through SISL.

HELP FUNCTIONS

Functions exist for visual presentation, for mathematical transformations, matrix computations, vector calculus and general equation solvers (direct and iterative methods).

BASIC ALGORITHMS

Basic algorithms for evaluation of position and derivatives and subdivision have been developed.

4.9.2.3 Extensions to SISL with impact on the SISL-STEP interface

SISL is a living product under constant development. New algorithms are added, existing algorithms are enhanced as e.g. those for handling NURBS. It is also likely that SISL will get a link to C++. Related to this an extended datastructure will be established. It seems opportune to use the Surface Application Protocol (Surface-AP) of STEP as a baseline for these extensions. This will ensure conformance between the SISL-format and the STEP-format.

This extended datastructure will, however, not include explicit topology; SISL is intended to remain a pure mathematical library. SISL can, therefore, only hold and reproduce by its pre-processor purely geometrically described surface models. The topic is discussed further in the requirement section below.

4.9.3 Requirements for the SISL-STEP interface

General interface requirements. Though the SISL-product is a mathematical library for surface modelling and not a CAD-system itself, it is necessary to enable communication between the SISL-representation of product shape and corresponding external representations. This is best done by implementing an interface between SISL and a standard exchange format. The format chosen for the CADEX project is STEP. The mechanism of model exchange, also called implementation form, is file based - in contrary to being e.g. a database solution.

Product model specification. Both information exchange and information sharing using STEP must be based on Application Protocols (APs). APs specify the product model domain that shall be communicated. Application Protocols must be selected for the SISL-STEP interface that are relevant for the SISL information model.

CADEX focuses on geometry data exchange; APs of corresponding scopes have been developed by the project. They are capable of representing CAD-models of types Boundary Representation (Brep), Constructive Solid Geometry, Surface, Wireframe, Compound Boundary Representation - when applied to mechanical design. Several of these APs offer different alternative model representations called Functional Levels (FL). FLs differ in their modelling capabilities.

SISL is designed to build the kernel of surface modellers. It is, thus, a natural choice to use the Surface-AP as an interface specification. SI supports all 3 Functional Levels of the Surface-AP:

FL1 - Geometrically bounded surface model

FL2 - Non-manifold surface model

FL3 - Manifold surface model.

Post-processors are developed for each of them. Only for FL1 (geometrically bounded surface model) a pre-processor is provided. As indicated above SISL can and will in the near future not hold explicit topology. Pre-processors are, therefore, not foreseen for SISL.

Customer oriented development. The SISL software respectively its predecessor APS-SS are part of several of the systems that are present in CADEX. It was a requirement to design the SISL-STEP interface in a way that minimizes overall development efforts, including potential efforts of users of APS-SS or SISL in developing their STEP processors.

Migration from APS-SS to SISL. The first processors developed by SI within CADEX were for SI's surface modelling software APS-SS. In course of the project period SI has released a new version of its surface modelling package. The old APS-SS software was replaced by SISL. The first two STEP processors are made for APS-SS according to the Sculptured Surface Application Protocol version May 1990 (as published in D2). A set of the four processors is based on SISL. These latter processors are conformant to the May 1992 versions of the AP 205. The total number of SI's STEP-processors amounts to six.

Software and hardware requirements. The SISL-STEP processors are written in ANSI C for a UNIX operating system. The CADEX software development guidelines have been followed. The CADEX Common Toolkit has been applied. The processors are implemented on Hewlett Packard 9000 workstations.

4.9.4 Architecture of the SISL-STEP processors

4.9.4.1 The CADEX common architecture

All CADEX partners agreed upon a common architecture for STEP pre- and post-processors. These architectures have been applied to all SISL-STEP processors. The figure 4.9.1 shows the typical structure of CADEX pre- and post-processors. Modules from the CADEX Common Toolkit, which are independent of any native CAD- or FEA-system, are shaded. The other modules have been developed for the APS-SS/SISL applications specifically.

The pre-processor architecture

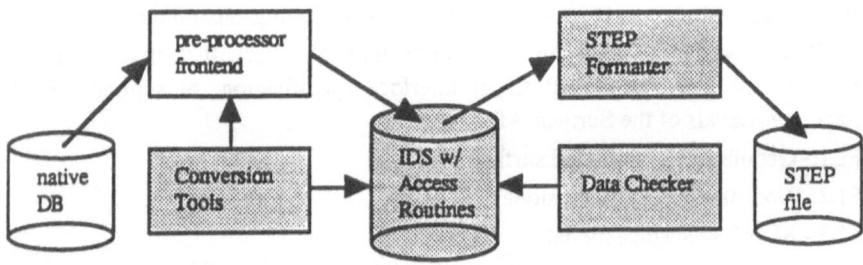

The post-processor architecture

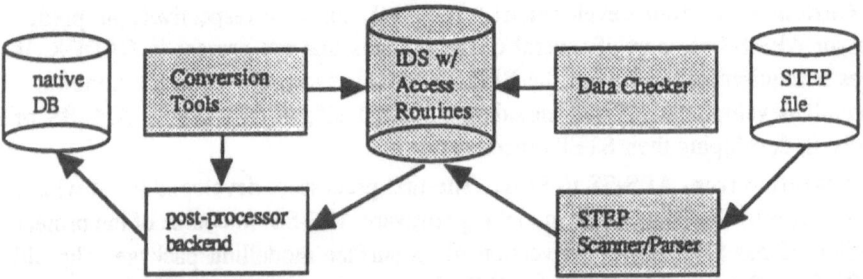

Fig. 4.9.1: Pre- and post-processors consist of system specific and system independent modules.

According to figure 4.9.1 the following common tools are involved in translating STEP models into SISL native models and vice versa:

Conversion Tools
- tools for converting from native to STEP representation and between different STEP-representations;

Intermediate Data Structure (IDS)
- in-core storage format based on the STEP specification;

IDS Access Routines
- set of subroutines manipulating the IDS;

Validation and Error Checker
- module for evaluating the contents of the IDS;

Formatter   - module for writing the IDS onto a STEP formatted file.

Pre-processing starts - as shown in the upper part of figure 4.9.1 - with the pre-processor frontend. The frontend scans the native model and calls appropriate IDS-routines to store the model in the Intermediate Data Structure (IDS). This process involves little or none conversions as the utilized Application Protocols provide a data structure similar to the one of SISL. The conformity of the generated IDS-model with the stated AP is validated by the Data Checker common software. After successful validation the STEP-Formatter writes the model in the STEP-format onto an ASCII-file; formatting is controlled by a configuration file.

A STEP-file is starting-point for post-processing (see lower part of figure 4.9.1). The Scanner/Parser common software reads the file into the IDS checking the file contents for syntax errors. The IDS-model is validated concerning AP-conformity. Depending on the type of post-processor more or less extensive conversions are done. Parts of these occur in a STEP-to-STEP manner, changing the contents of the IDS. Parts occur in the post-processor backend on the way of the model into SISL.

These two architecures are in principal independent of Application Protocols and Functional Levels. There are, however, AP dependent differences in the implementation of the architectures; the types of required conversions are different from processor to processor. These differences are described in detail below, when each processor is presented separately.

4.9.4.2 The native parts of the SISL-processors

Both pre- and post-processor have one native part each, the white boxes in figure 4.9.1: pre-processor frontend (short: frontend) and post-processor backend (short: backend) respectively. In contrary to the common tools these native parts are different for the different native systems that are involved in CADEX.

All SISL-processors are built around the same architecture. And according to the requirements stated above this architecture can be applied to the native parts of all processors of the APS-SS/SISL-systems family. Figure 4.9.2 below illustrates this relationship for the frontend. The backend process is obtained when reading the picture bottom-up.

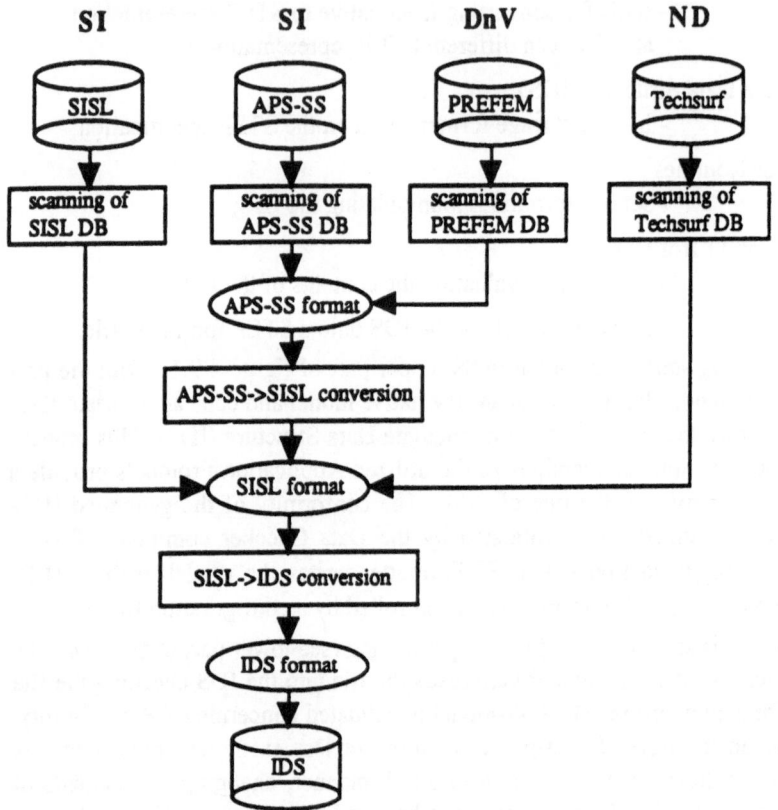

Fig. 4.9.2: Frontend-architecture for the APS-SS/SISL-systems family.

The figure shows databases, software (boxes) and data formats (ellipses). The architecture has been adopted and jointly developed by the Norwegian CADEX partners DnV, ND and SI.

It was decided on a stepwise conversion architecture for the involved systems (APS-SS/SI later SISL/SI, PREFEM/DnV, Techsurf/ND). Corresponding to figure 4.9.2 the frontends of the pre-processors consist of modules for:

- scanning of the native database (done by each partner independently);
- conversion to SISL-format;
- conversion to IDS-format;
- storage in IDS.

A common recursive algorithm has been applied by all the three partners for scanning the IDS - for details see below (figure 4.9.4). The conversion tools (APS

<-> SISL, SISL <-> IDS) were provided by DnV and SI and are now part of the CADEX Common Toolkit.

As already indicated above the SISL-processors vary in their use of conversion tools which is dependent on the type of product model to be converted. This topic is dealt with in detail per processor below.

4.9.4.3 General SISL-processor utilities

Front- and backend of the SISL-processors make use of common utilities of two categories:

1) entity relation tables;
2) error and log system.

Entity relation tables. When translating one native entity to one or several IDS-entities or vice versa, it is necessary for further processing to keep track of the relation between corresponding entities. A set of routines has been developed to store and access an unpredictable number of one-to-one or one-to-many relationships. The software is part of the CADEX Common Toolkit.

Error and log system. For the handling of error and log messages a system has been implemented based on the following requirements:

- integration of error messages from different sources (SISL, APS-SS, IDS, ...);
- error tracebility from log-file;
- all messages from ASCII-files;
- logging onto optionally user-defined or default files;
- logging of actions and of source and target entity names;
- log-summary/statistics after processing.

The system has been implemented in all SI and ND processors.

4.9.5 General information on the SISL-STEP processors

4.9.5.1 Available processors

The STEP pre- and post-processors that have been developed for SISL during the CADEX project support the following Application Protocols and Functional Levels.

Pre-processors:
Surface-AP
    FL1 - Non-topology (or implicit topology) surface model

Post-processors:
Surface-AP
    FL1 - Non-topology surface model
    FL2 - Non-manifold surface model
    FL3 - Manifold surface model.

On the pre-processing side only FL1 of the Surface-AP is supported. It has been argumented for this limitation above. However, it might be repeated that SISL has an internal datastructure consisting of geometry and implicit topology. This fact makes it natural to support this Functional Level, only. The three Functional Levels that are supported on the post-processing side require different types of conversions.

The total number of processors developed is six: Two pre-processors supporting the Surface-AP FL1 exist - one for the APS-SS system and one for the SISL-product; two post-processors supporting this Functional Level exist as well; finally for each of the other supported Functional Levels there is one SISL post-processor. Pre- and post-processors for the SISL-product are implemented in the "SISL demo system" which is presented below.

The APS-SS processors are built upon version 5.0 of the Common Toolkit, the SISL processors on version 6.1. The APS-SS processors have not been maintained throughout the project, i.e. they have not been updated to new versions of the toolkit and the APs. The SISL processors are up-to-date with the final version of the CADEX Common Toolkit. Thus they correspond to the formats specified in the May 92 version of AP 205. However, the entity types of the older version of AP 205 (May 90) are supported.

Before a description of the different processors is given, below the database of the APS-SS system, the datastructure of the SISL-product, and the SISL demo system are described shortly.

4.9.5.2 The data structures of APS-SS and SISL

The entity types of the APS-SS datamodel are shown in figure 4.9.3.

The following APS-SS entity types have been used for STEP-file processing:

PATCH: representation of a non-rational B-spline surface;

SURFACE: a set of PATCHes;

CURVE: representation of a non-rational B-spline curve;

CURVE SET: a set of CURVEs.

The following SISL entities are used by the SISL processors:

SISLPoint: coordinates of a cartesian point with arbitrary dimensionality;

SISLCurve: representation of a NURBS curve;

SISLSurf: representation of a NURBS surface;

SISLFace: representation of a trimmed NURBS surface, where the trimcurves are 2D NURBS curves in the paramter plane of the surface (note that this entity is not implemented at the moment this is written).

Other entities of the SISL datastructure will be of interest for future enhancements of the SISL-STEP interface:

SISLdir: direction cone of a SISLcurve/SISLSurf;

SISLbox: minimum surrounding box of a SISLcurve/SISLSurf;

SISLObject: contains one of SISLPoint/SISLCurve/SISLSurf;
SISLIntcurve: intersection curve;
SISLIntpt: intersection point;
SISLIntlist: list of intersection point;
SISLPtedge: intersection at edges;
SISLEdge: list of intersection at edges.

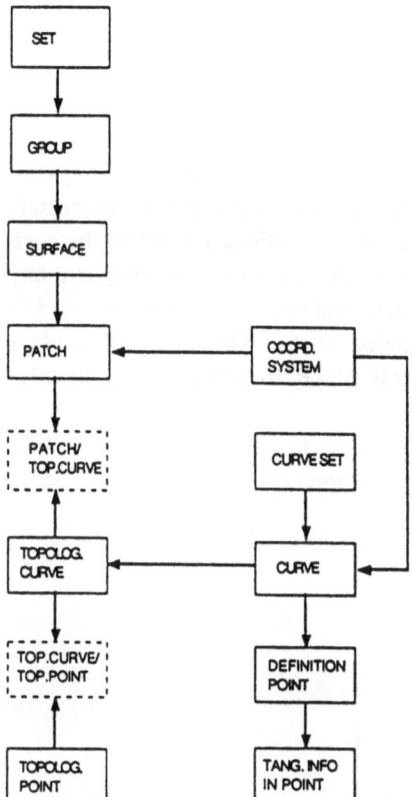

Fig. 4.9.3: The extended APS-SS datamodel

As mentioned above an extended SISL datastructure will be established. This datastructure will most likely include all subtypes of the STEP entities unbounded_curve and elementary_ surface.

4.9.5.3 Formats and conversions in SISL pre-processors

Two pre-processors have been developed, one for APS-SS, one for SISL. The necessary conversions are few because of the small number of basic geometric entities and the conformance between SISL and STEP formats. Some of the conversions are nevertheless complex, e.g. when rational SISLCurve/SISLSurf entities (NURBS) are converted to a non-rational representation on IDS-format. The corresponding mathematical tool exists in SISL, but is not part of the ordinary CADEX Conversion Tool Library.

The architecture of the pre-processor frontends are in principle identical. The native data structures are scanned for the entity types listed above in chapter "The data structures of APS-SS and SISL". They are converted "on the fly" to STEP conformant representations and put into the IDS. The target entities of these conversions are one of: cartesian_point, b_spline_curve, b_spline_surface or curve_bounded_surface.

4.9.5.4 Formats and conversions in SISL Post-processors

Post-processors supporting the Surface-AP FL1 exist for both the APS-SS system and the SISL-product. The other supported Functional Levels have post-processors only for the SISL-product. The number of conversion routines utilized by the post-processors is higher than for the pre-processors. The reason is obviously that any geometric entity covered by the supported Functional Levels must be converted to one of the four basic SISL entities mentioned above. Moreover, all explicit topology has to be converted to implicit topology, i.e. to geometrically bounded entities.

The overall architectures of the post-processor backends are the same. Before data is transferred from IDS to SISL all explicit topology must be converted to implicit topology.

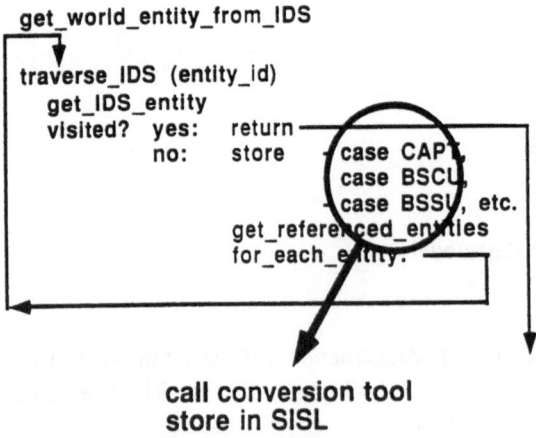

```
get_world_entity_from_IDS

traverse_IDS (entity_id)
   get_IDS_entity
   visited?  yes:    return
             no:     store    case CAPT,
                              case BSCU,
                              case BSSU, etc.
                    get_referenced_entities
                    for_each_entity.
```

**call conversion tool
store in SISL**

Fig. 4.9.4: The post-processor backend traverses the IDS recursively

Conversions of geometric STEP entities with both source and target entity being within the IDS are performed. Target entities of these conversions are one of:

cartesian_point, b_spline_curve, b_spline_surface or curve_bounded_surface. Moreover, the post-processor backends are all traversing the IDS in a recursive, bottom-up manner. Conversion to the relevant SISL entities are done "on the fly". In the "SISL demo system" the post-processors' backends contain the following functions:

- convert explicit to implicit topology (conversion between STEP-entities within the IDS);

- convert geometric entities (optional: conversion between STEP-entities within the IDS);

- recursive, bottom-up traversing of the IDS;

- convert to relevant SISL entities.

Figure 4.9.4 illustrates the functionality of the backend schematically as it has been utilized by the Norwegian partners.

4.9.5.5 The SISL Demo Sytem

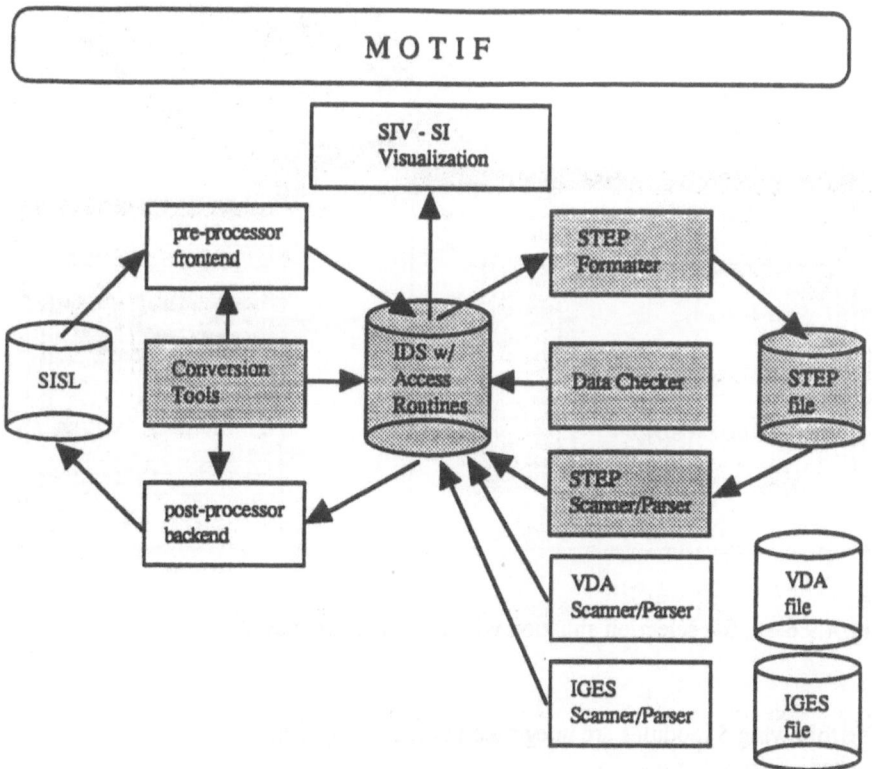

Fig. 4.9.5: The architecture of the SISL demo system

Figure 4.9.5 is a draft of the architecture of the system. The architecture is an extension of the general CADEX processor architecture shown in figure 4.9.1.

The set of SISL processors developed in CADEX has been embedded into a demonstration system, due to the lack of a user interface for SISL. This demo system provides user control mechanisms, that is menues for selecting functionality and model visualisation. There are no means for creating or manipulating geometry. Visualisation parameters, however, can be changed interactively.

The demo system is a MOTIF application. Figure 4.9.6 is a picture from a typical usage of the system showing two windows for user input and the display area.

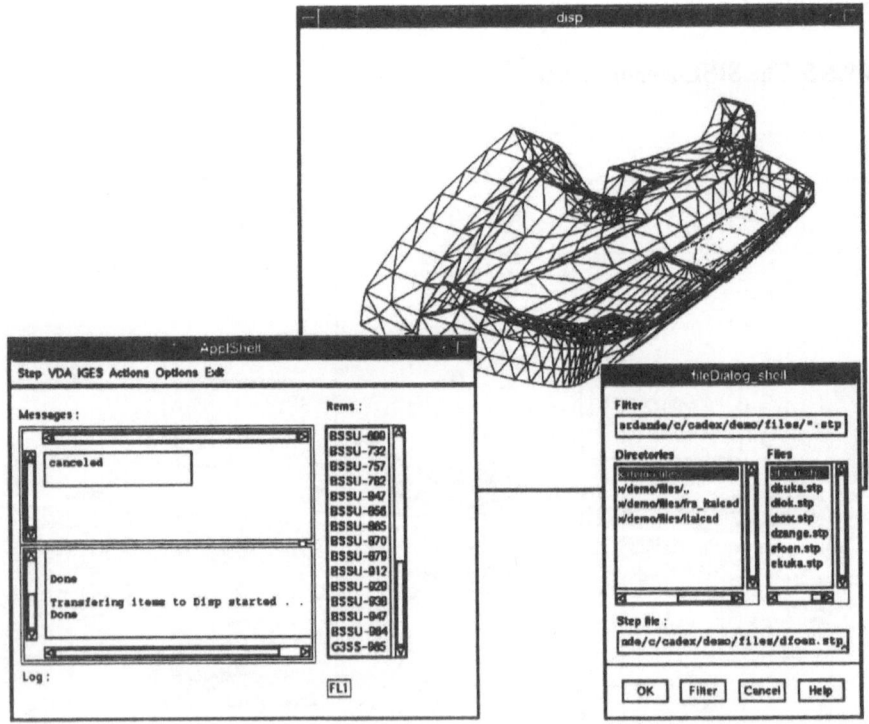

Fig. 4.9.6: A file selection situation with the SISL demo system

The following 5 modules are integrated in thedemo system:
- SISL (SI Spline Library)
- SIV (SI Visualisation)
- STEP <-> SISL processors

- VDA -> SISL processor

- IGES -> SISL processor

IGES, VDA and STEP files are read, converted and stored in the IDS. All IDS contents can be visualised, either model-wise ore entity-wise. Different conversions can be applied to the IDS contents such as

- FL3 -> FL2

- FL2 -> FL1

- rational -> non-rational splines

- conics -> b-spline curves

STEP files may be generated from the IDS independent of wether the original model was represented by IGES, VDA or STEP.

4.9.6 Processor details

4.9.6.1 Surface-AP, Functional Level 1, Pre-processors

Two pre-processors supporting the Surface-AP FL1 exist - one for the APS-SS system and one for the SISL-product. The difference lies in the frontend of the two processors. Figure 4.9.2 above illustrates this - the APS-SS frontend is scanning the APS-SS DB and the SISL frontend is scanning the SISL datastructure. The basic geometric entities of the two systems are B-spline curve and surface, thus the conversions needed are the same. An extra conversion from APS-SS- to SISL-format is needed for the APS-SS frontend.

The ordinary conversions from SISL- to IDS-format used by the pre-processors are:

CCSIPT	convert a SISLPoint to a cartesian_point on IDS-format;
CCSIBC	convert a SISLCurve to a b_spline_curve on IDS-format;
CCSIBS	convert a SISLSurf to a b_spline_surface on IDS-format;
CCSICS (*)	convert a SISLFace to a curve_bounded_surface on IDS-format.

Specifically for the APS-SS backend:

CSI_ASBC	convert a B-spline curve from APS- to SISL-format;
CSI_ASBS	convert a B-spline surface from APS- to SISL-format.

Other conversions used (availability restricted):

CSI_SIBC (*)	convert a rational SISLCurve to nonrational b_spline_curve on IDS-format;
CSI_SIBS (*)	convert a rational SISLSurf to nonrational b_spline_surface on IDS-format.

Testing. The APS-SS and SISL pre-processors have been tested in two ways: by file- cycle tests and by ordinary pre-processing of surface models. In the file-cycle tests files received from SNI were read in and then pre-processed. These files contain basically NURBS curves and surfaces and enabled us in particular to test the rational to nonrational conversions. Additionally, the pre-processors have been tested by generating STEP-files of APS-SS surface models. An example of these test cases is shown figure 4.9.7. The surface model of half of the plastic sledge consists of 36 B-spline surfaces (source format: VDA 1.0).

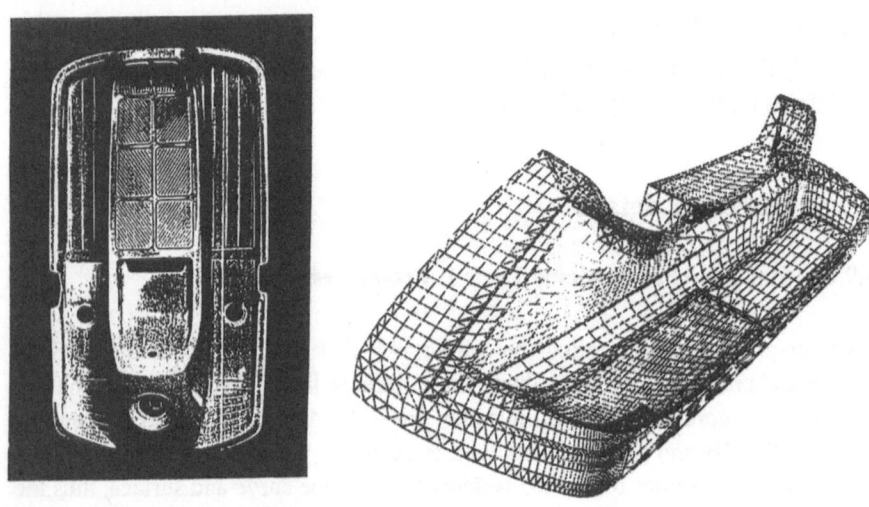

Fig. 4.9.7: Physical and data model of one of the test cases: a sledge.

4.9.6.2 Surface-AP, Functional Level 1, Post-processors

The difference of the two post-processors supporting the Surface-AP FL1 lies in their backends. Figure 4.9.2 illustrates this: the APS- SS backend is writing to the APS-SS DB and the SISL backend is writing to the SISL datastructure. An extra conversion from SISL- to APS-SS-format is needed for the APS-SS backend. The conversions from IDS- to SISL-format needed for the Surface-AP FL1 post-processors correspond to the ones for the pre-processors, of course converting in the opposite direction. The ordinary conversions from IDS- to SISL-format used by the post-processors are:

CCISPT   convert a cartesian_point on IDS-format to a SISLPoint;

CCISBC   convert a b_spline_curve on IDS-format to a SISLCurve;

CCISBS   convert a b_spline_surface on IDS-format to a SISLSurf.

   Specifically for the APS-SS backend:

CSI_SABC   convert a B-spline curve from SISL- to APS-format;

CSI_SABS   convert a B-spline surface from SISL- to APS-format.

Other conversions used (availability restricted):

CSI_ISBC   convert a rational b_spline_curve on IDS-format to non-rational SISLCurve;

CSI_ISBS   convert a rational b_spline_surface on IDS-format to non-rational SISLSurf.

The SISL demo system utilises the following conversions that are partly available in the Conversion Tools Library:

CCIPOC   convert a point_on_curve to a cartesian point on IDS-format (identifier level);

CCPOBC   convert a point_on_curve to a cartesian point on IDS-format (parameter level);

CCIPOS   convert a point_on_surface to a cartesian point on IDS-format (identifier level);

CCPOBS   convert a point_on_surface to a cartesian point on IDS-format (parameter level);

CCICBC   convert an "trimmed conic" to a b_spline_curve on IDS-format (identifier level);

CCPCBC   convert an "trimmed conic" to a b_spline_curve on IDS-format (parameter level);

CCILBC   convert a trimmed line to a b_spline_curve on IDS-format (identifier level)

CCPLBC   convert a trimmed line to a b_spline_curve on IDS-format (parameter level)

CCBPBC   convert a curve_on_surface to a b_spline_curve on IDS-format (parameter level)

CCBBPC   convert a (3D) b_spline_curve to a pcurve (represented as a 2D b_spline_curve) (parameter level)

CCPTBS (*)   convert a rectangular_trimmed_surface to a b_spline_surface on IDS-format (parameter level)

CCPEBS (*)   convert an elementary_surface to a b_spline_surface on IDS-format (parameter level)

CCPSBS (*)   convert a swept_surface to a b_spline_surface (parameter level).

(*) = not implemented within the CADEX project.

Testing. The APS-SS and SISL post-processors have been tested in two ways. By file- cycle tests and by ordinary post-processing of surface-models. In the file-cycle tests files received from SNI were used. These files contain basically NURBS curves and surfaces and enabled us in particular to test the rational to nonrational conversions. An example of these test cases is shown figure 4.9.8. The geometry, a torus, is shown both with a rational and a nonrational representation. The geometry tolerance used in the conversion is 0.001 relative to the SISLBox. The rational representation consists of totally 7x7 control points, while the nonrational representation needs 12x18 control points. Additionally, the post-processors have been tested by reading STEP-files and subsequent visual control.

Fig. 4.9.8: Source (rational, top) and target (nonrational, bottom) representation of a torus being post-processed.

4.9.6.3 Surface-AP, Functional Level 2

A post-processor supporting the Surface-AP FL2 exists only for the SISL- product. The main feature of this processor is the conversion from the STEP entity face_based_surface_model, with explicit topology, to the STEP entity geometric_3d_surface_set, with implicit topolgy. In addition to develop and utilize this conversion, most of the above mentioned geometry conversions are utilized. The only ordinary conversion from IDS- to SISL-format respectively between STEP-entities that have not been mentioned already are:

CS1221   convert a face_based_surface_model to a geometric_3d_ surface_set on IDS-format;

CSI_IISFA   convert a face from IDS to SISL format.

4.9.6.4 Surface-AP, Functional Level 3

A post-processor supporting the Surface-AP FL3 exists only for the SISL-product. The main feature of this processor is the conversion from the STEP entity shell_based_surface_model to the STEP entity face_based_surface_model, both with explicit topology. This conversion is succeeded by CCMFGS (see above). Most of the above mentioned geometry conversions are utilized here, too. The only ordinary conversion from IDS- to SISL-format respectively between STEP-entities not already mentioned is:

CS1321   convert a shell_based_surface_model to a face_based_surface_ model on IDS-format.

4.10 Siemens-Nixdorf-Informationssysteme

4.10.1 Overview

SNI's contribution to the CADEX STEP-processor development comprises the realisation of

- a pre- and postprocessor for the BRep Application Protocol (BRep AP). All 3 levels (facetted, ordinary and advanced Breps) have been implemented.
- a pre- and post-processor for the Surface Application protocol (SF AP). Only level 1 is supported.

An extensive test of these processors was possible by exchanging models with processors of other project partners. Figure 4.10.1shows the systems involved.

4.10.2 System and processor architecture

The processor can run either as a stand-alone module or as an integrated part of the SIGRAPH-CAD-3D system. SIGRAPH-CAD3D itself is a CAD-system, based on SIGRAPH WS30 workstations, mainly for design of mechanical parts, consisting of a basic 2D/3D system with all kinds of technical applications like sheet metal design, moulding, form features, etc. and interfaces to NC-applications as well as visualization systems. The rough system architecture is shown in figure 4.10.2.

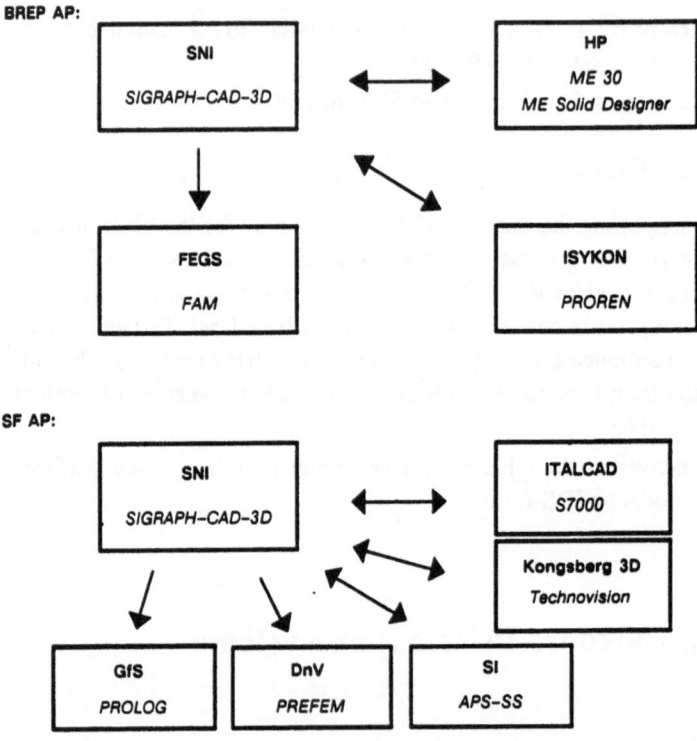

Fig. 4.10.1: Exchange between SNI's SIGRAPH-CAD-3D and other systems in CADEX

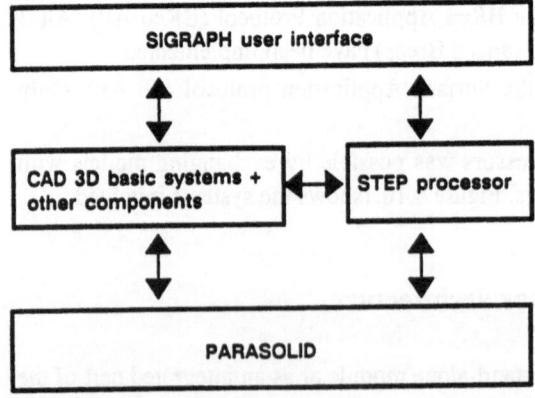

Fig. 4.10.2: Main components of the SNI system

The SIGRAPH-user interface is menu- and command-oriented. An example of invoking e.g. the STEP-preprocessor command via menu is given in figure 4.10.3. Several parameters give the user control over the command execution (type of application protocol, checking of elements in IDS, statistics file, approximation tolerance).

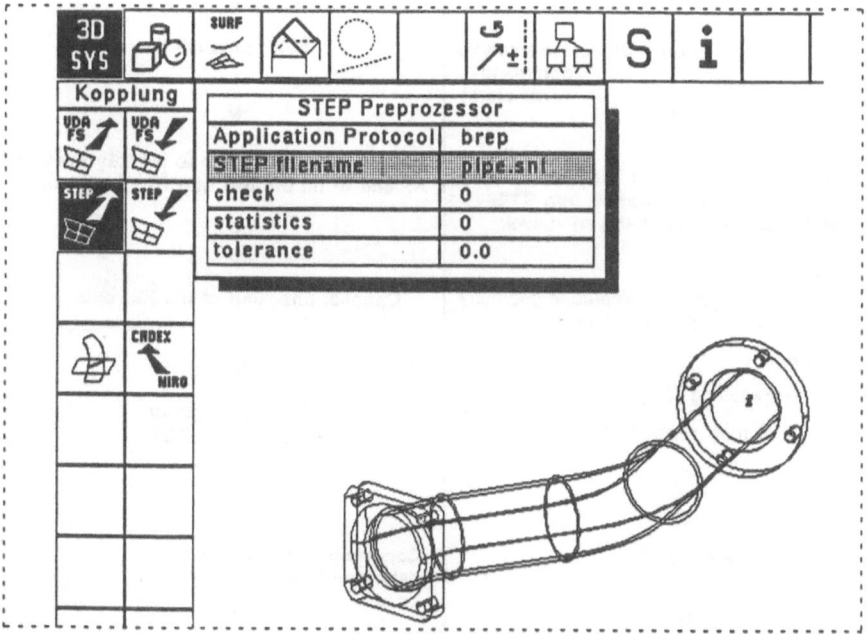

Fig. 4.10.3: SIGRAPH menu for STEP pre-processor

The processor is based on PARASOLID, which is a solid modeller with the following main properties:
- Exact Boundary Representation Modeller
- Fully integrated freeform surfaces (NURBS)
- Different kinds of BRep bodies -> wire, sheet and solid
- All topological and geometrical information within Parasolid

The basic "unit" of modeling is a body. The modeling operations available in Parasolid have to be performed on bodies or on topological subentities of a body. There are several body types which are allowed in the PARASOLID data structure. They enable the user to generate objects of different "complexity":
- Wire bodies which are a set of connected edges ("curve" bodies)
- Sheet bodies which consist of an open or a closed shell with wall thickness zero ("surface bodies")
- Solid bodies which occupy a continuous finite volume

All bodies have to be manifold, e.g. "T-sheets" or a combination of a wire and a solid in one body is not allowed.

The rough architecture of the processor itself and the use of the Common Toolkit is shown in figure 4.10.4, the details are given in a subsequent section.

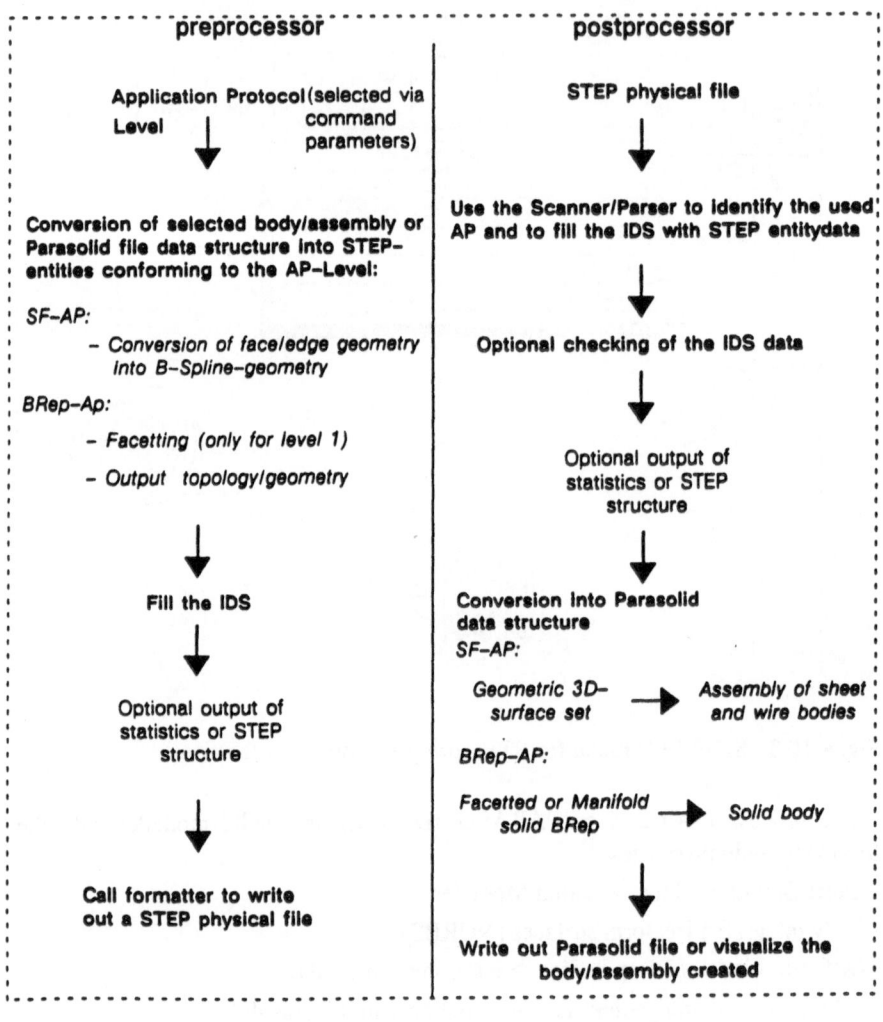

Fig. 4.10.4: Overview of actions in SNI's STEP processors

4.10.3 Processor for SF-AP

The processor for the SF AP covers only a part of the current level 1 scope. In fact only untrimmed curves and surfaces (B-SPLINE CURVE and B-SPLINE SURFACE entities) can be exchanged with the current implementation.

4.10.3.1 Preprocessor Parasolid --> SF-AP

The algorithm of the PARASOLID --> STEP preprocessor works as follows: Scan through the PARASOLID native model datastructure and find all top-level bodies and all bodies being instanced by top-level assemblies and sub-assemblies. Scan through all these found bodies and return their edges and faces.

For each edge/face found call a PARASOLID routine to output the corresponding curve/surface in B-Spline format. The output curve is bounded by the end points of the edge. The surface output is large enough to contain the face completely, that means the B-Spline surface is in general bigger (untrimmed).

The resulting data represents the originating curve/surface exactly if possible, but it may be necessary to approximate it. The representation of linear curves as non-rational B-Spline curves is exact, as well as that of circular and elliptical curves as rational B-Spline curves. Parametric curves are output exactly either as rational or as non-rational B-Spline curves depending on the PARASOLID curve type. Approximations are necessary in case of intersection curves. The representation of planar surfaces as non-rational B-Spline surfaces is exact, as well as that of cylindrical, conical, spherical and toroidal surfaces as rational B-Spline surfaces. Parametric surfaces are output exactly either as rational or as non-rational B-Spline surfaces depending on the PARASOLID surface type. Approximations are necessary in case of e.g. blend surfaces and offset surfaces.

The B-spline curves/surfaces are written into the IDS as B-SPLINE_CURVE/ _SURFACE entities together with its control points written as CARTESIAN_ POINTS.

For each Parasolid body a GEOMETRIC_3D_SURFACE_SET entity is created in the CADEX IDS referencing all B-Spline entities which have been generated before.

Finally the CADEX Formatter is called to write out the IDS to a STEP physical file.

4.10.3.2 Postprocessor SF-AP --> Parasolid

The algorithm of the STEP --> PARASOLID postprocessor works as follows:

Read in the STEP physical file by the CADEX Scanner/Parser software which fills the IDS. Optionally a check of the IDS by the CADEX IDS Data Checker can be performed afterwards.

If there is a top level entity of type GEOMETRIC_3D_SURFACE_SET generate a PARASOLID assembly which instances all data   (B-Spline-Curves/Surfaces) of the STEP file as parts. Then the entity of type GEOMETRIC_3D_SURFACE_SET is read in.

Read each B-Spline surface referenced by the GEOMETRIC_3D_ SURFACE_SET and generate a B-spline surface within PARASOLID. To be able to do this it is necessary to modify some of the attributes of the STEP B-Spline surface, for instance if some of the optional attributes are not filled but have the value "$", functions to provide default values for them have to be called. The cartesian points referenced by the STEP B-Spline surface have to be read in and their coordinate values have to be filled in one large array which will later be passed to the PARASOLID creation routine. This enables the postprocessor to try to generate a B-Spline surface in PARASOLID. If this geometric entity has been created successfully the processor tries to generate a sheet body which will be instanced by the PARASOLID assembly generated earlier.

A similar algorithm will be performed with the B-Spline curves in the GEOMETRIC_3D_SURFACE_SET:

For each B-spline curve referenced by the GEOMETRIC_3D_SURFACE_SET create a B-spline curve within PARASOLID, from which a wire body will be generated.

There is an important restriction for the import of data: PARASOLID is a solid modeller, so its topology checker doesn't allow storage of a face or an edge within the model, whose geometry is self-intersect ing, even if the creation of this surface or curve (pure geometry) has been possible within PARASOLID.

The next action is to generate a GEOMETRIC_3D_SURFACE_SET entity in the IDS, which has all the generated B-Spline surface and B-Spline curve entities as its defining attributes.

The postprocessor's work is finished with the storage of the created model in the Parasolid transmit file (standalone version) and the visualization of the generated bodies (CAD3D-system) resp..

For test purposes each sheet and wire body gets an attribute storing the entity id from the STEP file.

4.10.3.3 Test Experiences with SF-AP Processors

There have been no severe problems writing out our models into STEP SF-AP Implementation Level 1 format.

The only critical point however is that the STEP standard doesn't allow periodic B-Spline curves or surfaces, whereas Parasolid can hold such curve and surface types. Each periodic B-Spline entity has to be converted to a non-periodic one before being written to the IDS. This has to be done for instance with the periodic B-Spline surfaces representing complete analytical surfaces like cylindrical and spherical surfaces.

The figures in this section show some typical results of the SF-AP preprocessor (the models have been exported and imported again).

The figure 4.10.5 shows the output of a complete cone, i.e. the analytical surfaces and curves have been converted to B-Spline entities. The figure 4.10.6 is a 'surfsail' which consists of triangular and rectangular surface patches.

The figures 4.10.7 to 4.10.10 show some models which have been imported from the systems Technovision and S7000.

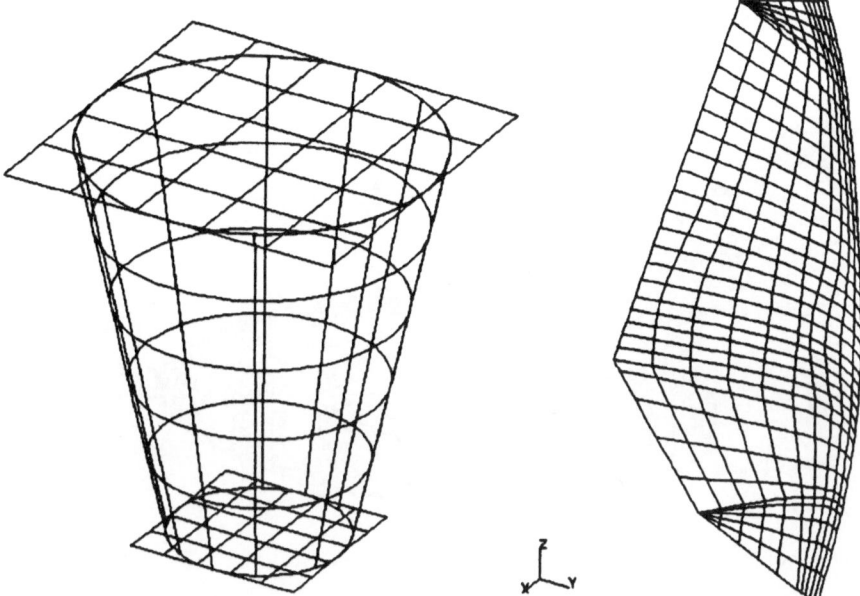

Fig. 4.10.5: Export of a 'cone' via SF-AP

Fig. 4.10.6: Export of a
'surfsail' via SF-AP

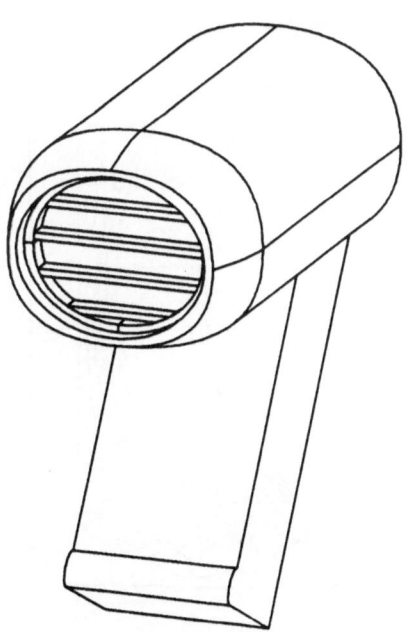

Fig. 4.10.7 Export of a 'hairdryer' via SF-AP

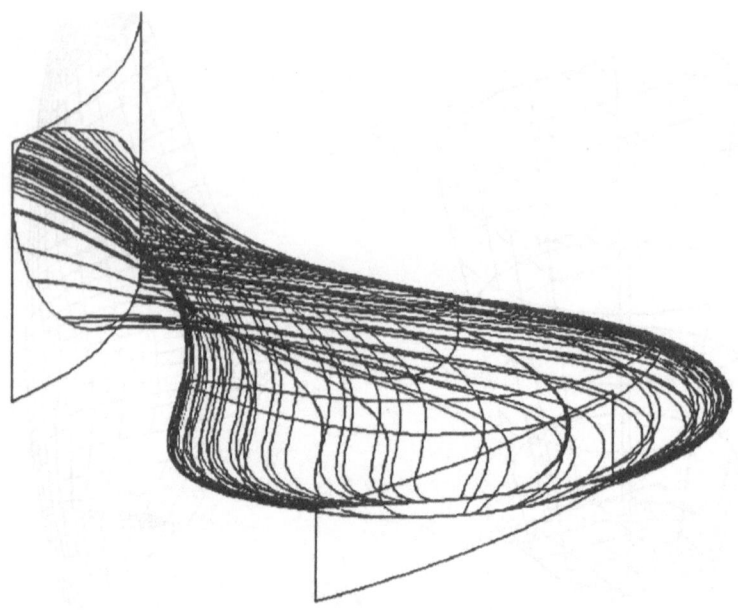

Fig. 4.10.8: Import of a 'propeller blade' from Technovision (ND) via SF-AP

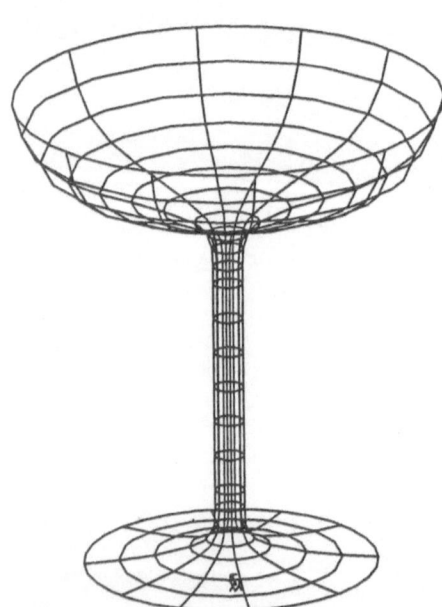

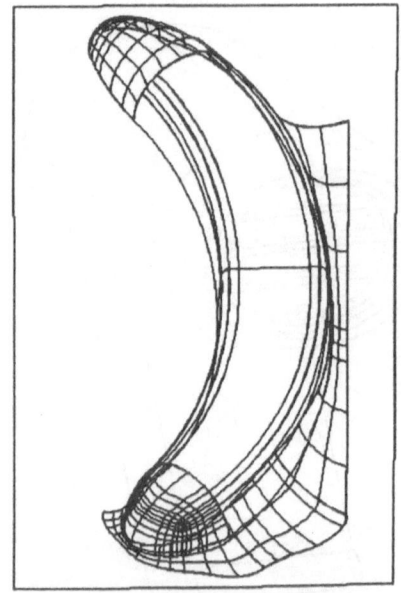

Fig. 4.10.9: Import of a 'glass' from
S7000 (Italcad) via SF-AP

Fig. 4.10.10: Import of a 'banana'
from Technovision (Norsk Data) via
SF-AP

In some cases the creation of the entire Parasolid model was not possible. The import of certain geometries was stopped with the error message "Geometry fails to pass checks". This indicates that Parasolid was not able to handle this specific entities. The most likely reason is that the rejected surface or edge is self-intersecting which we could prove in some cases. In general however it is very difficult to detect the area which causes the problems.

4.10.4 Processor for BRep-AP

4.10.4.1 Preprocessor Parasolid --> BRep-AP

The algorithm of the Parasolid --> STEP preprocessor works as follows:

Scan through the Parasolid native model datastructure and find all top-level solid bodies and all solid bodies being instanced by top-level assemblies and sub-assemblies. The conversion of a body is dependent on the selected level.

Facetted BRep (Level 1):

Parasolid bodies, which can contain any kind of surfaces, have to be converted into facetted bodies. This is done by a facetter, which is inside Parasolid and is mainly used for graphical purposes (shading, etc.). For every body the facetter generates a set of planar facets (faces of facetted body), which are described by bounding polylines. For each face the polyline and the easily derived planar surface are written to the IDS. Of course nonplanar faces have to be approximated. The approximation tolerance can be given by the user.

Ordinary and advanced BRep (Level 2/3):

In this case the data structure of Parasolid can be mapped entirely to the BRep AP. There are bodies, shells, faces, loops, edges and vertices as topological entities and surfaces, curves and point as geometrical entites. If necessary flags are stored with these entites indicating an orientation, i.e. LOGICAL_STRUCTURE entities don't exist in Parasolid.

There are only few differences between the data structures of STEP BRep-AP (Level 2/3) and Parasolid:

In Parasolid there are edges without bounding vertices, i.e. so-called ring edges. Before the edge can be exported to BRep-AP a vertex has do be defined splitting this edge.

On a complete torus and on a complete sphere there is no loop in Parasolid. Before the face can be written to BRep-AP a loop has to be defined.

The BRep-AP allows only non-rational B-Spline entities, so true rational B-Spline entities have to be approximated.

The general idea of the BRep-AP pre-processor is to scan each body in the bottom-up way related to the entity types, i.e. start with the points and vertices, go on with edges, loops, surfaces, faces, shells. The LOGICAL_STRUCTURE entities are generated in the IDS immediately after the 'non-logical-structure' entities. As a consequence only backward references are used in the process writing entities into the IDS.

The unique entity identifier which is used in the IDS is stored as an attribute of the Parasolid entity. This enables entities which are handled later in the data structure hierarchy to know the IDS name of the referenced entity.

The last step of the preprocessor is to call the CADEX Formatter software which generates a STEP physical file.

4.10.4.2 Postprocessor BRep-AP --> Parasolid

The algorithm of the STEP --> Parasolid postprocessor works as follows:

Read in the STEP physical file by the CADEX Scanner/Parser software which fills the IDS. Optionally a check of the IDS by the CADEX IDS Data Checker can be performed afterwards.

The back-end of the BRep postprocessor consists of four steps:

In the first step the topology of the STEP BRep body is imported into Parasolid. Due to the fact, that the topology structure of a facetted BRep "ends" with the face entities, it is necessary to derive edge and vertex entities from the POLYLOOPs in the STEP file in order to fill up the Parasolid topology structure entirely.

In the 2nd step geometry associated to each topological element is attached to each Parasolid face, edge and vertex.

In the next step Parasolid tries to mend the body. It attempts to recalculate edge and vertex geometry so that the body confirms to Parasolid's accuracy requirements assuming that the imported surfaces are accurate. As the geometry that was originally attached to the faces of the body is used in this calculation, success will depend on the accuracy of the initial data.

If these three steps succeeded, in the last step a check of the generated Parasolid model is performed. If it passes this check full modelling functionality on the imported BRep model can be guaranteed.

For test purposes each Parasolid entity gets an attribute storing the corresponding entity id from the STEP file.

The postprocessor's work is finished with the storage of the created model in the Parasolid transmit file (standalone version) and the visualization of the generated bodies (CAD3D-system) resp..

4.10.4.3 Test experiences with BRep-AP processors

SNI's BRep postprocessor could successfully import a lot of models. Test cases came mostly from ME30/Solid Designer of Hewlett-Packard and PRO*CAD.

The following areas of problems were identified:

If the imported geometry is not accurate enough modifications of attached geometrical entities have to be done. As long as the surface geometry which is the basis for necessary recalculations in SNI's post-processor has high accuracy, the resulting imported model will be very similar to the original. If the surface geometry however isn't accurate enough the import of the model might be stopped because of an unexpected topological situation. This can be the case, if e.g. a planar surface should meet a cylindrical surface tangentially and the data are too inaccurate for Parasolid to find the intersection curve (see also the chapter on test results).

Another problem occured in cycle tests with bodies containing true rational B-spline-surfaces. In the current AP the entity in the STEP-file has to be non-rational, so approximations were necessary when exporting the model. During import the high accuracy of geometry data needed by Parasolid could not be guaranteed, so the mending process often fails. Of course the approximation could be made with a smaller tolerance, but this ends up with more complex approximating B-spline-surfaces.

Parasolid doesn't have POLY_LINE, PARABOLA and HYPERBOLA entities, i.e. they have to be recalculated. The idea is to climb in the IDS data structure starting from the curve up to the two surfaces defining the intersection curve, calculate the intersection of the corresponding Parasolid surfaces and attach the correct portion of the intersection curve to the Parasolid edge.

The figures 4.10.11 and 4.10.12 show some BRep models which have been imported from Hewlett-Packard's system ME30 and Solid Designer resp., figure 4.10.13 shows a model exported to SolidDesigner.

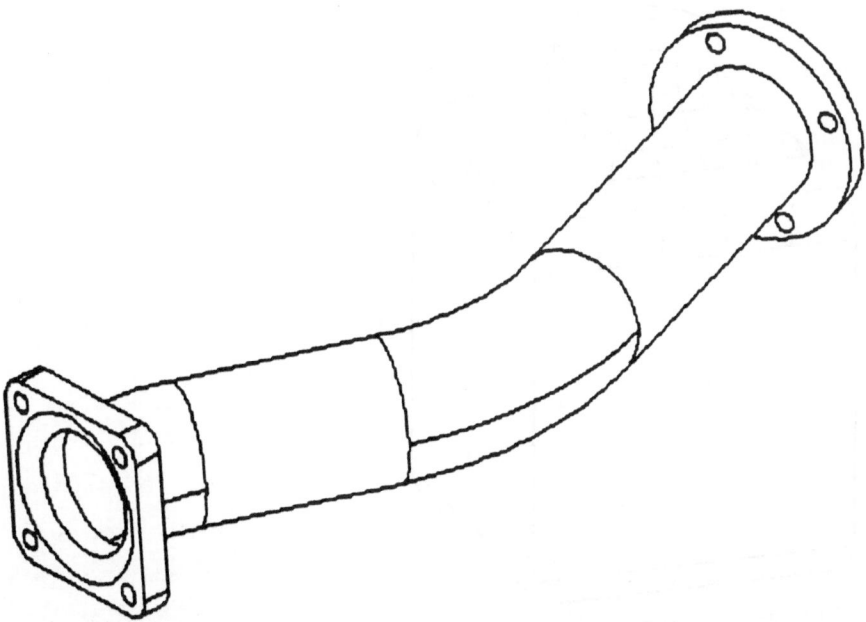

Fig. 4.10.11: Import of a test part from ME30 (Hewlett-Packard) via BRep-AP (MODEL7 of FIAT library)

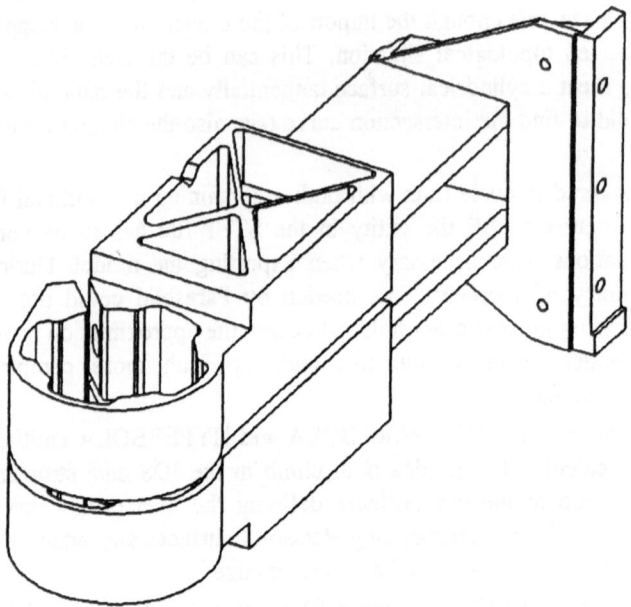

Fig. 4.10.12: Import of a test part from ME30 (Hewlett-Packard) via BRep-AP

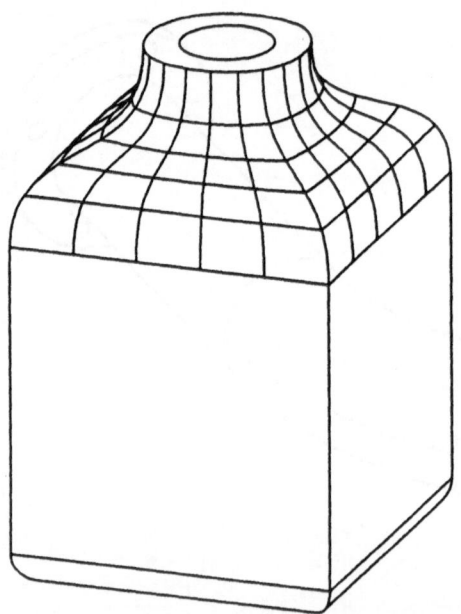

Fig. 4.10.13: Export of a test part with freeform surfaces to ME Solid Designer (HP) via BRep-AP

4.10.4.4 Exchange of models with NIRO project

Based on the BRep-AP Level1 it was also possible to exchange models between SIGRAPH-CAD3D and systems involved in the ESPRIT project NIRO. Only small addings to the processor have to be made due to the fact that CADEX and NIRO used different STEP dialects.

In fact we could import models from CATIA, ROBCAD and KISMET. On the other hand we could export BRep-models as facetted models to these systems.

In figure 4.10.14 a robot from REIS is shown, which was created in CATIA, consisting of an assembly of about 10 facetted bodies.

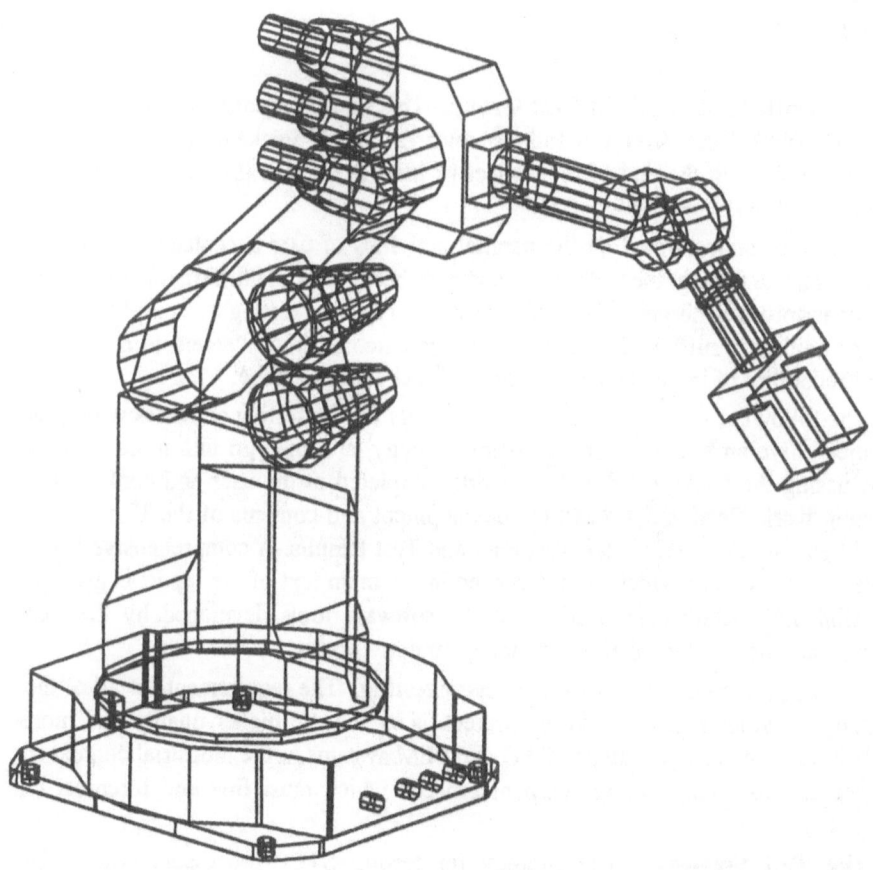

Fig. 4.10.14: Import of a test part (Reis robot) via BRep-AP (facetted) from NIRO-project

5. Test and Validation

5.1 Introduction

The work of the CADEX Test Group. This section reports the work done by the CADEX Test Group members towards Work Package 6 on Test and Validation, up to the end of the project in June 1992. It is the final report of the group.

Work has concentrated on the parallel activities of user acceptance testing and conformance testing, the two being separated because of the different nature of the tasks involved. The group see these different types of testing as complementary, since neither is sufficient in isolation to guarantee an improvement in the quality and reliability of industrial engineering product data exchange.

The following sections state the requirements for the testing activities within the project, give an overview of the testing strategy, and then go into more detail in discussing the background and the work completed in the user and conformance testing work. Section 5.4 detail the development and contents of the Test Library and Section 5.5 describes the Test Plan and Test Results. A comprehensive list of relevant documents which are referenced in the main text of the report is given in section 5.6. Detailed descriptions of the software tools developed by the Test Group are included in Section 3 on the Common Toolkit.

The requirements for STEP processor testing. The requirement for all of the testing activities performed in the project, is to achieve higher quality and more reliable data exchange between CAD and FEM systems in the industrial context. It is the end user engineering companies needs which must first and foremost be satisfied.

The Test Strategy. The strategy for testing STEP processors within the CADEX project is to perform some conformance testing work where possible, and some user acceptance testing. The importance of addressing both types of testing has already been stated, and it is to the projects credit that efforts in both directions are underway at such an early stage in the development of the standard.

To reflect this strategy, it was agreed that resources would be most effectively deployed if conformance testing were to be the primary responsibility of

CADDETC, and that user acceptance testing be the primary responsibility of the users themselves, ie. BMW and FIAT [16]. This reflects both the interests and the expertise of the partners concerned. Having said that however, each has supported the other as fully as possible with individual tasks when required, and the group continue to co-ordinate their work and liaise at project Working Group meetings.

The Test Methodology. The general methodology of testing is the same for both user acceptance testing and conformance testing. It requires the specification of:

- the types of tests to be performed
- the type of data required for each type of test
- the test methods or procedures
- the evaluation or verdict criteria
- the required software tool support

However, there are significant differences in the order in which these activities are carried out for the two types of testing. In particular, for user acceptance testing, the test models were identified first and then evaluation criteria were defined. For conformance testing on the other hand, the approach adopted is to define within the standard the list of things to be tested for and then to derive appropriate test cases from this list. The list of what needs to be tested for is itself derived from the information models with the AP documents.

The following sections detail the work performed in each of these areas.

5.2 User acceptance testing

5.2.1 The user roles in CADEX

Definition. User acceptance testing is defined as the process of testing a processor for acceptance by a user according to his specified criteria in his working environment when applied to his products.

Motivation. BMW and FIAT elected to take prime responsibility for the use acceptance testing aspects of the CADEX Test Group work because they are the people who will be affected most by the quality of the STEP processors developed by the CAD and FEM system vendors. It is therefore in their interests to ensure that their requirements are clearly stated and that the processors developed satisfy those requirements.

The motivation for performing user acceptance testing is to give more confidence that a processor or pair of processors performs to user specified criteria, above the level of confidence that can be gained from performing conformance testing in isolation.

Influence of BMW and FIAT. The major influences in this area of work within CADEX comes from the users of processors, BMW and FIAT. BMW in particular have a substantial amount of experience in performing testing of IGES processors within their industrial environment, and that experience has been used to develop the testing methods, test cases and software tools for user testing of STEP processors within CADEX.

5.2.2 Types of tests

The material described here has been contributed by BMW. It is essentially a transcription of [13].

Figure 5.2.1 summarises the possible ways for the types of tests described below. Basic tests for stability, maintenance, portability and system support are not within the scope of the CADEX user acceptance testing work.

1. Inter-system testing without localization of errors

   This involves the testing of a pair of processors to establish if together they can accomplish a user required data exchange.

2. Inter-system testing with localization of errors

   This test involves analysing the STEP file produced by the pre-processor in the test. Figure 5.2.2 illustrates this process.

3. Cycle and file cycle test of processors without localization of results

   These tests assess the overall performance of a system's pre and post-processors. Figures 5.2.3 and 5.2.4 illustrate these processes.

4. Single processor testing of pre-processors.

   This is covered by conformance testing described below.

5. Single processor testing of post-processors.

   This is covered by conformance testing described below.

5.2.3 Test data

In order to support the tests described above, several types of test data are required:

- Inter-system tests require models of production parts in the native format of the systems being tested.
- Cycle tests require models in the native format of the system being tested, produced from a model generation description.
- File cycle tests require STEP format files produced from a model generation description.

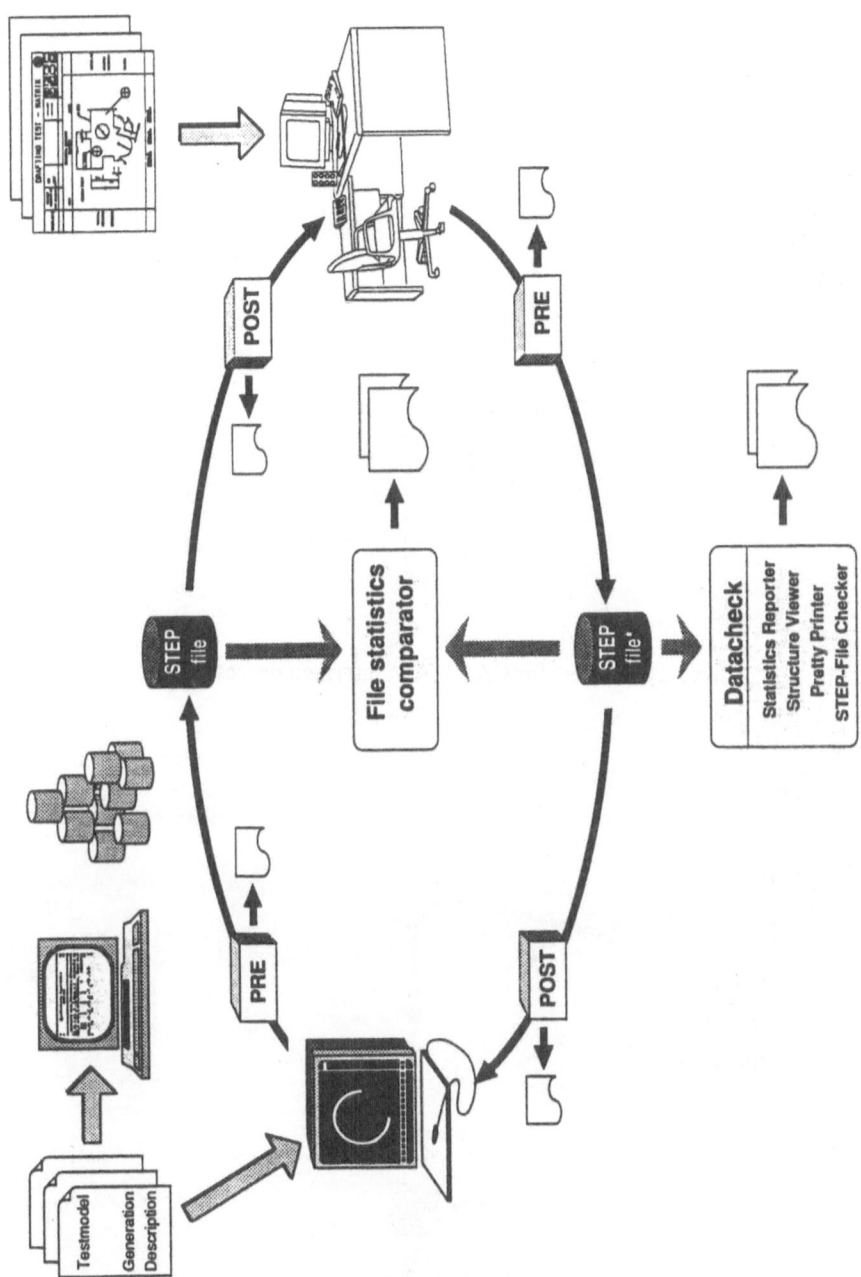

Fig. 5.2.1: Test methodology summary for CAD/CAM interface processors

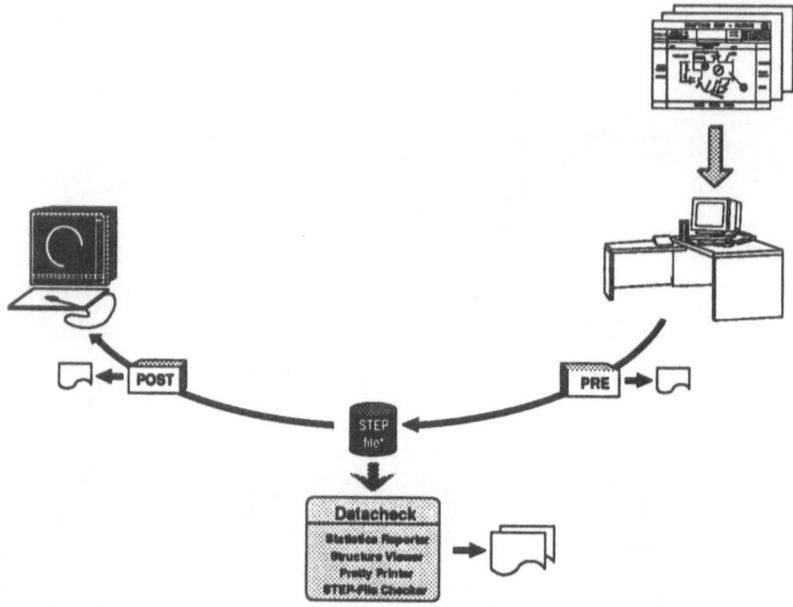

Fig. 5.2.2: Inter-system test for CAD/CAM interface processors

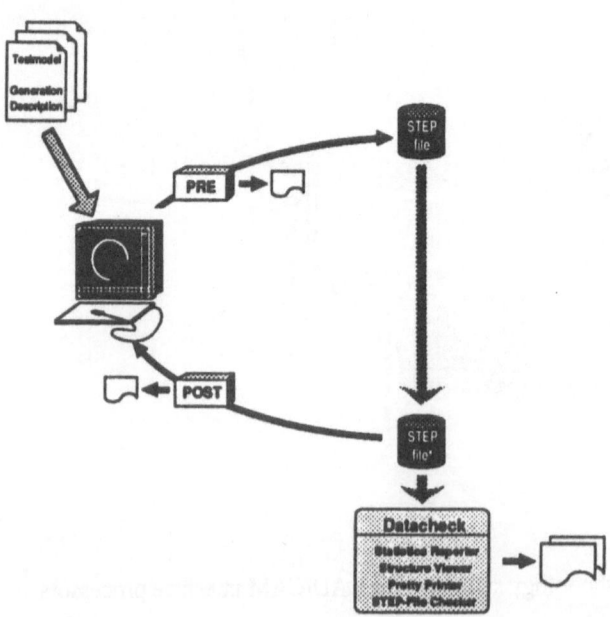

Fig. 5.2.3: Cycle test for CAD/CAM interface processors

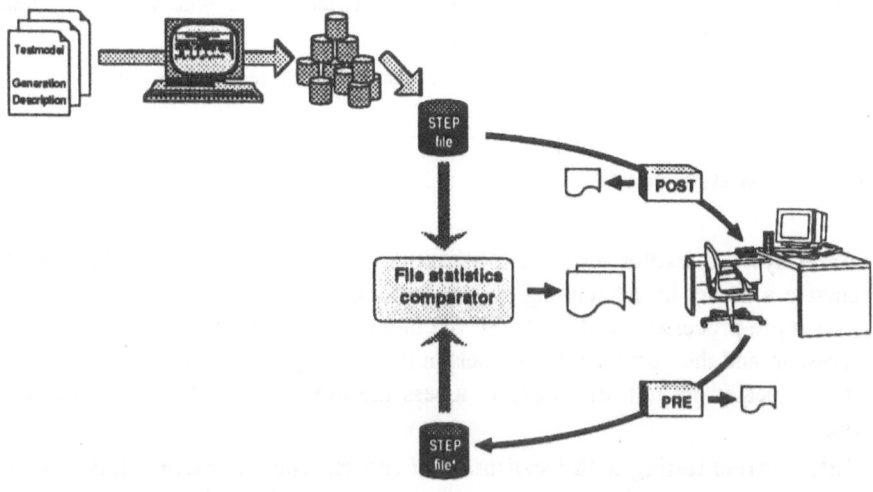

Fig. 5.2.4: File cycle test for CAD/CAM interface processors

Test Methods and Data Overview

test goal	test method	test data
exchangeable data set between system A and B	inter system test	system specific data test matrices and productive models
processor capacities of system X	cycle test	system specific data test matrices
processor capacities of system X	file cycle test	STEP data test matrices
preprocessor test	comparison of native and STEP data	simple system specific data
postprocessor test	comparison of native and STEP data	simple STEP data
syntax, semantic and statistic check of a STEP file	syntax and semantic check statistic evaluation	no specific STEP data

Fig. 5.2.5: Test methods and data

Cycle and file cycle test data should contain test matrices of all the entity types specified by the standard so that a quick verification of whether the system supports the standard or not, can be made.

Figure 5.2.5 summarises the types of test data used for each type of test.

5.2.4 Test procedures

Inter-system testing without localization of errors. The procedure is to generate a model in the sending system, process that into a STEP file using the systems pre-processor, read the STEP file into the receiving system using its post-processor, and then compare the models in the sending and receiving systems. The test is repeated in both directions to assess the overall compatibility of the two systems.

Inter-system testing with localization of errors. The procedure in this type of test is the same as for the previous one except that the STEP files produced by the system pre-processors are analysed also.

Cycle and file cycle testing without localization of errors. The first of these tests consists of defining a model in the CAD system, generating a STEP file from this, reading the STEP file back into the system being tested, and comparing the two CAD models. The second test involves starting with a STEP file created from a test model generation description, reading this into the system, generating a second STEP file from the system, and then comparing the two STEP files.

Evaluation criteria. Data transfer evaluation criteria, ie. what needs to be measured during the testing procedures, have been developed for each model. Criteria such as visual appearance, the evaluated volume of the component (for CSG and B-rep APs), and the values of critical model parameters have been included. These are detailed in [19] and for this report included in section 5.4.

5.2.5 Software tool support

Implicit in the performance of all of these tests, is a requirement for software tool support for the analysis and comparison of STEP files. This includes:
- a STEP file statistics reporter
- a STEP file statistics comparator
- a STEP file structure viewer
- a STEP file checker
- a STEP file editor

These software tools are described in Chapter 3.

5.3 Conformance testing

5.3.1 Introduction and definition

CADDETC was elected to take prime responsibility for the conformance testing aspects of the CADEX Test Groups work because of its background in this area through ongoing work with IGES conformance testing and the leading role which it is playing in STEP through ISO TC184/SC4/WG6 (Conformance Testing Procedures).

CADDETC has carried out some background work in gaining an understanding of what work has been and is currently being done under ISO TC184/SC4/WG6 and in the CTS2 project, and has disseminated this to other CADEX partners. This was essential before deciding what work should be done using CADEX resources.

[2] defines what conformance testing means with respect to STEP, and describes the overall concepts of what is proposed for STEP conformance testing methods and a framework for their implementation.

[2] defines conformance testing as: The testing of a candidate product for the existence of specific characteristics required by a standard; testing the extent to which an IUT (implementation under test) is a conforming implementation.

5.3.2 Motivation for performing conformance testing

The motivation for performing conformance testing is to increase the confidence that data exchange between two implementations of a given standard will work, by ensuring as far as practically possible, that the the two systems concerned conform at least to the requirements specified in that standard.

The principle objectives of specifying these within the standard are to achieve:

- repeatability of results
- comparability of results
- auditability of results

It was stated in [1] that as far as the Test Groups efforts in the CADEX project are concerned, there was no suggestion that complete conformance testing will be carried out. However, the importance of conformance testing, as illustrated by the fact that there is a separate Working Group (WG6) under ISO TC184/SC4, and a class of parts devoted entirely to this topic, clearly meant that the project had a responsibility to perform some work in this area.

5.3.3 Influence of the CEC funded CTS2 project

Much of the original work in developing conformance testing procedures and methods has come from a CEC funded CTS2 project, the full title of which is:

"Conformance Testing Services 2 (project number 15): Establishment Of Harmonised Conformance Testing Services For CAD/CAM Systems Data Exchange Interfaces"

The project is managed by CADDETC and also involves partners from Germany, France and Sweden. The objective is to establish harmonised conformance testing services for IGES, SET Schematics, VDA-FS and VDA-IS in the short term, and for STEP as and when that standard is published. To this end, common testing procedures and methods have been developed, including test suites and a supporting software architecture. The methodology has much in common with that used for conformance testing of GKS, MAP and programming language compiler products. The commercial IGES conformance testing service at CADDETC has been operating since the second quarter 1991 and a DXF conformance testing service has also recently been launched.

The work of the project is detailed in the four Technical Reports produced so far [6], [7], [8] and [36], all of which have already been made available to CADEX partners. All of this work has been reported to ISO TC184/SC4/WG6, and much of it has been adopted as a basis for specifying STEP conformance testing procedures and methods [2]. A presentation of the project was given to CADEX Test Group members at the Test Group meeting held in Leeds on November 16th 1990 [15].

5.3.4 Types of tests

Conformance testing corresponds directly to detailed single processor testing as enumerated under section 5.2. The fundamental principle is that only a single processor is tested at any one time so that problems encountered are known to be in that processor. In other types of tests described above, pairs of processors are involved, and it may not be easy to isolate errors.

One strategy for testing processors could therefore be to perform conformance testing prior to user acceptance testing so that there should be little doubt about how each individual processor handles the relatively simple conformance test cases in isolation. Knowing the results from these tests will help to isolate errors when more complex tests are carried out.

5.3.5 AP conformance requirements

Much of the work described in this and the next section is essentially AP work and could have been reported under Work Package 2 or Work Package 7. However, because it is directly related to testing, it is included here under the report of the Test Group (Work Package 6).

The first requirement which has to be satisfied for conformance testing to be possible is that a list of statements specifying what is to be tested for, has to be provided. In STEP, this is in the form of a list general conformance requirements which are common to all APs, and a detailed set of Test Purposes for both pre and post processors which are specific to individual APs. Test Purposes are described later.

By June 1991, it was becoming clear that the issues surrounding conformance requirements and test purposes were not understood to the same extent by all

parties involved in AP development and testing and therefore a meeting was held between the Test Group and the CADEX AP owners to discuss them in some detail. Extracts from the minutes of that meeting [24] are reproduced below:

5.3.5.1 Extracts from the Oslo Test Group meeting minutes (June 1991)

The meeting was initiated by BMW in order to expedite the completion of the conformance testing aspects of the CADEX APs which are currently being forwarded to ISO for inclusion in STEP Version 1.0, ie. the B-rep and Surface APs.

CADDETC therefore arranged the meeting, to include the Test Group and the relevant AP owners, and carried out some preparation work based on the Part 201 meetings in Leeds in June and discussions with SI on Part 205 with the CTS2 project. This formed the basis for a substantial part of the debate during the meeting.

Some papers which emerged from the Part 201 meetings in June in Leeds were copied and distributed by CADDETC prior to the meeting. These included:

- Draft High level conformance requirements
- Draft AP structure diagrams for Part 201.
- Draft Test Purposes for a subsection of Part 201.

Subsequent to the Part 201 meetings, SI had discussions with the CTS2 project partners both in Leeds and in Brussels at a CTS2 project meeting. The CADDETC participants prepared AP Structure diagrams for each functional level of Part 205, and these too were copied and distributed prior to the meeting.

Discussion started around the draft High level conformance requirements document. Specific points discussed were:

1.) The meaning of the word implementation at the end of the first sentence requires clarification. Is this, in the case of CADEX, the implementation of the IDS which mirrors the AP EXPRESS schemata, or the implementation of the CAx system, or both, or neither ?

In the case of CADEX, it is not difficult to forsee a situation where the common tool element of a system (or all the systems for that matter) could be tested and seen to conform with what is currently specified in the APs. This would not help the cause of data exchange between two systems however !

There seemed to be consensus that the minimum set of operations which need to be performed in a conforming system should be specified in the AP in some way. Being able to perform boolean operations on a CSG tree was cited as a possible example for the CSG AP. Taken to its logical conclusion, adding methods to the entity definitions of an EXPRESS version of the ARM would seem to be the only way of formally specifying such functionality. The reintroduction of the business rules section of the AP would also go some way to satisfying this requirement.

2.) There was substantial debate about the issue of substitution/mapping/ conversion of entities. The reason for this appearing is that without such a statement, vendors will be free to map entities from the AP to whatever is convenient for their particular CAx system. This could and often does lead to some loss of information and a change in the number and types of operations which a user can perform on the converted entity, between one system and another. This is not desirable from the users point of view.

The counter argument was that existing systems do not implement AP schemata directly in their data bases and that it is not realistic for every CAx system vendor to do so, which is the implication behind the statement if a vendor is to be able to claim conformance with the standard. Further more, acceptable data exchange is still possible without this requirement, depending on what is required to be transmitted and what is required to be done with the data at the receiving system, if the capabilities of the two systems is known. This is also true for IGES !

However, STEP is not intended to be a standard which preserves the individual quirks of each CAx system and the current requirement to go through a long and tedious process of establishing if one system can transfer data to another. It is intended to be a standard which improves the speed, quality and reliability of industrial engineering product data exchange in the future. It was strongly suggested that this was the dominant requirement.

3.) There was a similar debate about the completeness issue. The reason for this clause appearing is so that if a vendor claims that his system conforms with an AP, then a user should be able to judge the functionality of that system and others with which he wishes to transfer data, from the AP rather than having to delve into the precise representations and functionalities of each system. This is obviously closely related to the conversion issue above. If completeness is not required the whole philosophy of APs is fairly redundant and users may be no better off than they are with IGES.

Dividing the APs into levels of functionality as already exists is one way of reducing requirements on vendors and clarifying their conformance claims. This would also probably be acceptable to users.

It was agreed that post-processors should be able to gracefully handle all AP entities without the system crashing. The wording of the clause is still very vague however. What exactly does process mean ? In the case of CADEX, the scanner/parser will recognise each entity type in a physical file and store it in the IDS, but the back end of each system post processor and the system itself may just throw some entities away, or convert them.

For pre-processors, there was a lot of support for the idea that completeness was not required. A pre-processor should be classed as conformant if the files which it produces contain only valid AP entities. This is a reasonable argument if the assumption is made that a user has performed the process of establishing all his future requirements for data exchange from such a CAx system. It was strongly suggested that this is an impossible task, if a system claims conformance with an AP, then the user should be safe in the assumption that he can exchange data with other systems claiming the same conformance.

It was suggested that it should be stated in the APs that if a 1:1 mapping exists between AP entity and CAx system entity, then that mapping should be used as the default in the CAx system. Further more, if a 1:1 mapping does not exist, the vendor must specify the CAx system specific entity corresponding to the AP entity, and the algorithm for performing the conversion which has been implemented in the system. Or does this information go in the Protocol Implementation Conformance Statement (PICS) ?

4.) There was agreement that entities not in an AP should not be generated by a pre-processor, and that post-processors should be able to recognise and report such entities.

5.) There was agreement that conformance with the requirements of the implementation form should be stated, ie. in the case of CADEX, Part 21 and the physical file mappings in the APs.

6.) The consensus of the meeting was that the clauses regarding completeness and conversions should remain, but that the way in which results of a conformance test are reported and therefore the claims which vendors could make, should reflect something akin to levels of conformance:

- Level 1 - systems which implement APs directly in their database, and provide the complete functionality specified with the AP entity definitions
- Level 2 - systems which implement all AP entities and corresponding functionality, but convert some of them
- Level 3 - systems which lose some entities and/or functionality

These would be based on tables which detail the conformance of the system with the AP. The vendor would fill in each table, and one result of the conformance testing process would be for the test laboratory to also fill in the tables as a cross check:

- Table 1 - list of AP entities which are directly supported in the system
- Table 2 - list of AP entities which undergo a mapping in the system
- Table 3 - list of AP entities which are lost in the system
- Table 4 - list of AP entities not available in the system (for pre-processors only)

Presenting the information in this way would enable users to quickly judge whether two systems can communicate with one another, regardless of their performance in a conformance test.

In conclusion, it is safe to say that a substantial leveling in understanding was achieved amongst all concerned. The main points which emerged were:

- STEP should strive to be better than current standards and specifications. This necessarily means more stringent conformance requirements with respect to completeness and conversions, which users must exert pressure to achieve.
- If conformance requirements are not specified in the APs, they cannot be tested for. All APs are currently lacking in this respect. The process of industrial validation of APs MUST take the testing aspects seriously.
- Better communications between ISO TC184/SC4/WG6 and the AP owners/projects would avoid many of the mis-understandings which were apparent at the beginning of the meeting.

With respect to this last point, CADDETC agreed to arrange for WG6 representatives to meet with CADEX representatives at the Sapporo meeting in July 1991 to take the discussions further.

5.3.5.2 AP conformance requirements work since the Oslo meeting

Since the Oslo meeting in June 1991, CADDETC has puts its emphasis on contributing to this work because without it, the CADEX APs will not progress as quickly within ISO as they should. In the long term this will be to the detriment of the standard as a whole since if it cannot support relatively simple APs dealing with geometry, there will be very little incentive for industry to adopt it in preference to other standards and specifications such as IGES, VDA-FS and VDA-IS, and SET.

To this end CADDETC participated in the joint US/CEC Workshop on Manufacturing Technologies in July 1991 in Berlin in order to represent the CADEX project and establish a dialogue with the US representatives working on AP integration within PDES, Inc. and the relevant ISO committees. The following section is an extract from [27], the report from the Berlin workshop.

5.3.5.3 Extracts from the Berlin CEC/US workshop report

This is a brief summary of the relevant aspects of the joint workshop between US and CEC representatives to discuss the possibility of closer collaboration between the people involved in research, development and standardisation activities related to manufacturing technologies.

The workshop took place at IPK in Berlin from June 29th to July 2nd 1991. It was a follow on from a previous workshop held in Daytona Beach on January 28th to 31st 1991. Willy Van Puymbroeck (CEC) and Howard Bloom (NIST) co-chaired the Workshop.

The official report from the workshop was made available from the CEC.

Workshop structure. The first morning was devoted to a plenary session which detailed the overall objectives and general administrative matters. The workshop then divided into three parallel tracks for the remainder of the first day and the subsequent two days:

- Product data sharing
- Enterprise integration frameworks
- Open systems architecture

Brad Harris (CADDETC) attended the first of these three tracks as a CADEX representative from June 29th to June 31st. Subsequent days were for meetings of the Action Offices involved to discuss and document the overall outcome.

Objectives. The objectives of the workshop were cited as being to:

- Assess inter project collaboration since the last workshop
- Assess the common technical ground between participants
- Formulate appropriate concrete short term collaboration plans
- Formulate long term collaboration plans

Short term was taken as including collaboration between existing projects and starting in the 6 - 12 month timeframe, long term could include new projects and would not start for at least 12 months.

The objectives of any proposed collaboration would be to:

- Define global requirements
- Pool resources
- Build early consensus
- Expedite rapid and extensive testing of newly developed technology

The guiding principles in proposing collaboration were stated as:

- No cross funding
- Comparable resources, balanced benefits
- Consortia of equal weight
- Industrial relevance
- Value added international co-operation
- Involvement of enterprises of all sizes

Product data sharing track. The Action Officers for this track were Reiner Anderl (RPK Karlsruhe) and Bob Kiggins (General Manager, PDES, Inc.). A list of the attendees during the afternoon of the 29th and the 30th. is given below.

The format of this track for the for the first day was a series of presentations from participants representing various CEC and US projects. The projects presented were:

- Horst Nowacki - NEUTRABAS
- Brad Harris - CADEX
- F Vernadet - CIM-OSA
- Wim Gielingh - IMPPACT
- Rolf Schmidt - CACID
- Reiner Anderl - VIMP
- Bob Kiggins - PDES, Inc. general information, RAMP, CALS and the STEP
  Implementors workshop held in May 1991
- Bill Conroy - IGES/PDES Organisation (IPO)
- Steve Ryan - PDES, Inc. AP development
- Reiner Reschke - PDES, Inc. software prototyping

Following these presentations, the Action Offices suggested that the strategic goals for collaboration in this area were to:

- Focus on the critical technically complex problems associated with modelling and managing engineering product data
- Establish a critical mass of expertise for solving the problems
- Establish a broader consensus on the proposed solutions

There was then a collaborative brain storming to determine just what the critical technical issues and barriers to progress were. These were then grouped and each participant asked to prioritize them and state whether they considered them to be short term or long term issues. The categories and the most important issues, in order, were:

- Methodologies (AP integration; Integration methods; AP structure and contents)
- Testing (Inter-operability; Conformance; Performance)
- Implementation (SDAI; Toolkit for application developers; Quality EXPRESS tools)
- Future STEP developments (EXPRESS language; STEP interface to CIM-OSA and Multi-functional aspects of products and features; Description languages for manufacturing processes)
- AP related (Complete existing APs, specifically CADEX APs, and CDIM B4 / IMPPACT work; NIDDESC/NEUTRABAS work; Sheet metal AP)
- Information dissemination (Common workshops; Improved mutual review process; Education and awareness)
- Relationship with other standards (IRDS, EDI, SQL, SGML, ODA...)

It was evident from this that APs, integration/qualification and testability of implementations were high on the list of short term critical issues for STEP.

The requirement for more people, particularly in Europe, to understand and contribute to the integration activities was voiced several times. How this is to be achieved is still an open question.

The third day was then devoted to specific bi-lateral discussions on selected projects. The following section gives an overview of the discussion between Brad Harris, on behalf of CADEX, and Steve Ryan.

CADEX / US collaboration. The critical issue as far as CADEX was concerned, was to try and push the development of the CADEX APs (204 and 205) through the ISO integration and qualification process more quickly than had somehow been agreed at the Sapporo meeting. My understanding was that the agreement there was to delay the date for CD issue of these documents until late 1992, well after the end of the CADEX project. This did not seem reasonable, and therefore a lengthy discussion was held as to how this situation may be changed.

The second issue was that of harmonising/integrating 203, 204 and 205. This obviously has some overlap with the CADEX Mechanical Design AP idea.

It was clear that some work had already been done on evaluating the common ground between 203 and 205 (document by Bill Anderson, PDES, Inc.). There was a fairly clear requirement to look at if and how the other geometry based CADEX APs could be integrated with 203 in order to increase its domain and usefulness. In particular, the addition of a B-rep capability would be highly desirable.

It was strongly suggested by Brad Harris that the main priority was to address the individual APs first, before considering the bigger picture. When improved industrial data exchange of geometry based information between dissimilar CAD systems using physical files, was proven, then addressing bigger APs implemented in databases could be more confidently considered. Without this first step, there could be little confidence that STEP version 1.0 will actually be any better from a users point of view than IGES !

There was insufficient time to discuss the CADEX common toolkit in any detail

in relation to SDAI. A project based document review of SDAI would be a useful input to ISO TC184/SC4/WG7. Reaching consensus on this document should be another high priority for all those implementing STEP based processors.

The result of the discussion with Steve Ryan was that a draft of a proposed joint work item between CADEX and PDES, Inc. was written, the text of which is shown below in the form specified by the CEC.

The immediate actions are listed, but it will require some effort on the part of CADEX to maintain the issue on the PDES, Inc. and ISO agendas. If CADEX does not do this, it may well be neglected by others ! The project is in a strong position to do this through its funding for attendance at ISO meetings, and the fact that the CADEX APs have already been discussed to some extent with the 203 project. There are also very good technical reasons for being persistent in pushing these APs.

Harmonisation of CADEX and PDES, Inc. APs - 203, 204, 205. Significant resources have already been invested in developing the CADEX APs, STEP Parts 204 and 205 - B-rep and Surface geometry respectively. This include implementation of prototype processors by several European vendors. There is a clear industrial requirement for these APs, and they need to be progressed more quickly within ISO than is currently planned.

The domain of their information content also overlaps with Part 203 significantly, on which substantial resources have also been invested. There are other CADEX APs covering Wire Frame, CSG and Non Manifold B-rep geometry which may also be important either individually or in the context of Part 203.

The key technical issue is AP integration methodology. This project must use the methodology (planned to be developed elsewhere), to integrate these existing APs.

Objectives:

The first objective is to progress the APs 204 and 205 to DIS status within ISO before the end of the CADEX project (June 1992).

The second objective is to ensure compatibility and integration of Parts 204 and 205 with Part 203 prior to the end of the Part 203 balloting process (Nov. 1991).

The third objective is to prove the AP integration methodology.

The international collaboration is required to co-ordinate the activities and ensure consensus on the APs to be included in STEP version 1.0 and the AP integration methodology.

No IPR problems are foreseen.

Proposed project:

- To exchange current AP documents.
- To document and resolve the technical issues between the APs and with the AP integration methodology.
- To drive Parts 204 and 205 through the ISO integration and qualification processes.
- To evaluate the feasibility of integrating Parts 203 and 204.

Deliverables:

- Issues document on the existing APs and the integration methodology.
- Qualified and integrated Part 204.
- Qualified and integrated Part 205.
- Part 203/204 integration evaluation document.
- Perhaps Users and Implementors Guides for the resulting APs.

Proposed participants:

- 203 - Steve Ryan, Mitch Gilbert (?) (PDES, Inc.)
- 204 - Werner Weick (PROCAD), Ray Goult (?) (Cranfield)
- 205 - Jochen Haenisch (SI)
- Wire Frame, Non Manifold B-rep - Jon Aas (FEGS)
- Testing - Brad Harris, ++ (CADDETC)
- Methodology - Mark Palmer (?) (NIST)

Immediate actions:

Jon Aas to distribute all CADEX AP documents to Steve Ryan asap.

Steve Ryan to distribute PDES, Inc. AP 203 to Jon Aas for further distribution within CADEX asap.

Document issues against 203 as part of its CD balloting process.

Steve Ryan and Y Yang to discuss the 203/204/205 integration issues and the feasibility of resolving them in the short term, at their meeting in August, and report the results to CADEX asap.

5.3.6 AP test purposes

The following paragraph is reproduced from the D5 deliverable report of January 1991:

"Some of the CADEX partners responsible for contributing to STEP AP specifications have provided draft conformance requirements and test purposes as part of their work. The work of the Test Group in conformance testing is predicated on those conformance requirements and test purposes already being available. Further discussion of these is not within the scope of the Test Group work described here."

Since that report was written, there has been a considerable amount of work put into the development of APs within the CADEX project by several partners. Unfortunately, there have also been several changes with respect to the STEP resource models, AP Guidelines, and integration requirements, upon which the APs are based.

The result has been that the AP models have been continually changing over this period and the way in which test purposes should be written has been evolving also. Consequently it has been impossible to consider the development of test cases for conformance testing of the CADEX APs. Despite this, the subject was discussed at the CADEX Oslo meeting in June 1991 also and the following section is a further extract from the minutes of that meeting.

5.3.6.1 Further Extracts from Oslo Test Group meeting minutes (June 1991)

In discussing the test purposes required for each AP, the work done in the Part 201 meetings was offered as an example of what was required. In preparing these test purposes, the Part 201 project found it very useful to draw a structure diagram which essentially represents the structure of the ARM on a single sheet of paper. The purpose of it is to identify all the possible paths through the model to the low level entities, ie. where a choice has to be made or an option is available when instanciating the model. These occur when an attribute type is OPTIONAL, where a referenced entity is a SUPERTYPE of other entities, or where an attribute is an ENUMERATION or SELECT type. The CTS2 participants from CADDETC produced similar diagrams for each functional level for Part 205 prior to the meeting.

The idea is to enumerate all options within the model which need to be tested. This is NOT all combinations. Combinations are specifically not tested for.

For each test purpose, an (abstract) test case would be written from which could be generated a reference STEP physical file and from which could be derived a model generation description for each CAx system under test [8].

Each test case would specify verdict criteria. In this way, the simple functions of the system under test would be evaluated for conformance with the standard. The rationale behind this approach is that only when there is confidence that a system can meet the simple requirements of the standard, should more complex user acceptance tests be performed which evaluate the performance of a system in an industrial context and with live data, so that aspects such as execution time, robustness and the capability of dealing with user specific data sets can be judged.

Against this argument is the fact that this approach requires a large number of test purposes in the AP document, and a corresponding large number of test cases in the 300 part documents.

It was suggested that a single test case could be designed to elicit all that is required by containing an instance of each of the optional characteristics of the ARM. This may indeed be possible, but the size and complexity of such a model would almost certainly make it impossible to debug if a problem was found, since this would in fact be testing combinations, not options. There is nothing to prevent such large and complex models being designed, indeed the FIAT test cases are an attempt at designing representative industrial test cases, but these will be used for inter-operability, cycle and file cycle testing, not for conformance testing.

The consensus was therefore that there seems to be no other alternative to having many small test purposes and test cases if the conformance testing process is to cover the conformance requirements methodically and completely.

In addition, there was consensus that test purposes should be specified in terms of the AIM rather than the ARM. However, if the ARM's were specified in EXPRESS, the view may change ? This is not such a contentious issue.

Completion of this work necessarily requires a fully expanded and attributed form of the relevant AIM's.

5.3.6.2 The Leeds workshop in December 1991

Subsequent to the Berlin workshop, further AP discussions took place at the ISO meeting in Houston in October based around a CADDETC plan to expedite the progress of Parts 204 and 205, and a CADEX/ISO workshop was organised for December to continue this work.

This workshop was arranged to take forward the work of detailing the AP information models for Parts 204 and 205 and thereby lay the foundation for developing test purposes. The remit of this meeting was to develop AIM's for Parts 204 and 205 (and perhaps contribute to 206), and to document the Application Interpreted Constructs (AICs) which resulted from this work and the consideration of the integration of these CADEX APs with AP 203 from PDES Inc. Key ISO Integration and AP co-ordination people from the US attended, together with CADEX AP representatives. The intention was to be to able to complete this integration work subsequent to this meeting if necessary, without having to relay on US people travelling to Europe and/or CADEX people travelling to the US.

The minutes from this meeting are available separately [28]. The main conclusion from the meeting was that CD versions of Parts 204 and 205 could be released by July 1992, six months earlier as planned at the Sapporo meeting in June 1991, and all efforts have been put into bringing this forward further.

Unfortunately, the APs have not yet achieved CD status. This is due to many factors which space does not permit discussion of here. However, the pace of development of the APs within the ISO arena lends much justification to the decision to invest effort in this area; without it, achieving CD status would be much further away than it is now.

The CTS2 project has liaised with CADEX in terms of helping to develop Test Purposes for one or more APs as has the CADDETC DTI project.

5.3.7 Test data and procedures

Once the test purposes for each AP have been developed, a suite of abstract test cases, one test case per test purpose, will be written. The individual test cases may be grouped and ordered in a logical hierarchical structure. The abstract test cases will be defined in a formal language which will be specified in [4].

As part of its contractual commitments to the CEC, the CTS2 project is currently planning to start the development of an abstract test suite for a STEP AP, possibly a CADEX based AP. The objective of this work package is to evaluate the CTS2 testing methodology developed for IGES, SET Schematics VDA-FS and VDA-IS, with respect to STEP.

However, the time scales are such that it has been impossible for this test suite to be completed before the end of the CADEX project.

Test methods for STEP processors are described in an abstract way, there being a single method described for each of the implementation forms of STEP APs (Physical File, Working Form, DBMS, IKBS). These methods will be detailed in a subsequent document in the class devoted to conformance testing, [5], although each method will necessarily be consistent with the framework described in [2].

The conformance assessment process consists of testing an implementation of a given AP against the requirements specified in that AP. [3] details the process and gives the responsibilities of the test laboratory and client. The following is a brief summary:

- preparation for testing - selecting the test method and test suite
- performing the test - selecting and parameterising the abstract test cases, giving an executable test suite, and observing the results obtained.
- analysing the results - assigning a verdict to each test.
- Writing the conformance report - including a summary and the detailed results.

It can be seen that the process is potentially complex and that a large volume of work needs to be done to test even the simplest of APs. This is due to the complexity of the standard itself rather than any inherent flaws in the testing methodology, which simply states what is required to systematically check all testable aspects specified in the standard.

There was no commitment by any CADEX partner to develop a conformance testing service for STEP processors within the framework of the project. It has been impossible to perform any informal conformance testing of processors developed within the project because of the delays described above in reaching consensus on APs and the derivation of test purposes. These factors have been outside the control of CADEX partners, and in the case of CADDETC, has required that work be done on AP development activities rather than test case development.

5.4 The FIAT test library

This part of the document reports on the work done by FIAT within Work Package 6 - Test and Validation. The emphasis was on User Acceptance Testing, and the main target was to support the CADEX partners with a test library based on "industrial" models, fully explained in a system independent model description. From this one, every partner could be able to produce STEP files to be used for performing different tests according to the methodology defined elsewhere.

The chapter consists of three sections: the first deals with the choice of models to be tested, the second contains the descriptions of the models and the third describes the data evaluation criteria.

The activities are described comprehensively in the reports [40], [41], [19].

5.4.1 Choice of the models

In February 1991, FIAT defined a set of nine industrial test cases to be used for system testing [40]. The following approach has been adopted:
1. identification and analysis of an initial population, meaningful sample of the company reality (172 drawings)
2. extraction of a still representative sub-sample (25 drawings)
3. geometric evaluation of the sub-sample
4. choice of a final set of industrial test cases (9 drawings).

Note that it was tried, as far as possible, to make any consideration for defining the sample from a "user point of view" and not from a "CAD expert point of view".

5.4.1.1 Identification and analysis of the initial population

The sample population was chosen among drawings representative for parts really designed by IVECO (FIAT's truck division), and it was formed by 172 drawings. The parts (assemblies and components) came from different engineering departments, concerning areas like:
- mechanics (engines, frames, gear boxes, shafts, etc.)
- carbody
- electric layout
- pneumatic layout.

The sample has been chosen in compliance with the statistical distribution of the size of the drawings in the company (see figure 5.4.1).

Drawing percentage

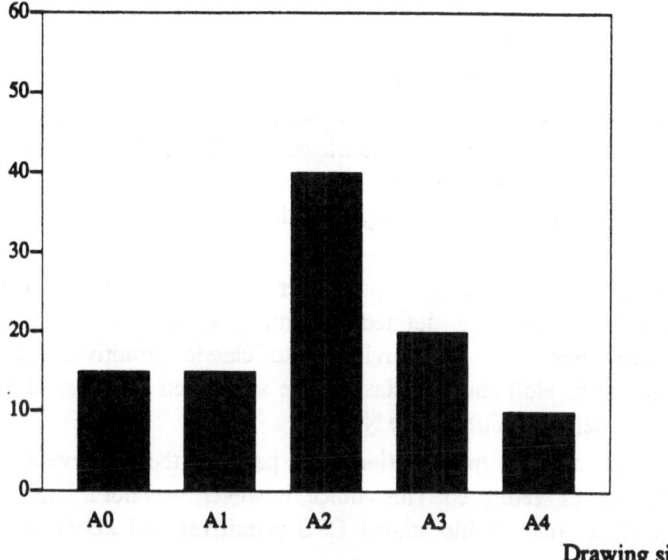

Fig. 5.4.1: Distribution of size in 172 drawings

The analysis has been made according to drawing size, type of execution (i.e. manually or CAD), geometric complexity (intuitively, from the shape) and engineering type of the part. The concept of "intuitive" geometric complexity was purely subjective, as it was based on human sensibility and intuition, i.e. getting a look to the shape and not thinking in CAD entity terms. The possible values of it were low, medium and high. The concept of engineering type shared the sample according to the different engineering classes, identified before.

5.4.1.2 Extraction of the sub-sample

The rules used for the extraction of the sub-sample were based on the following three steps:

1. examination if the part was an elementary component or an assembled one, choosing the former. This step reduced the starting sample to 74 drawings;
2. classification according to part design classes. The parts have been grouped into similar classes (e.g. console, exhaust pipe, cog wheel, gear box, etc.), then one drawing per class has been chosen, as far as possible.
3. consideration on drawing size and on "intuitive" geometric complexity. About the former, it was tried to choose a set respecting the statistical distribution. About the latter, it was tried to cover all degrees of "intuitive" geometric complexity (low, medium and high).

Finally, a sub-sample of 25 drawings was obtained.

5.4.1.3 Geometric evaluation of the sub-sample

The sub-sample was analysed and evaluated taking into account two aspects:
- geometric complexity of the part
- mathematical representation of the part.

The definition of the complexity and its measure was an arbitrary and subjective choice, but reasonably rational. The geometric complexity of a part was defined as the set of the solid primitives (i.e. box, cylinder, etc.) and the operations (i.e. union, intersection, duplication, mirroring, etc.) theorically needed for creating the part itself.

As measure of the complexity, a parameter named Numerical Geometric Complexity (N.G.C.) has been defined, assuming values depending on the primitives. These ones have been divided into classic primitives (like box, cylinder, sphere, torus, etc.) and non-classic (like sculptured surfaces, B-splines, etc.), associated to defined values of the N.G.C.

For giving a mathematical representation of the part and, then, for evaluating the N.G.C, CSG primitives were used. This choice, however, was not a limitation; in fact, there is a tight relationship among CSG primitives and 2D/3D non-CSG entities, as we can see:

```
Primitives        <-> 2D entity      <-> 3D entity
box               line                  line
wedge             line                  plane
cylinder          circle                cylindrical surface
cone              conics                conic surface
torus             (not used)            circular fillet
sculpt. surface   param. curve          param. surface
```

So, it was intuitive to assign an N.G.C. value to each primitive: for the classic ones the values were chosen related to the degree of the corresponding mathematical representation; for the non-classic ones, the values were assigned intuitively in the following way:

```
Primitives                          N.G.C.
box/wedge                           1
cylinder/cone                       2
torus                               4
surface of extrusion                5
surface of revolution               6
swept surface                       7
spline/interpolation surf.          8
poles surf. (Bezier, B-splines)     10
```

Therefore, the N.G.C. of a part was defined as the sum of the N.G.C.s of the primitives used for their generation. The results are listed in figure 5.4.2. For a more detailed description see Part II.5 in [40].

number / object	size	CAD	compl exity	box (1)	cyl (2)	con (2)	wed (1)	tor (4)	extr sf(5	revo sf(6	swep sf(7	spli sf(8	pole sf(10)	NGC
1 plate support	A4	-	low	1	2	2	.	.	.	.	.	.	.	9
2 nozzle holder	A2	-	med	.	7	9	.	3	.	.	.	.	.	44
3 gasket	A0	-	med	5	18	.	.	.	.	.	.	.	.	41
4 water pump holder	A2	-	med	2	18	.	.	.	.	.	.	.	.	38
5 fastening bracket	A2	-	low	7	3	.	.	.	1	.	.	.	.	18
6 support union	A2	x	med	4	3	.	.	.	.	.	.	.	.	10
7 plateband f.support	A1	x	med	6	4	.	.	.	.	.	.	.	.	14
8 plate holder	A1	-	low	4	4	2	.	1	.	.	.	.	.	20
9 upper element	A3	-	low	6	2	1	.	1	.	.	.	.	.	16
10 brake disc	A0	x	med	.	24	4	.	1	.	.	.	.	.	60
11 disc wheel	A1	x	med	.	13	14	.	.	.	.	.	.	.	54
12 air cleaner	A0	-	med	4	6	.	.	2	1	.	.	.	.	29
13 support	A1	-	low	12	6	.	.	.	.	.	.	.	.	24
14 gaskets	A2	-	low	4	7	.	.	.	.	.	.	.	.	18
15 driving pulley	A2	x	low	.	8	2	.	.	.	.	.	.	.	20
16 stirrup	A2	x	low	3	6	.	1	.	.	.	.	.	.	16
17 shaft	A1	-	med	.	14	4	.	.	.	.	.	.	1	46
18 differential pinion	A1	-	med	.	9	8	.	.	.	1	.	.	2	60
19 plate support	A1	-	low	4	4	.	.	.	.	.	.	.	.	12
20 stirrup	A2	x	high	4	3	.	.	.	.	.	.	.	.	10
21 gas outlet line	A0	-	high	2	7	.	.	.	2	.	1	.	.	33
22 alternator support	A1	x	med	6	6	.	.	.	.	.	.	.	.	18
23 compartment cover	A1	-	high	10	28	.	.	.	.	.	.	.	.	66
24 bracket	A2	-	high	7	8	.	.	1	.	.	.	.	.	27
25 soundproof protects	A0	x	high	8	3	3	.	.	.	.	3	2	1	67

(NGC values given in parentheses)   Average: 30.8

Fig. 5.4.2: Characteristics of the sub-sample

5.4.1.4 Choice of the final set

This sub-sample was still too large. The number of models was planned to be around ten. It was not regarded to be necessary to have a great number of models with a similar complexity.

The previous sub-sample has been reduced to the final one according to the following rules:

- draw the distribution of the sub-sample of the 25 drawings versus the N.G.C. (see figure 5.4.3)

- analyse the distribution, choosing those drawings having the values of the N.G.C. around the peaks and covering all specified geometric entities.

More precisely, the final sample was constituted by:

1 drawing for $10 <=$ N.G.C. $<= 17$
3 drawings for $18 <=$ N.G.C. $<= 20$
2 drawings for $30 <=$ N.G.C. $<= 50$
3 drawings for $51 <=$ N.G.C. $<= 67$

This was assumed to be a representative subset of models for use in a CAD project.

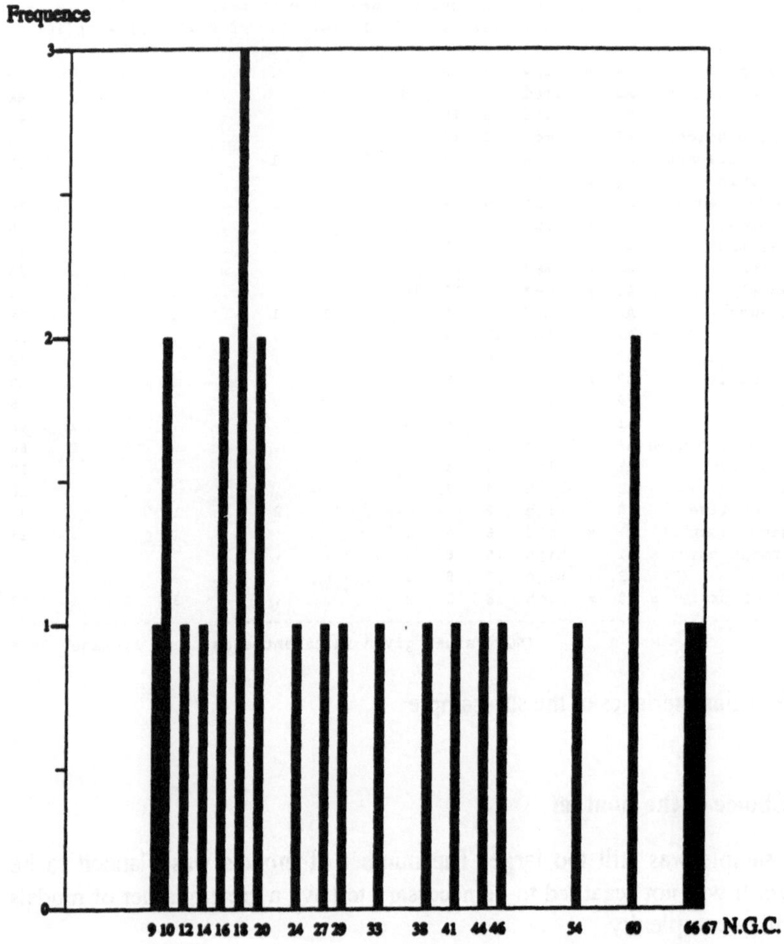

Frequence

Fig. 5.4.3: Distribution of NGC in 25 drawings

These industrial test cases were (characteristics see figure 5.4.4):
- Model 1: Nozzle holder
- Model 2: Fastening bracket
- Model 3: Support union
- Model 4: Plate holder
- Model 5: Gaskets
- Model 6: Differential pinion
- Model 7: Gas outlet line from blower
- Model 8: Timing system compartment cover
- Model 9: Soundproof protections

number object	size	CAD exity	compl	box (1)	cyl (2)	con (2)	wed (1)	tor (4)	extr sf(5	revo sf(6	swep sf(7	spli sf(8	pole sf(10)	NGC
1 nozzle holder	A2	-	med	.	7	9	.	3	.	.	.	.	.	44
2 fastening bracket	A2	-	low	7	3	.	.	.	1	.	.	.	.	18
3 support union	A2	x	med	4	3	.	.	.	.	.	.	.	.	10
4 plate holder	A1	-	low	4	4	2	.	1	.	.	.	.	.	20
5 gaskets	A2	-	low	4	7	.	.	.	.	.	.	.	.	18
6 differential pinion	A1	-	med	.	9	8	.	.	.	1	.	.	2	60
7 gas outlet line	A0	-	high	2	7	.	.	.	2	.	1	.	.	33
8 compartment cover	A1	-	high	10	28	.	.	.	.	.	.	.	.	66
9 soundproof protects	A0	x	high	8	3	3	.	.	.	.	3	2	1	67

(NGC values given in parentheses)   Average: 37.3

Fig. 5.4.4: Characteristics of 9 industrial test cases

5.4.2 Model generation description (MGD)

In June 1991, FIAT described the nine chosen models in a system independent form [19]. The following approach has been adopted:

1. identification of Application Protocols (APs) entities
2. classification of MGDs according to the APs (21 in total)
3. definition of the model generation schemas
4. MGD of the test cases.

5.4.2.1 Identification of the entities

In this activity, the documents [11], [12], [42], [43], have been used for getting the table of STEP geometric and topological entities included in the various CADEX APs. The MGDs were described according to the entities listed in this table.

5.4.2.2 Classification of MGDs according to APs

In this step, the selected components have been divided into two classes, according to their shapes: open parts (i.e. sheet metal part) and closed parts (i.e. mechanical part). The former are represented using Surface and Wire Frame Aps, the latter using B-rep, CSG and Wire Frame APs. B-rep and Wire Frame APs have been divided into analytic and non analytic.

A matrix for matching the geometry of each of the selected parts to the various CADEX APs has been drawn (see figure 5.4.5). If the component can be modeled using a specific AP, an MGD has been defined.

The above considerations brought to explode the 9 models into 21 MGDs.

Model number	B–Rep		Constructive Solid Geometry	Sculptured Surfaces	Wireframe	
	analytic	non–analytic			analytic	non–analytic
Model 1 NGC = 44	x		x		x	
Model 2 NGC = 18				x		x
Model 3 NGC = 10				x	x	
Model 4 NGC = 20				x	x	
Model 5 NGC = 18				x	x	
Model 6 NGC = 60		x				x
Model 7 NGC = 33		x	x			x
Model 8 NGC = 66	x		x		x	
Model 9 NGC = 67				x		x

Fig. 5.4.5: AP of FIAT test cases

Model number	7
Part name	Gas outlet line from blower
Topology class	closed part
Geometry representations	B–rep / non–analytical CSG Wireframe / non–analytical
General characteristics	– carbody part – constituted by 3 parts : first flange second flange outlet body

Fig. 5.4.6: Model identifier card

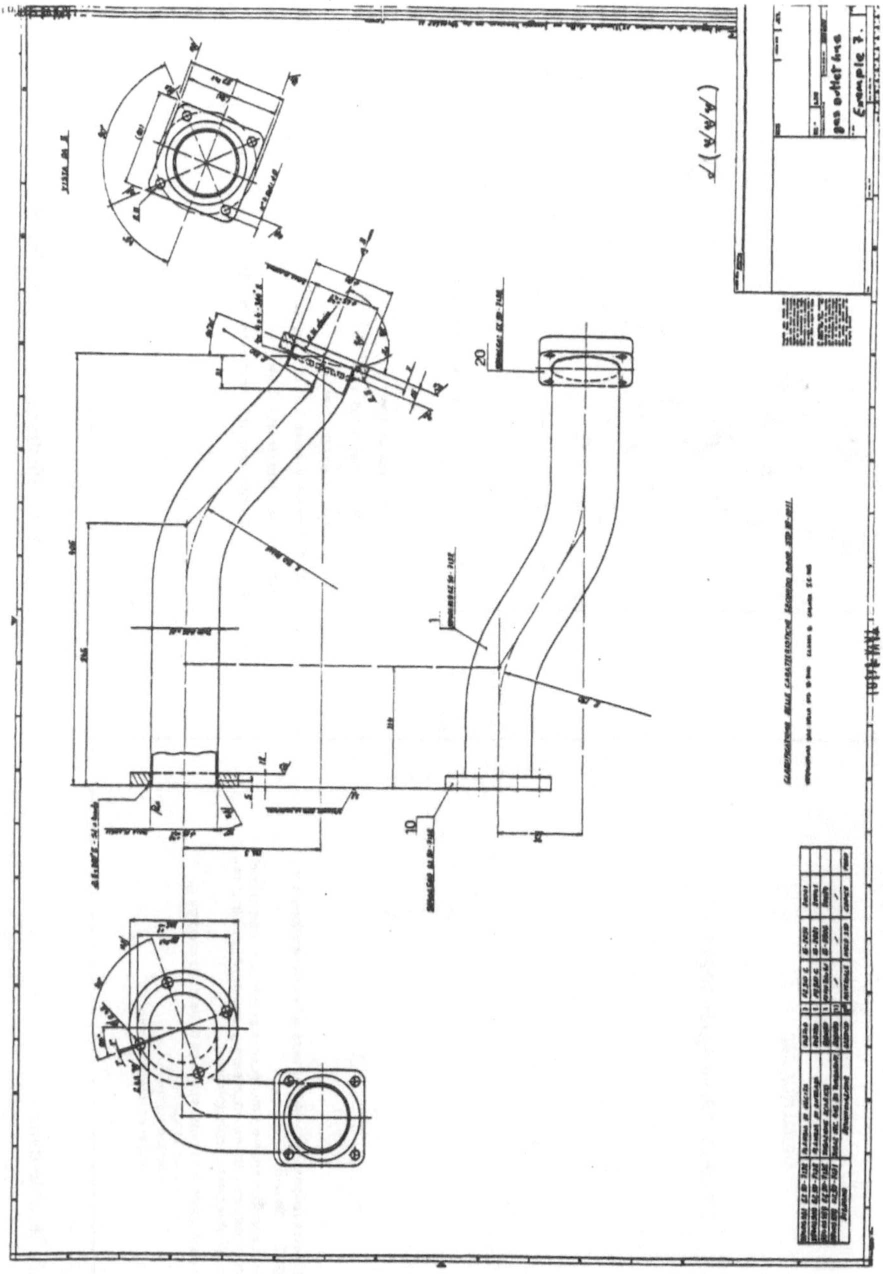

Fig. 5.4.7: Source engineering drawing

Model Number 7

Csg Representation Description

- the construction is reported in the attached illustrations from page 2 to page 5
- in the description some simplifications have been introduced in comparison with the engineering drawing (i.e. minor radius fillets were not considered)
- the description has been split into three parts regarding:
  - first flange
  - second flange
  - central part (outlet body)

Fig. 5.4.8: A note card

Second flange description

2.1) Cylinder diameter 9 mm.

2.2) Cylinder diameter 105 mm.

2.3) Cylinder diameter 65 mm.

2.4) Difference between 2.1, 2.2 and 2.3

2.5) Extruded line length 12 mm.

2.6) Solid swept area of linear extrusion of 2.4

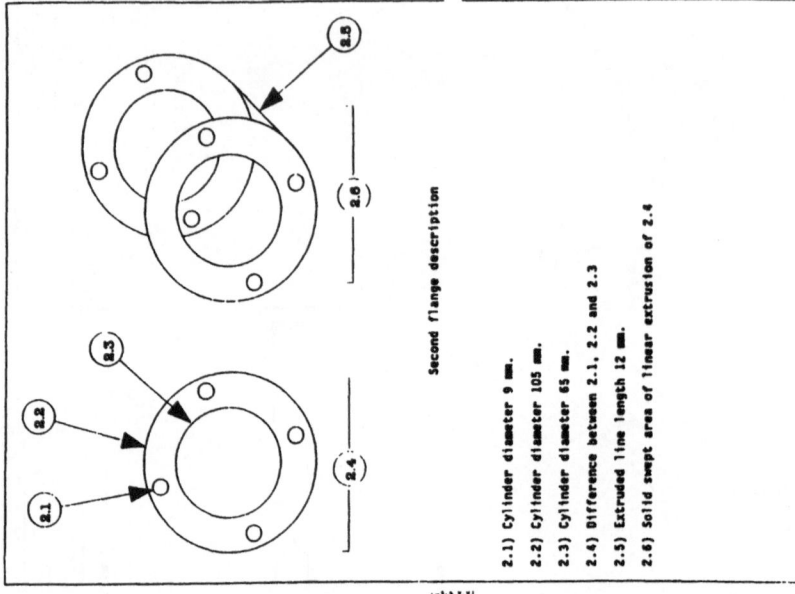

Fig. 5.4.9: Illustration for generating the model

5.4.2.3 Definition of the model generation schemas

Each description of the models has been set up according to the following schema:
- an identifier card, resuming the main part characteristics (see figure 5.4.6) - an engineering drawing of the part (see figure 5.4.7)
- for each AP representation: a "note" card and more illustrations, describing the possible steps for building the model, using the corresponding AP representation (see figures 5.4.8 and 5.4.9).

5.4.2.4 MGDs of the test cases

So far, according to the identified schemas, it was produced an MGD package for each of the 21 cases identified as before. Note that a big effort has been done for producing illustrations and comments, in order to explain through sequences of logical operations and crucial dimensions the modeling of the part. The above quoted figures 5.4.6, 5.4.7, 5.4.8, 5.4.9 refer to the so called Model 7.

5.4.3 Data transfer evaluation criteria

In January 1992, FIAT issued [41], as the first revision of [19], including:
- modifications and corrections according to the feedbacks received from CADEX partners
- evaluation criteria on model transfer
- "zoomed" engineering drawings
- some explicative drawings with axis position, and outlining critical dimensions.

In comparison with [19], this version [41] contained for each of the nine selected cases, an evaluation criteria card (see figure 5.4.10). In this one, what needs to be measured during and after the transfer has been put in evidence. The card has been divided into two main parts:
- visual evaluation, i.e. the result that must be provided by the systems for visually checking the data transfer (perspective drawing, orthogonal views and plane sections)
- numerical properties, i.e. the result that must be provided by the systems for computing mass properties (volume, centre of gravity and moments of inertia) and detected critical dimensions.

It should be noted that these criteria have been stated from a "strict" user's point of view and not from a CAD expert's point of view; in fact, they have been set up with the help of an expert designer of IVECO engineering department.

MODEL 7 *Gas outlet line from blower*

	Graphic types	Y/N	Notes
	Visual Evaluation		
PERSPECTIVE Drawing	wireframe	Y	Drawings related to the two pipe front flanges
	hidden lines	N	
	isoparametrics	Y	Surface representing the pipe
ORTHOGONAL Views	x–y plane	Y	
	y–z plane	Y	
	x–z plane	Y	
Plane SECTIONS	x = const.	N	
	y = const.	N	
	z = const.=0 z = const.=75	Y	Two needed sections for controlling all main dimensions (diameters and length)
Others			

	Numerical Properties	
Types	Y/N	Notes
Centre of gravity	Y	
Dimensions	Y	Pipe diameter and length; dimensions of the flanges
Mass	Y	
Moments of Inertia	Y	
Others		

Fig. 5.4.10: Evaluation criteria card

5.5 Test plan and results

This chapter reports on the plan of the processor testing provided by CADDETC and the results that, according to [44] and [45], have been sent to FIAT from different partners.

5.5.1 Test plan

5.5.1.1 General test files for all partners

According to the meeting held in Aachen 12th September 1990 [46], FIAT collected the test data sent periodically by CADEX partners and distributed every

new version of this library on each project board meeting. FIAT updated the test
data to be compatible with each current processor versions.

The latest and up to date list (June 1992) of these STEP files is:

Filename	Length	Date	AP	Owner
APOLLO.SS	16548	05-06-92	Surface	Siemens Nixdorf
ARM.STE	587068	05-06-92	Brep	Hewlett Packard
ASCON.STE	86731	05-06-92	Brep	Hewlett Packard
ASPIS.STE	153680	05-06-92	Brep	Hewlett Packard
BLOCK.PRE	6234	05-06-92	Surface	Siemens Nixdorf
BOGEN.NF	3605	05-06-92	Brep	GfS
BOTTLE.PRE	34482	05-06-92	Surface	Siemens Nixdorf
BOTTLE_S.STE	29483	05-06-92	Surface	Det Norske Veritas
CHAIR.NF	245549	05-06-92	Surface	SI
CONE_1.PRE	4887	05-06-92	Surface	Siemens Nixdorf
CONE_2.PRE	3662	05-06-92	Surface	Siemens Nixdorf
CRANFLD.NF	72695	05-06-92	Brep	
CSGTEST.NF	4826	05-06-92	CSG	Disel/GfS
CUBE.NF	6035	05-06-92	Brep	Hewlett Packard
CURSUR.NF	16045	05-06-92	Surface	SI
CYLINDER.PRE	6091	05-06-92	Surface	Siemens Nixdorf
EXSI1.NF	11046	05-06-92	Surface	SI
EXSI2.NF	9852	05-06-92	Surface	SI
EXSI3.NF	12330	05-06-92	Surface	SI
FBSM_FI1.NF	87629	05-06-92	CBrep	FEGS
FOEN.SNI	116976	05-06-92	Surface	Siemens Nixdorf
GELENK.NF	62794	05-06-92	CBrep	GfS
HOOD.NF	54222	05-06-92	Surface	SI
INJECTOR.NF	4932	05-06-92	CSG	DISEL
KANNE.PRE	97225	05-06-92	Brep	Procad/Isykon
KEGEL.PRE	3618	05-06-92	Brep	Procad/Isykon
KLINKE.NF	54544	05-06-92	Surf/brep	Siemens Nixdorf
MAINHOUS.NF	175951	05-06-92	Brep	Hewlett Packard
MERKER.PRE	55676	05-06-92	Brep	Procad/Isykon
P14.NF	3189	05-06-92	Brep	Siemens Nixdorf
P140.NF	257367	05-06-92	Brep	Siemens Nixdorf
PANEL.NF	167446	05-06-92	Surface	Siemens Nixdorf
PART7.STE	47298	05-06-92	Brep	Hewlett Packard
PRISM.NF	25903	05-06-92	CBrep	FEGS
QUADER.NF	7488	05-06-92	Brep	Procad
QUADER.PRE	6782	05-06-92	Brep	Procad
QUADZYL.PRE	9466	05-06-92	Brep	Procad
QUZYKEKU.PRE	13352	05-06-92	Brep	Procad
RUNDQUAD.PRE	15218	05-06-92	Brep	Procad
SBWMGE20.STE	200635	05-06-92	Wireframe	FEGS

SBWM_FI1.STE	73560	05-06-92	Wireframe	FEGS
SLEDGE.NF	63776	05-06-92	Surface	SI
SOCKET.PRE	78540	05-06-92	Surface	Siemens Nixdorf
SPHERE.PRE	3485	05-06-92	Surface	Siemens Nixdorf
TEAPOT.PRE	254266	05-06-92	Surface	Siemens Nixdorf
TEST63.NF	15335	05-06-92	CBrep	GfS
TORUS.PRE	4689	05-06-92	Surface	Siemens Nixdorf
VASE.PRE	13393	05-06-92	Surface	Siemens Nixdorf
WBEI.NF	2905	05-06-92	Wireframe	GfS
WBOGEN.NF	2415	05-06-92	Wireframe	GfS
WEXSI1.NF	4633	05-06-92	Wireframe	GfS
WFTEST.NF	2887	05-06-92	Wireframe	GfS
WKLINKE.NF	27098	05-06-92	Wireframe	GfS
ZYLINDER.NF	3565	05-06-92	Brep	Procad
ZYLINDER.PRE	3567	05-06-92	Brep	Procad
-----	------			
58	3352192			

With either this collection or the FIAT industrial test cases, every partner could perform cycle and file cycle tests. Inter-system testing could be done according to a test plan proposed by CADDETC [9] and described in section 5.5.1.2.

5.5.1.2 Test plan with industrial test cases

The test plan specified which data exchange tests should be carried out, between which companies, using which models and which APs, with regard to the list of processors in the Technical Annexe. Where it was stated that processors could not be available until the end of the project (according to the latest list from GfS), no tests were possible and no results reported.

Model numbers referred to those contained in FIAT report [3]. The types of tests were as specified by BMW in the corresponding chapters.

Where it was stated that no models were available for testing, it was because FIAT don't have such models in their industrial environment (for example, B-rep level 1). Where it was stated that a partner cannot expect to receive models of a given type from other partners, they can only perform cycle and file cycle tests, not inter-system tests.

The following table suggests a test plan for industrial test STEP files library.

Det Norske Veritas Research

Write models 2 and 9 using Surface AP level 2 to test pre-processor and send to GfS.

Expect to receive models 2, 3, 4, 5 and 9 from Isykon, Italcad, ND, Procad, SI and SNI using Surface level 1 AP to test post-processor.

Expect to receive No models for Surface level 2 AP to test post-processor.

DISEL

Write models 1, 7 and 8 using CSG AP.

Expect to receive models 1,7,8 from ITALCAD to test the postprocessor.

FEGS

Write all models using Wireframe AP to test pre-processor and send to GfS.

Expect to receive models 1 and 8 from HP, Isykon, Procad and SNI using B-rep AP level 2 to test post-processor.

Expect to receive all models from GfS using Wireframe AP to test post-processor.

Gesellschaft für Strukturanalyse

Write all models using Wireframe AP to test pre-processor and send to FEGS.

Expect to receive models 1 and 8 from HP, Isykon, Procad and SNI using B-rep AP level 2 to test post-processor.

Expect to receive models 2, 3, 4, 5 and 9 from Isykon, Italcad, ND, Procad, SI, and SNI using Surface AP level 1 to test post-processor.

Expect to receive models 2 and 9 from DnV using Surface AP level 2 to test post-processor.

Expect to receive No models for Surface AP level 3 to test post-processor.

Expect to receive all models from FEGS using Wireframe AP to test post-processor.

Hewlett-Packard

Write models 1 and 8 using B-rep level 2 AP to test pre-processor and send to FEGS, GfS, Isykon, Procad and SNI.

Expect to receive models 1 and 8 from Isykon, Procad and SNI using B-rep level 2 AP to test post-processor.

ISYKON

Write models 1 and 8 using B-rep level 2 AP to test pre-processor and send to FEGS, GfS, HP, Procad and SNI.

Write models 2, 3, 4, 5 and 9 using Surface AP level 1 to test pre-processor and send to DnV, GfS, Italcad, ND, Procad, SI and SNI.

Expect to receive models 1 and 8 from HP, Procad and SNI using B-rep level 2 AP to test post-processor.

Expect to receive models 2, 3, 4, 5 and 9 from Italcad, ND, Procad, SI, and SNI using Surface AP level 1 to test post-processor.

ITALCAD

Write models 2, 3, 4, 5 and 9 using Surface AP level 1 to test pre-processor and send to DnV, GfS, Procad, ND, Isykon, SI and SNI.

Expect to receive models 2, 3, 4, 5 and 9 from Isykon, ND, Procad, SI, and SNI using Surface AP level 1 to test post-processor.

Kongsberg 3D Partner (Norsk Data)

Write models 2, 3, 4, 5 and 9 using Surface AP level 1 to test pre-processor and send to DnV, GfS, Procad, Italcad, Isykon, SI and SNI.

Expect to receive models 2, 3, 4, 5 and 9 from Isykon, Italcad, Procad, SI, and SNI using Surface AP level 1 to test post-processor.

PROCAD

Write models 1 and 8 using B-rep level 2 AP to test pre-processor and send to FEGS, GfS, HP, Isykon and SNI.

Write models 2, 3, 4, 5 and 9 using Surface AP level 1 to test pre-processor and send to DnV, GfS, Italcad, ND, Isykon, SI and SNI.

Expect to receive models 1 and 8 from HP, Isykon and SNI using B-rep level 2 AP to test post-processor.

Expect to receive models 2, 3, 4, 5 and 9 from Italcad, ND, Isykon, SI, and SNI using Surface AP level 1 to test post-processor.

Senter for Industriforskning

Write models 2, 3, 4, 5 and 9 using Surface AP level 1 to test pre-processor and send to DnV, GfS, Procad, Italcad, Isykon, ND and SNI.

Expect to receive models 2, 3, 4, 5 and 9 from Isykon, Italcad, Procad, ND, and SNI using Surface AP level 1 to test post-processor.

Siemens Nixdorf Informationssysteme

No models available for testing B-rep level 1 AP pre-processor and post-processor.

Write models 1 and 8 using B-rep level 2 AP to test pre-processor and send to FEGS, GfS, HP, Isykon and Procad.

Write models 6 and 7 using B-rep level 3 AP to test pre-processor.

Write models 2, 3, 4, 5 and 9 using Surface AP level 1 to test pre-processor and send to DnV, GfS, Italcad, ND, Isykon, SI and Procad.

Expect to receive models 1 and 8 from HP, Isykon and Procad using B-rep level 2 AP to test post-processor.

Expect to receive No models for B-rep AP level 3 to test post-processor.

Expect to receive models 2, 3, 4, 5 and 9 from Italcad, ND, Isykon, SI, and Procad using Surface AP level 1 to test post-processor.

This plan is also shown in figure 5.5.1. The results of the tests described here and those on other tests made on the STEP file library, can be seen in the following sections, dedicated to particular partner's contributions.

		B-rep 1	2	3	to: from:	Surface 1	2	3	to: from:	Wire Frame 1	2	3	to: from:	CSG	to: from:
DISEL	send													1, 7, 8	
	receive													no one	
DNV	send						2, 9		GfS						
	receive					2, 3, 4, 5, 9	no one		Isykon, Italcad, ND, Procad, SI, SNI						
FEGS	send									all	all	all	GfS		
	receive		1, 8		HP, Isykon, Procad, SNI					all	all	all	GfS		
GfS	send									all	all	all	FEGS		
	receive		1, 8		HP, Isykon, Procad, SNI	2, 3, 4, 5, 9 (–)	2, 9 (*)	no one	Isykon, Italcad, ND, Procad, SI, SNI (–) DnV(*)	all	all	all	FEGS		
HP	send		1, 8		FEGS, GfS, Isykon, Procad, SNI										
	receive		1, 8		Isykon, Procad, SNI										
ISYKON	send		1, 8		FEGS, GfS, HP, Procad, SNI										
	receive	2, 3, 4, 5, 9 (–)	1, 8 (*)		DnV, GfS, Italcad, ND, Procad, SI, SNI (–) HP, Procad, SNI (*)	2, 3, 4, 5, 9			Italcad, ND, Procad, SI, SNI						
ITALCAD	send					2, 3, 4, 5, 9			DnV, GfS, Isykon, ND, Procad, SI, SNI						
	receive					2, 3, 4, 5, 9			Isykon, ND, Procad, SI, SNI						
ND	send					2, 3, 4, 5, 9			DnV, GfS, Isykon, Italcad, Procad, SI, SNI						
	receive					2, 3, 4, 5, 9			Isykon, Italcad, Procad, SI, SNI						
PROCAD	send		1, 8		FEGS, GfS, HP, Isykon, SNI	2, 3, 4, 5, 9			DnV, GfS, Italcad, ND, Isykon, SI, SNI						
	receive		1, 8		HP, Isykon, SNI	2, 3, 4, 5, 9			Italcad, ND, Isykon, SI, SNI						
SI	send					2, 3, 4, 5, 9			DnV, GfS, Isykon, Italcad, ND, Procad, SNI						
	receive					2, 3, 4, 5, 9			Isykon, Italcad, ND, Procad, SNI						
SNI	send	no one	1, 8 (*)	6, 7	FEGS, GfS, HP, Isykon, Procad (*)	2, 3, 4, 5, 9			DnV, GfS, Isykon, Italcad, ND, SI, Procad						
	receive	no one	1, 8		HP, Isykon, Procad	2, 3, 4, 5, 9		no one	Isykon, Italcad, ND, SI, Procad						

Fig. 5.5.1: Plan for exchange of test models

5.5.2 Test results of Gesellschaft für Strukturanalyse

5.5.2.1 Ranges of test models in GfS

Being a vendor of a finite element system, GfS is extremely interested in testing models. Therefore GfS has repeatedly asked all other partners to send them CADEX STEP files. While GfS was committed to write Compound_Brep and Wireframe files, their STEP postprocessor is able to read Solid Breps and Surfaces as well, thus covering the ISO Application Protocols 204/205/206 plus the CADEX Compound_Brep AP.

During the project GfS has tested each file they have received from any partner. CSG files could only be passed through the Common Tools part of the GfS processor, because the GfS modeler cannot handle CSG. All other files were processed further and submitted to

- storage, checking, manipulation

- conversion

- mapping to native data

- if possible writing back to a STEP file.

GfS has received STEP files mainly from 4 different sources:

1. the FIAT test file library (its main purpose was to test processors)

2. files created by the STEP preprocessor of GfS from models designed in the native system or by modifying any existing data

3. a set of (wireframe) files created by FEGS with their program FAMEX

4. files of the 9 test cases (worked out by the test group) and distributed by various partners

A few more files were exchanged for additional test purposes. Many files belong to more than one group.

5.5.2.2 Experiences with models in the FIAT test file library

Most of these files are not large, some are made to test the behaviour of software in the Common Tool Kit and in the system- specific parts of the processors. Prior to performing the tests all files were converted to the format prescribed by the latest version of APs (1992) and demanded by Revision 6.0 of the Common Tool Kit.

The following table gives some basic data about the tests. Performance was measured on a workstation Apollo DN3000 (Model 3010A) with operating system Aegis 10.3, main memory 8 MB and processor 68020. Times are in seconds of elapsed time, file sizes in kB.

The test procedure comprises always a full cycle. In most cases it begins with an existing STEP file. This is read into the data base (IDS) of the STEP processor. Then the model is converted to native representation and can be displayed in the modeler of the FEM system. Except from the manipulations described below no changes and no meshing are done there. The modeler hands over the data again to the STEP processor, where they are converted to STEP representation and stored in the IDS again (the data base was initialised before). Then the data are written to a STEP file using the formatter. The resulting file is compared with the original one.

In some cases (when using models created by the local system) the cycle may begin at another point. In case of losses the time for the formatter was measured again with the original model.

The following table shows the times used for the scanner/ parser, for the formatter, for the postprocessor backend (converting from STEP to native representation) and for the preprocessor frontend (converting from native to STEP representation). The column 'comparison' gives an indication of the correctness (explained below).

Filename	AP	size	sca/par reading time	formatter writing time	write to native data	read from native data	comparison after cycle
block.pre	SF	6	4	3	24	9	ok
bogen.nf	BR	4	4	3	22	7	equiv
bottle.pre	SF	35	6	9	24	13	ok
chair.nf	SF	237	61	74	50	39	ok
cone_1.pre	SF	5	2	3	20	9	ok
cone_2.pre	SF	4	2	3	21	9	ok
cranfld.nf	BR	71	17	17	46	14	not all loops
cube.nf	BR	6	2	3	20	6	equiv
cursur.nf	SF	16	6	5	21	11	ok
cylinder.pre	SF	6	2	2	21	9	ok
exsi1.nf	SF	11	6	5	23	7	equiv
exsi2.nf	SF	10	4	5	22	7	equiv
hood.nf	SF	54	13	16	25	15	ok
kanne.pre	SF	95	20	39	28	20	ok
kegel.pre	BR	4	2	1	20	6	equiv
klinke.nf	BR	53	13	37	37	11	not all loops
mainhous.nf	BR	170	36	28	178	20	not all loops
merker.pre	BR	54	13	34	32	11	not all curves
p14.nf	BR	4	1	2	20	6	equiv
p140.nf	BR	246	52	126	31	19	not all loops
panel.nf	SF	162	38	43	36	30	ok
quader.nf	BR	8	2	3	21	6	equiv
quader.pre	BR	7	2	3	21	6	equiv
quadzyl.pre	BR	9	3	4	20	6	not all loops
quzykeku.pre	BR	13	3	5	22	7	not all loops
rundquad.pre	BR	15	4	7	22	7	equiv
sledge.nf	SF	63	15	20	27	16	ok
socket.pre	SF	77	18	21	30	22	ok
sphere.pre	SF	4	1	1	20	9	ok
teapot.pre	SF	246	52	63	47	42	ok
torus.pre	SF	5	2	3	23	9	ok
vase.pre	SF	13	4	3	21	10	ok
wbei.nf	WF	3	1	3	20	6	not all curves
wbogen.nf	WF	3	1	1	20	6	equiv
wexsil.nf	WF	5	2	3	20	7	equiv
wftest.nf	WF	3	1	2	20	6	not all curves

```
wklinke.nf    WF   26      8       7       28      9       equiv
zylinder.nf   BR    4      1       2       21      6       equiv
zylinder.pre  BR    4      1       1       20      6       equiv
              comparison:
ok     = same number of entities before and after cycle, same model
equiv  = same model, logical structures removed, some redundancy
            in points and topology, model is still equivalent.
```

The wireframe models were partly generated by converting existing models of higher dimensionality. This is done by the GfS STEP processor by removing all faces and shells and collecting the remaining edges in an edge_based_wireframe model. If edge_loops are kept, also a shell_based_wireframe_model can be generated.

All models could be read without difficulties. Surface models are transferred to the native system and then back to STEP again without any loss. If the models have no topological information, the geometric sets have to undergo a topology generation before meshing (creating finite elements from the CAD entities). Brep models cause a few difficulties, if they use particular analytical curves (parabola, hyperbola) that are presently unavailable in the native system. Large loops and inner loops cause loss in mapping to the native format. Therefore the cycle test was not successful for some Brep models. As far as only logical structures were lost or only inefficiencies were shown (e.g. in not using one cartesian point for several purposes), the model was regarded as being equivalent to the original model.

5.5.2.3 FEM modeler files

The modeler in GfS's FEM system is a Compound Brep modeler. It always works with topology (i.e. no geometric sets), but allows for non-manifoldness of topology. Therefore faces can be shared by adjacent bodies and a model can consist of many bodies. Models can be one-dimensional, two-dimensional, three-dimensional or mixed.

From the internal models, the STEP preprocessor of GfS can generate the following STEP models:

only curves, no edges   -> geometric_3d_curve_set *)
only surfaces, no faces   -> geometric_3d_surface_set *)
only edges, no faces   -> edge_based_wireframe_model
edge_loops, no faces   -> shell_based_wireframe_model
with faces, no shell   -> face_based_surface_model
faces and open_shell   -> shell_based_surface_model
only one closed_shell   -> manifold_solid_brep
several closed_shells   -> body_based_solid_model

*) geometric sets occur only when using imported models, further processing evoques a topology generation

Tests were performed with the following models:

Filename	AP|size	|sca/par reading time	formatter writing time	write to native data	read from native data	comparison after cycle
test63.nf	CB 15	4	4	21	7	ok
haus.nf	CB 142	53	28	43	22	equiv
schnecke.nf	SF 196	50	46	157	27	ok
part1_wfsf	SF 75	37	47	117	15	equiv
gelenk.nf	CB 61	13	13	31	12	ok
gelenk4.nf	CB 198	48	38	36	27	ok
wheel.nf	CB 116	26	27	32	28	ok
hex2.nf	CB 43	11	10	23	13	ok
prism.nf	CB 25	9	7	22	8	ok
wfpoly7.nf	WF 122	24	24	28	22	ok
wbottle.nf	WF 164	38	30	27	17	ok

These models are well appropriate for the GfS modeler and its STEP processor. The models in the SF_AP and WF_AP files have both Level 2 (i.e. non_manifold topology). Therefore they could be written according to the Compound_Brep AP as well. In order to show better the dimensionality, the processor automatically choses the Surface or Wireframe AP when appropriate (see table above).

5.5.2.4 Wireframe files written by FEGS's program FAMEX

A set of files was sent from FEGS to GfS in order to test Wireframe processors. Partly they describe models of the 9 industrial test cases described by the CADEX users group. They cover all three levels of the wireframe AP:

- geometric_3d_curve_set (trimmed analytical curves without topology)
- edge_based_wireframe_model (trimming done by edges collected in one or more connected_edge_sets)
- shell_based_wireframe_model (additional edge_loops are given to enable creation of faces)

Filename	AP = WF|size	|sca/par reading time	formatter writing time	write to native data	read from native data	comparison after cycle
fbsm_fimod1.nf	85	23	15	104	14	equiv
sbwm_fimod1.step	71	19	14	99	13	equiv
sbwm_gehe20.step	193	50	34	162	16	equiv
g3cs_fimod3_surf	14	4	5	22	10	not trimmed cu
ebwm_fimod4_cbr	202	52	39	304	25	equiv

Some shell_based_wireframe_models were converted to surface AP (level 2, i.e. face_based_surface_model) by generating faces from the given edge_loops. These models are described in the previous subchapter. Some examples are:

```
   name of WF file          name of Surface file
sbwm_spiral_shell.step -> schnecke.nf
sbwm_fimod1.step       -> part1_wfsf.nf
```

5.5.2.5 Industrial test cases

As far as these models were available to GfS they have been read and converted.
Depending on source system and AP a varying number of entities is transferred in
a full cycle. Difficulties were encountered with faces having inner loops or loops
consisting of too many edges. In the present version of GfS's modeler the curves
hyperbola and parabola are not representable. B-spline_curves and -surfaces are
representable and can also be written back to STEP format. For meshing purposes
they are converted to piecewise low degree polynomial splines.

Filename	AP	size	sca/par reading time	formatter writing time	write to native data	read from native data	comparison after cycle
part7.ste	BR	47	11	15	40	10	not all loops
part7.sni	SF	96	24	27	32	21	ok
part74.nf	CB	76	28	36	30	27	not all loops
sbwm_fimod4_	WF	353	89	59	507	25	equiv
part5.sni	SF	117	28	29	31	24	ok
part6.sni	SF	75	20	19	27	19	ok
part1.sni	SF	93	25	25	32	21	ok
sbwm_f3_brep	WF	42	11	11	34	9	equiv

5.5.2.6 Conclusions from GfS tests

The performance measured in elapsed time is not in all cases linearly dependent
from the size of the file. There is always a large range in which the points in a
size/time diagram lie. For this particular computer system the following estimates
can be given:

t_p = time for parser in s
t_f = time for formatter in s
t_w = time for conversion to native in s
t_r = time for conversion from native in s
s = file size in kB

$$t_p = 1 + 0.23 * s \qquad (t_p/s = 0.21 \ldots 0.48)$$
$$t_f = 1 + 0.37 * s \qquad (t_f/s = 0.17 \quad\;\; 0.64)$$
$$t_w = 21 + 0.235 * s \qquad (t_w/s = 0.13 \ldots 1.03)$$
$$t_r = 8 + 0.125 * s \qquad (t_r/s = 0.06 \ldots 0.34)$$

The correctness of models was checked by using the file comparator and (in
addition:) graphical displays. Furthermore the converter and the modeler give
messages, if an entity cannot be processed properly. The file checker could not be
used, because it was incompatible with the used version of the Application
Protocols and the other modules of the Common Tool Kit.

In general the ability of the processors to transfer and convert all kinds of
models could be shown. Shortcomings of native systems (e.g. discarding logical
structures) have effects on cycle tests, but with a few exceptions all models are

valid: They can be used in the local system and can be transferred to other systems, where they can be used correctly, too.

The test result shows an excellent functionality and a satisfying performance.

5.5.3 Test results of Siemens Nixdorf

SNI has concentrated its work on the models in FIAT test library. The following models have been exchanged via B-rep AP and Surface AP resp.:

MODEL 1 Nozzle holder (B-rep AP)

MODEL 5 Gasket (SF AP)

MODEL 7 Gas outlet line from blower (B-rep AP)

MODEL 8 Timing system compartment (B-rep AP)

General route of importing a B-rep model:

1) Fill the IDS by the scanner/parser

2) Create the topology structure in Parasolid

3) Attach geometry to the topological elements

4) Mend the body (try to fulfill Parasolid's accuracy requirements)

5) Check for consistency of topology and geometry

In the following paragraphs the experiences with the four models are reported.

MODEL 1:

The model has been imported from HP's ME 30 and the PROCAD system. In every case the creation of a valid solid model in Parasolid via the route mentioned above was possible. The body could also exported to these systems without any problems.

MODEL 5:

The model has been imported successfully as a set of untrimmed surfaces. This fact made it very difficult to regenerate a solid body from the STEP file. Exporting these surfaces was possible without any problems.

MODEL 7:

The model has been imported from HP's ME 30 system. The creation of a valid solid model in Parasolid via the route mentioned above was possible. For the reason that Parasolid allows edges without any vertex the topology of the toroidal surface in the model could be simplified after importing. The body could also be exported without any problems.

MODEL 8:

This model, imported from ME 30, was the only one of these four models which caused some slight problems when importing into Parasolid. The original model couldn't be mended successfully. This means that the received data have been too inaccurate for Parasolid, so that sometimes a recalculation of edge geometry failed. (Parasolid takes the surface geometry as master geometry for an imported model and recalculate the edge geometry by intersecting the surfaces. The problem occurs on up to ten edges and was caused by a:

- non-intersection of a planar and a cylindrical surface

- non-intersection of two cylindrical surfaces

on edges where the two surfaces should meet tangentially. The result of import is a so-called inaccurate body where some of further modelling operations will fail.

Different attempts have been made to avoid this problem:

- Transforming the original model to a different space position

- Scaling down

- Increase the number of significant digits in the STEP file

The result has been that the number of "problematic edges" has been changed but in the end there was no full success.

The problem will be solved in later versions of Parasolid, because this version will allow inaccurate data in order to have valid solid models where further modelling will be possible.

Summary: Most of the tested models have been exchanged successfully. Problems as mentioned above depend mainly on the modeller itself (strong accuracy requirements). Also incorrect STEP files (wrong entities) have sometimes been the reason for exchange failure.

In the end, the use of a B-rep AP for exchanging data between solid modelling systems will be a very good means.

5.5.4 Test results of Senter for Industriforskning

A test plan had been set up concerning all CADEX partners who developed STEP-processors. This plan could not be followed by SI for the following reasons.

- The relevant files were not available, and they could not be produced by SI. The SI product SISL is a library of mathematical functions for surface modelling. It is not a complete CAD-system. There is, therefore, no input module for the creation of geometry though the mathematical functions for creating B-splines exist. Those have been utilised by the SISL-STEP processors. SI could, thus, not generate the models requested by the test plan themselves, but was dependent on input from the project.

- The project as a whole put more effort into the development of Common Tools as originally foreseen. At SI both processor development and testing suffered from this change.

- The basis for the CADEX-developments changed in the course of the project. STEP proved to be far less stable than expected. Adaptions to Common Tools and processors took more resources than planned. As being the last in the chain of software development, testing did not get the necessary attention.

Nevertheless the SI-processors were tested, though not in an as systematic way as desirable. The project did not generate many pure surface models; many Brep-models were, however, converted by SNI to SS-AP FL1 models. Therefore, most of the files we tested came from SNI. The following files were exchanged with DnV, GfS, ITALCAD, ND, PROCAD and SNI:

pre-processed file	origin	FL
chair.nf	VDA	1
cursur.nf	VDA	1
hood.nf	VDA	1
panel.nf	VDA	1
sledge.nf	VDA	1
e30splc.nf	VDA	1
splcurv.nf	IGES	1
splsurf.nf	IGES	1

post-processed file	origin	FL
blade.nf	ND	1
bottle.nf	SNI	1
bottle_level2.nf	DnV	2
cylinder.nf	SNI	1
exsi4.nf	SI	2
exsi5.nf	SI	2
exsi6.nf	SI	3
fiat_p7.nf	SNI	1
foen_ss	SNI	1
foen_dnv_ss	DnV	1
hood_level2.nf	DnV	2
kanne.pre	PROCAD	1
klinke.nf	SNI	1
sledge_level2.nf	DnV	2
socket.nf	SNI	1
surfsail.nf	SNI	1
teapot.nf	SNI	1
torus.nf	SNI	1
wiggle.nf	ND	1

In processor development and testing SI focused on SS-AP FL1. Processors for FL2 and FL3 were added at a late stage in the project and were mainly tested by manually edited files and a very few and simple models pre-processed by DnV. Preprocessing of FL2 and FL3 models is hardly supported by the project. No problems occured in interpreting these non-manifold and manifold surface models. As SISL cannot represent topology, the models were converted to geometrically bounded surface models and with good results.

Most of the above listed files are conforming to FL1. Our experiences with those can be summarised as follows:

- Initially neither our processors nor the STEP-files received were fully AP-conforming. This was especially valid for STEP header information. As soon as a common understanding of the SS-AP FL1 was established these problems disappeared.
- SI had no major problems in reading FL1-files. Even big models as the original foen.nf were processed properly. Files pre-processed by SI were not always successfully read by other partners. This was mainly due to limited functionality of the receiving system. Self-intersecting geometry and degenerated b-spline surfaces e.g. caused errors in post-processing.

5.5.5 Test results of Kongsberg 3D Partner A.S.

Kongsberg 3D Partner has received four of the test cases, defined by the test and validation group in the CADEX project, from SNI. These are test models number 1, 5, 6 and 7. Originally these models contain very little sculptured geometry. SNI has however converted the analytical geometry to b-splines and written it to STEP file using their SS-AP level 1 preprocessor. This then enables SS-AP level 1 post-processors to read the files. In some case, however, it is quite artificial to represent the geometry by b-splines without trimming.

Writing the models to STEP files again gives the same results as with the postprocessing, except for the fact that all faces represented in TECHPRO must have boundaries, so each of the b-spline surfaces postprocessed will automatically get 4 b-spline curves along the minimum and the maximum parameter values of the surfaces as boundaries. These curves will be additional b-spline curves on the STEP file when the model is preprocessed.

TECHPRO is still a prototype modeler without the functionality necessary for creating the CADEX test models in the system. Therefore we had to rely on receiving the test models from other partners. The files shown in the preprocessing table below are made in the surface modeler Techsurf and imported into TECHPRO via a dedicated link.

Results from the postprocessing:

```
Model number                            :    1
Part name                               :    Nozzle holder
Sending Company                         :    Siemens Nixdorf
Number of geometric_3d_sets in file     :    1
Number of b_spline_surfaces in file     :    34
Number of b_spline_curves in file       :    34
Errors detected during processing       :    0


Model number                            :    5
Part name                               :    Gaskets
Sending Company                         :    Siemens Nixdorf
Number of geometric_3d_sets in file     :    1
```

```
Number of b_spline_surfaces in file  :     36
Number of b_spline_curves in file    :     128
Errors detected during processing    :     0


Model number                         :     6
Part name                            :     Differential pinion
Sending Company                      :     Siemens Nixdorf
Number of geometric_3d_sets in file  :     1
Number of b_spline_surfaces in file  :     33
Number of b_spline_curves in file    :     48
Errors detected during processing    :     0


Model number                         :     7
Part name                            :     Gas outlet line from blower
Sending Company                      :     Siemens Nixdorf
Number of geometric_3d_sets in file  :     1
Number of b_spline_surfaces in file  :     32
Number of b_spline_curves in file    :     55
Errors detected during processing    :     0
```

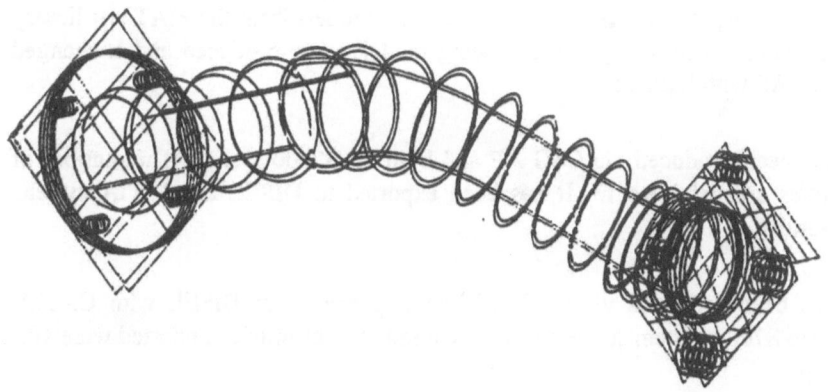

Fig. 5.5.2: Test model 7: Gas outlet line from blower (sufaces not trimmed)

A number of other files were exchanged with SI, DnV, SNI and ITALCAD:
  Preprocessing:

```
file                  origin
---------             ------
blade                 Techsurf/Techpro
wiggle                Techsurf
drill                 Techsurf
```

```
demo                    Techsurf
banana                  Techsurf
And: All postprocessed files.
```
   Postprocessing:
```
file                    origin
---------               ------
chair.nf                SI
cursur.nf               SI
hood.nf                 SI
panel.nf                SI
sledge.nf               SI
foen                    ITALCAD, SNI
teapot                  SNI
bottle                  SNI
surfsail                SNI
vase                    SNI
```
All the tested files were processed with success.

5.5.6 Test results of Italcad

In the following the experiences of Italcad with models from the FIAT test library are described. Specifically, the following models were produced and exchanged via CSG AP with DISEL:

Model 1:

   It has been produced via CSG AP and built on S7000 system. The number of primitives created were 13. It has been exported to DISEL on CATIA system successfully.

Model 7:

   It has been produced via CSG AP and imported from DISEL with CATIA system to S7000 system successfully. The number of primitives imported were 40.

Model 8:

   It has been produced via CSG AP and built on S7000 system. The number of primitives created were 98. It has been exported to DISEL on CATIA system successfully.

5.5.7 Test results of Procad and Isykon

5.5.7.1 Test methods during the development phase

For the first versions of our pre- and postprocessor we have done cycle-tests in our CAD system. Using the scanner/parser software this test method could mainly test the syntax of the generated STEP files. Geometry and Topology could only be checked for completeness but not for correctness. With more stable versions of the processors we began to exchange models with several partners. There we had good results with the exchange of simple Brep models like cubes and cylinders with HP and SNI, and SS models with GfS.

5.5.7.2 Testing the preprocessor

At the end of the development phase we have implemented all the CADEX test tools in our postprocessor. These test tools were used to check our generated STEP files for correct syntax and topology. As we still had some problems with defining the orientation of edges in a loop, we missed functions in the test tools which check these topological entities. Therefore we had to put a lot of effort in checking our own STEP files. Next we have done more detailed testing by sending more complex STEP models to the partners with Brep and SS postprocessors receiving related feedback on errors by SNI and HP, which enabled us to find the last bugs in our preprocessor. In the last step we decided to use the analytic brep models #1 and #8 of the FIAT test library, which we generated in our CAD system and sent them as STEP files to all partners.

Example of successful transfers:

- Brep model MODEL1 from PRO*CAD/PROREN to ME30 and Sigraph-CAD3d
- Brep model MODEL8 from PRO*CAD/PROREN to ME30 and Sigraph-CAD3d
  - see figure 5.5.3
- SS model KANNE from PRO*CAD/PROREN to SISL and PROLOG - see figure 5.5.4.

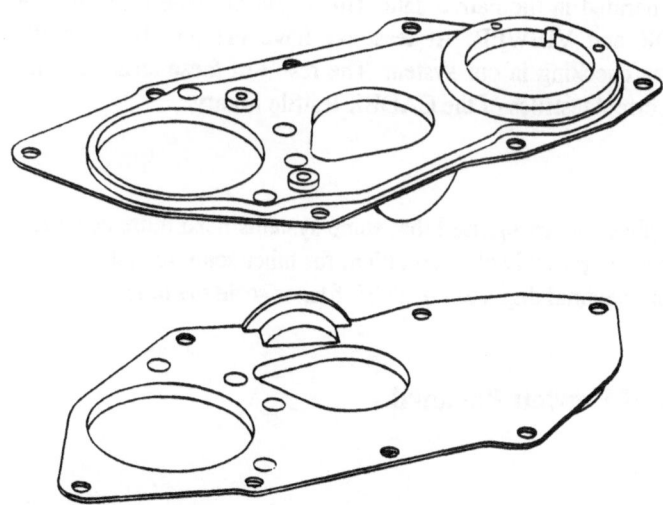

Fig. 5.5.3: Industrial model 8

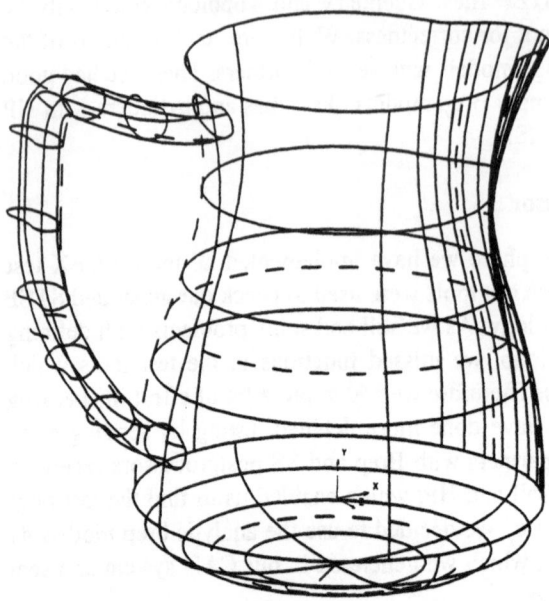

Fig. 5.5.4: Model "Kanne"

5.5.7.3 Testing the postprocessor

A detailed testing of the Brep-, CSG-, and SS-postprocessor was also done by data transfers with several partners. For this work we used all the CADEX test tools. A testfile running through parser and IDS-checker without any errors was regarded to be a correct STEP file. Next step was to compare the number and types of STEP entities with those generated in the native data. This could be done by using the STEP-COMPARATOR and VIEWER. At least we have the possibility to do topology and geometry checking in our system. The result of these tests was that we could process all correct testfiles of the CADEX testfile library.

5.5.7.4 Problems

At the end of the test phase we recognised that some systems need more accuracy than we have in our system, possibly also a problem for other sending systems. We had either to write more decimal digits to the STEP file, or scale the model.

5.5.8 Test results of Hewlett Packard

5.5.8.1 Introduction

At HP there existed a need for a test environment for three Processors: ME30 Preprocessor and a bidirectional STEP link from and to the new product HP-PE/SolidDesigner.

Because the special situation not only to develop new post- and pre-processors for an existing CAD system but for a new system under development it was not possible to adopt the test concept proposed by WP6 of the CADEX project without any changes. The concept of WP6 bases on statistical analysis of industrial CAD models. The method described in [1] allowed the selection of a small but characteristic sample of models. So the successful transfer of the test models allows a qualification of the processors.

5.5.8.2 HP's test concept

The main problem of using the models described in [3] in an environment where CAD system and processor are developed in parallel is that the test models allow only a final "stop or go" test which can at first used in a late stage of both projects. So HP was forced to put more effort on partial testing of the processors which had to be harmonised with the actual status of the CAD system itself. The following concept allows both to evaluate post- and pre-processors, but a working solution in the "inner loop" within HP does not guarantee conformance to the CADEX AP's because it is not possible to avoid related bugs in the post- and the pre-processor which may compensate each other. So a part of the test models were distributed to other partners. Because HP was not member of WP6 no special time was allocated to build a systematic and useful test database for all partners.

HP classified the models their processors must handle into the following groups. Although some groups seem to represent more "simple" models, the following enumeration should not be interpreted as a hierarchy for succeeding tests.

1. One simple geometric surface type per file with minimal topology but no surface singularities. Surface singularities are more difficult to handle for a modeler and may need "degenerated topology".
2. One simple geometric surface type per file with minimal topology but including surface singularities. Examples for this group are cones with apex or toroidal surfaces as apples and lemons.
3. Models containing intersections between two geometric surface types. For HP's final test of their processors a complete test matrix will describe most of the possible intersection types. Note that there may be more than one test case for a given surface-surface intersection because there are different types of resulting curves possible. The test cases should cover the complicated cases in respect to HP's CAD system.
4. Models containing complex topology but only simple geometry. A model for this group may contain only straight lines and planar surfaces but a lot of such entities distributed in the object space. This group is not only interesting for performance testing but also allows the test of the stability of the processors considering static and dynamic memory allocation and deallocation.
5. Transfer of models created after rotation in 3D space to detect implicit dependencies on the position of the geometry referring to the position of coordinate system of the modeler.

6. Models with reduced accuracy.
7. Realistic test models derived from "industrial" test cases. These models should contain a lot of different geometry, complex intersections and complex topology. For this test HP uses the results of WP6.

5.5.8.3 Test plan

The quality of the processors can be evaluated by analysing the quality of transmitted models from the groups described above. To test the real contents of the STEP file no recalculation of analytic intersections was performed. In the post processor the geometry was created exactly with the same parameters as used in the STEP file. The ability of the SolidDesigner to recalculate surface-surface intersections was only used for the STEP POLY_LINE entity. To evaluate transmitted models three successive tests can be distinguished.

1. Visual tests: the models should be turned in 3D space and compared with drawings or by direct comparison between different modelers.
2. Body checker call in the receiving system to detect accuracy problems not detected in visual tests.
3. Modeler Operation like punch, stamp, bore or blend as well as boolean operations to demonstrate that the imported model can be used in the receiving system as a native model.

5.5.8.4 Test results

Most of the time spent in testing HP's processors was used to create and transmit models of the classes 1-3 and 5. In addition to the visual tests after a successful import with the post-processor the STEP files themself were analysed both to detect possible problems as well as to evaluate the quality of the exported files. Class 4 is mostly related with performance and stability of the processors. These tasks will not be done before the end of the CADEX project. Realistic models of different contents including complex geometry and topology have been exchanged between HP, PROCAD and SNI. In addition, a lot of the models created by HP's modelers are contained in the collection of test models maintained by FIAT.

From the FIAT Test Library itself 3 of the 4 BREP test models were used to demonstrate the processors.

Model 7 was generated and distributed to the other CADEX partners. It describes a "gas outlet line from blower" consisting in HP's version of toroidal, spherical, cylindrical and planar surface types but only analytic surface-surface intersection tracks. The original model from the test library included a non-analytic surface component. Therefore it is planned to redesign model 7 as soon as the STEP processors can handle BREP level 3 data.

Model 1 describes a "nozzle holder" and was originally designed by PROCAD, transmitted to HP PE/SolidDesigner and rewritten as a STEP file. It consists of a collection of conical and cylindrical surfaces to bound a plate. Beside the visual test it was possible to manipulate the model, e.g. a polygonal hole was punched into its cylindrical outer cover. Despite of these manipulations the SolidDesigner body checker was not satisfied with the model. The detected problems need additional evaluations.

Model 8 consists of conical and cylindrical surfaces. As with Model 1 it was possible to work with Model 8 using local transformations. But some operations fail in areas where the body checker detects inconsistencies.

5.5.8.5 Conclusions

In general the detected problems can be classified in three groups:

1. Different constraints between topology and geometry in the sending and the receiving modeler. For instance prob-edges must be translated correctly. Another problem can be caused by (geometric) surface singularities which may need a topological representation in some systems but not in others. This is an issue which should be included in future versions of the BREP AP: what kind of topology should be generated in the case of curve and surface singularities?

2. A STEP face provides an optional parameter for one outer boundary. The definition of a loop type can be different in different systems. There may be no problem with planar surfaces but for instance if a cylinder topology definition is used there exist a need to represent two outer loops in the file. So a recalculation of the STEP file information may be necessary.

3. Different accuracy in modelers may cause problems for a receiving system with a higher accuracy. If intersection curves and points have to be recalculated the original analytic representation of such an entity may be lost. A result of that can be that for instance the labeling of the radius of a circular arc is difficult in the recalculated intersection. The accuracy of the transmitted STEP files is also an issue to be handled by the BREP AP.

5.5.9 Test results of DISEL

The models used in the test processors have been produced by FIAT.

For the CSG processors we used the models 1,7 and 8. The models were designed in our CATIA system and sent to ITALCAD, who postprocessed the three models without errors. In his turn, ITALCAD produced the same models and sent them back to be postprocessed by us. All models were imported succesfully.

For Wireframe models we received the models 1,3,4 and 5 from FEGS and other three differents models not included in FIAT's document. All models were successfully postprocessed. However, some problems were detected referred to the models tolerances: two different points in physical file were the same if the distance tolerance specified in the STEP file was considered. Moreover we have found serious problems when we tried to change the CATIA tolerances referred to point and curve discrimination tolerances.

Since FIAT could not provided models to test Brep processors level 1, we generated models converting the CSG models 1,7 and 8 in faccetted brep models, but we found important problems with the size of the models. For example, a CSG model composed by 5000 entities, in the brep facetted version included more than 50000. That volume of information could not be neither preprocessed nor postprocessed, because it collapsed the memory. The models were split into several small parts, each of them were successfully processed.

To test the surfaces processors we received models from PROCAD and SI. All of the models were postprocessed successfully. The preprocessors were tested internally.

5.6 Bibliography

[1] ESPRIT project 2195 - CADEX - Interim Report; A Haub, H Helpenstein (Editors); 20th August 1990

[2] STEP Part 31 - Conformance Testing Methodology And Framework: General Concepts; ISO TC184/SC4/WG6 Document Number N10; J Owen; 15th November 1990

[3] STEP Part 32 - Conformance Testing Methodology And Framework: Requirements On Testing Laboratories And Clients For The Conformance Assessment Process; ISO TC184/SC4/WG6 Document In Preparation

[4] STEP Part 33 - Abstract Test Suites For ISO 10303 Application Protocols; ISO TC184/SC4/WG6 Document In Preparation

[5] STEP Part 34 - Abstract Test Methods For ISO 10303 Implementation Form; ISO TC184/SC4/WG6 Document In Preparation

[6] Technical Report Number 1 - Interim Report on Global Test Concepts (tasks A--E); CTS2 Project Document Number CTS2/89/A/005/c, Revision 2

[7] Technical Report Number 2 - Specification of Harmonised Conformance Testing Methodology and Software Architecture (tasks F--I, L--N, V--W); CTS2 Project Document Number CTS2/90/A/006/c, Revision 3

[8] Technical Report Number 3: Abstract test cases: Rationale, definitions and examples (Tasks Q, L--O); CTS2 Project Document Number CTS2/91/A/001/c, Revision 2

[9] IDS Data Checker Requirements Specification; CADEX Document Number TESCC01P; X Ni; 11th December 1990

[10] IDS Data Checker Functional Specification; CADEX Document Number TESCC02P; X Ni; 11th December 1990

[11] Application Protocol For The Data Transfer of STEP B-rep Models Via A Physical File (Version 2.0); CadEx Document Number BREPR12P; W Weick; 11th September 1990.

[12] Application Protocol For The Data Transfer of STEP Sculptured Surface Models Via A Physical File (Version 2.0); CadEx Document Number SS_SI08P; J Haenisch, P Evensen; 3rd October 1990.

[13] Test And Validation Of STEP Processors; CADEX Document Number WG-BM12C; W Kerschbaum; 9th December 1990

[14] General Model Description From FIAT; CADEX Document Number WG3FI03C; D Gerbino; 8th January 1991

[15] Minutes Of The CADEX WG3 Meeting Held In Leeds On November 16th 1990; S B Harris; CADDETC Document Number RO/90/0005 Version 1.1; 27th November 1990

[16] Minutes Of The CADEX WG3 Meeting Held In Turin On December 13th 1990; S B Harris; CADDETC Document Number RO/90/0008 Version 1.0; 18th December 1990

[17] ESPRIT project 2195 CADEX - Deliverable 5: Report Of The Test Group Work; S B Harris; CADDETC Document Number RO/91/0002 Version 1.0 (CADEX Document Number WG3CC01C); 18th January 1991

[18] Interim Report Of The CADEX Test Group - June 1991; S B Harris; CADEX Document Number WG3CC02C; 2nd August 1991

[19] Model Generation Description of "Industrial" Test Cases; CadEx Document Number WG3FI05C; G F Facciano, D Gerbino; June 1991.

[20] ESPRIT project 2195 CADEX - Deliverable 5: Report Of The Test Group Work; S B Harris; CADDETC Document Number RO/91/0002 Version 1.0 (CADEX Document Number WG3CC01C); 18th January 1991

[21] IDS Data Checker Functional Specification; CADEX Document Number TESCC02P; X Ni; 18th July 1991.

[22] IDS Data Checker Software Guide; CADEX Document Number TESCC08C; X Ni; 24th July 1991.

[23] Minutes Of The CADEX WG3 Meeting Held In Bochum On April 25th 1991; S B Harris; CADEX Document Number WG3CC03C; 10th May 1991.

[24] Minutes Of The CADEX WG3 Meeting Held In Oslo On June 24th and 26th 1991; S B Harris; CADEX Document Number WG3CC04C; 28th June 1991.

[25] Minutes Of The CADEX WG3 Meeting Held In Madrid On September 19th 1991; S B Harris; CADEX Document Number WG3CC05C; 25th September 1991.

[26] Minutes Of The CADEX WG3 Meeting Held In Munich On November 14th 1991; S B Harris; CADEX Document Number WG3CC06C; 26th November 1991.

[27] Report of the joint CEC / US workshop on Manufacturing Technologies; S B Harris; CADDETC Document Number RO/91/0018; 9th August 1991.

[28] Joint ISO and CADEX AP Integration Meeting: Minutes; R J Goult, A Mckay, S Arabshahi; December 11th - 13th 1991.

[29] Software for screen management on different hardware; W Kerschbaum; CADEX Document Number CT_BM14C; 19th February 1991.

[30] Requirement for error handling in common tools; W Kerschbaum; CADEX Document Number CT_BM16C; 6th March 1991.

[31] Structure viewer for STEP files; W Kerschbaum; CADEX Document Number WG3BM18R; 23rd April 1991.

[32] Statistics report for STEP files; W Kerschbaum; CADEX Document Number WG3BM19R; 23rd April 1991.

[33] Statistics comparator for STEP files; W Kerschbaum; CADEX Document Number WG3BM21R; 21st June 1991.

[34] Technical description of the STEP Statistics viewer and the STEP file comparator; W Kerschbaum; CADEX Document Number WG3BM23R; 31st July 1991.

[35] Technical description of the STEP file structure viewer; W Kerschbaum; CADEX Document Number WG3BM24R; 31st July 1991.

[36] Technical Report Number 4: Final report on global test concepts (Tasks A-E); CTS2 Project Document Number CTS2/91/A/008/c; 20th December 1991.

[37] STEP Part 42 - Shape representation; R J Goult; January 1991

[38] STEP Part 21 - Clear text encoding of the physical file; J V Maanen; October 1990

[39] STEP Part 11 - EXPRESS reference manual; May 1990.

[40] Analysis and Choice of a Set of Industrial Test Cases; CadEx Document Number WG3FI04C; G F Facciano, D Gerbino; February 1991.

[41] Model Generation Description of "Industrial" Test Cases; CadEx Document Number WG3FI06C Revision 1; G F Facciano, D Gerbino; January 1992.

[42] Application Protocol For The Data Transfer of STEP Wire Frame Models Via A Physical File (Version 1.3); J Aas, C Miles; 30th July 1990.

[43] Application Protocol For The Data Transfer of STEP CSG Solid Models Via A Physical File (Version 1.0); CadEx Document Number CSGDI08P; F Gonzalez; 3rd July 1990.

[44] Minutes of Torino WG meeting 7th May 1992; CadEx Document Number WG_FI07C; D. Gerbino.

[45] CADEX Status Report 14th February 1992

[45] Minutes of Aachen WG meeting 12th September 1990; CadEx Document Number WG_GF30R; H. Helpenstein.